P9-AOG-370

Effects of Petroleum on Arctic and Subarctic Marine Environments and Organisms

Volume I

NATURE AND FATE OF PETROLEUM

Effects of Petroleum on Arctic and Subarctic Marine Environments and Organisms

Volume I

NATURE AND FATE OF PETROLEUM

EDITED BY

Donald C. Malins

Environmental Conservation Division
Northwest and Alaska Fisheries Center
National Marine Fisheries Service
National Oceanic and Atmospheric Administration
U.S. Department of Commerce
Seattle, Washington

ACADEMIC PRESS, INC., New York San Francisco London 1977
A Subsidiary of Harcourt Brace Jovanovich, Publishers

ACADEMIC PRESS RAPID MANUSCRIPT REPRODUCTION

ACADEMIC PRESS, INC.
111 Fifth Avenue, New York, New York 10003

United Kingdom Edition published by
ACADEMIC PRESS, INC. (LONDON) LTD.
24/28 Oval Road, London NW1

Library of Congress Cataloging in Publication Data

Main entry under title:

Effects of petroleum on arctic and subarctic marine environments and organisms.

Includes bibliographies and index.
CONTENTS: v. 1. Nature and fate of petroleum in arctic and subarctic marine environments.
1. Oil pollution of the sea–Arctic regions. 2. Oil spills–Environmental aspects. 3. Aquatic animals, Effect of water pollution on. I. Malins, D. C.
GC1085.E34 363.6 77-6693
ISBN 0–12–466901–8 (v. 1)

PRINTED IN THE UNITED STATES OF AMERICA

Contents

CHAPTER 3

ALTERATIONS IN PETROLEUM RESULTING FROM PHYSICO-CHEMICAL AND MICROBIOLOGICAL FACTORS

Neva L. Karrick **225**

Contributors

DONALD W. BROWN
ROBERT C. CLARK, JR.
NEVA L. KARRICK
WILLIAM D. MacLEOD, JR.

Environmental Conservation Division
Northwest and Alaska Fisheries Center
National Marine Fisheries Service
National Oceanic and Atmospheric Administration
U.S. Department of Commerce
2725 Montlake Boulevard East
Seattle, Washington 98112

Preface

The sinking of the *Argo Merchant* off Nantucket Island on 21 December 1976, one of a series of recent tanker mishaps, resulted in several million gallons of Bunker C fuel oil entering the marine environment. This intrusion of petroleum posed a threat to a host of marine organisms, some of which were important commercial species. The well publicized mishaps underscored and reemphasized a long-standing concern about the impact of petroleum transport and drilling operations on the marine environment.

The first step in addressing the problem should consist of an up-to-date and comprehensive evaluation of what is known about the subject. The information should provide an understanding of the chemical and physical properties of petroleum, together with data on possible sources, inputs, weathering, and sinks. Then it should provide an evaluation of data relating to whether petroleum pollution significantly alters the biological processes of the many diverse organisms that comprise the ecosystems. No attempt had been made in the past to provide this information through the publication of a comprehensive treatise. The May 1973 Workshop on "Inputs, Fates, and Effects of Petroleum in the Marine Environment" at Airlie, Virginia, and the subsequent 1975 report by the National Academy of Sciences provided an overview of the subject. The substantial body of evidence accrued in the last few years was largely reported in review articles and symposia proceedings and covered only limited aspects of the problem.

In two volumes of the present work, we have attempted to compile and evaluate current knowledge on the effects of petroleum on arctic and subarctic marine environments and ecosystems. Clearly, the potential for petroleum pollution has increased dramatically in these areas and is expected to increase even more as petroleum exploration is extended more northerly and into deeper areas of the continental shelf. We are cognizant of various definitions of Arctic and Subarctic; however, for convenience we have taken a rather liberal view in defining these regions. Arctic waters are considered to be those regions covered by extensive sea ice at least part of the year, excluding shallow bays or inlets. The southern boundary of sea ice is shown in Figure A. Subarctic waters are considered to be essentially the zone between the southern limit of sea ice and the southern limit of the 3°C-water isotherm at 100 meters depth (about 40°N latitude in the Pacific and the northern edge of the Gulf Stream in the Atlantic); the southern boundaries become diffuse near the continental margins (Fig. A).

Volume I deals with the nature and fate of petroleum in the marine environment, thereby setting the stage for Volume II, which covers the biological effects of petroleum, specifically, alterations in life processes and in community structures. The volumes are intended to provide scholarly and cogent discussion of the various effects of petroleum on the arctic and subarctic environments. The discussions are directed toward research workers, environmentalists, petroleum industry managers and executives, and public officials concerned with the many aspects of petroleum pollution of marine waters. The contributors were encouraged to provide a balanced perspective, coupled with cogent speculation, on subjects that are often areas of lively controversy.

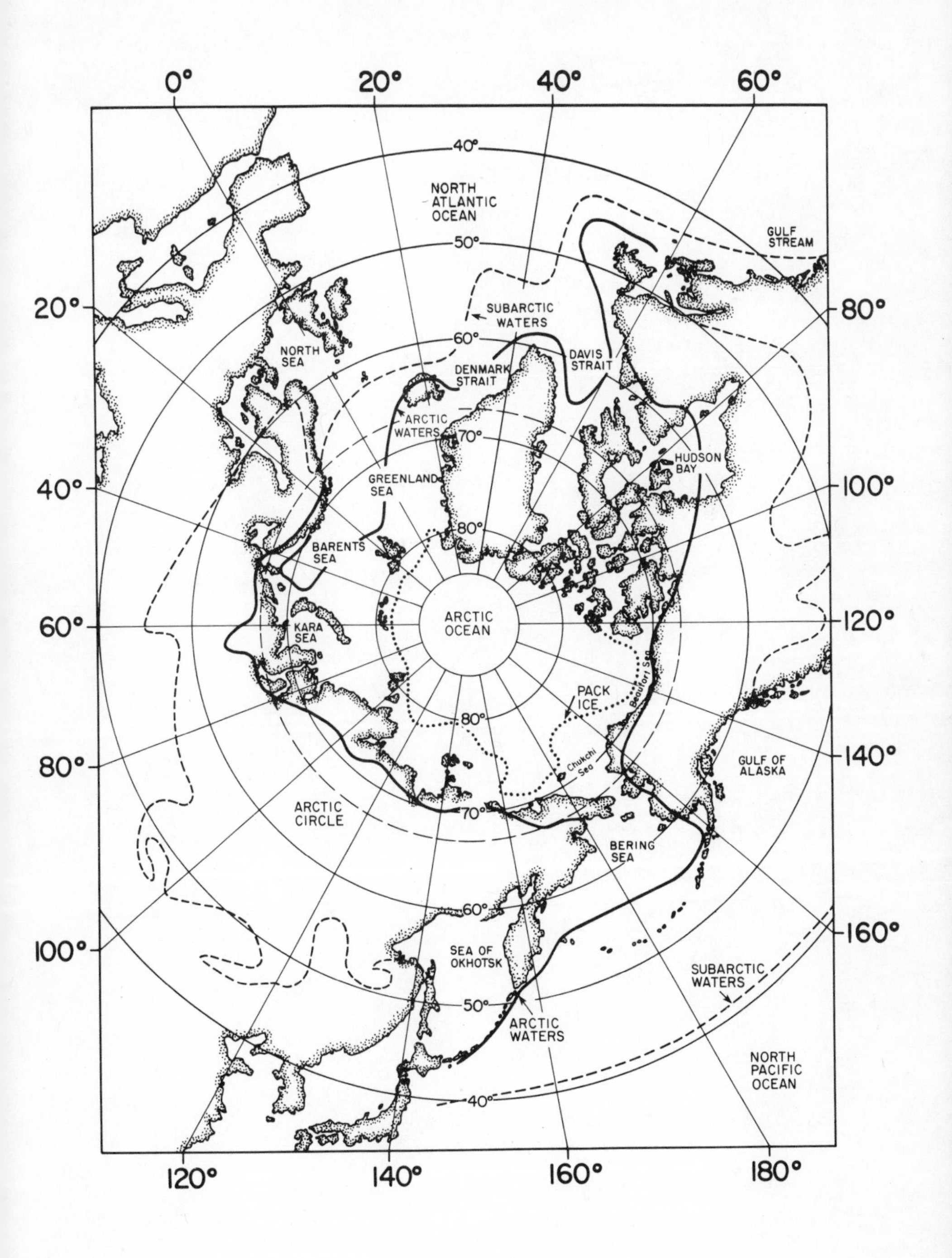

0°
20°
40°
60°
80°
100°
120°
140°
160°
180°
NORTH ATLANTIC OCEAN
GULF STREAM
SUBARCTIC WATERS
NORTH SEA
DENMARK STRAIT
DAVIS STRAIT
ARCTIC WATERS
GREENLAND SEA
HUDSON BAY
BARENTS SEA
ARCTIC OCEAN
KARA SEA
PACK ICE
Beaufort Sea
Chukchi Sea
GULF OF ALASKA
ARCTIC CIRCLE
BERING SEA
SEA OF OKHOTSK
SUBARCTIC WATERS
ARCTIC WATERS
NORTH PACIFIC OCEAN

FIG. A

Acknowledgments

Preparation of this book was encouraged and partially supported by the Outer Continental Shelf Environmental Assessment Program, Environmental Research Laboratories, National Oceanic and Atmospheric Administration, with funding furnished by the Bureau of Land Management, U. S. Department of the Interior.

We are grateful to our colleagues at the Northwest and Alaska Fisheries Center, Department of Oceanography at the University of Washington, and other institutions for help and advice during preparation of this book and for review of drafts of the manuscripts. A special note of appreciation is due to several individuals: Mr. Maurice E. Stansby for guidance in production and in obtaining literature through retrieval services and personal contacts; Mr. Frank Piskur for his dedicated and conscientious efforts in editing manuscripts and in organizing materials; Mr. James Peacock, Ms. Carol Oswald, and Mr. Steve Jensen for preparation of figures and illustrations and design of cover; Ms. Gail Siani for her tireless efforts in typing drafts and for her fine contribution to the organization of materials; and Mrs. Isabell Diamant for her typing skill and review of literature references.

We thank those publishers and authors for graciously granting permission to reprint original figures and tables from the referenced articles. Especial indebtedness is due the following specialists who kindly reviewed drafts of the manuscripts:

JACK W. ANDERSON, Battelle Pacific Northwest Laboratories, Sequim, Washington

WALTER J. CRETNEY, Environment Canada, Victoria, British Columbia, Canada

MELVIN W. EKLUND, Northwest and Alaska Fisheries Center, Seattle, Washington

JOHN W. FARRINGTON, Woods Hole Oceanographic Institution, Woods Hole, Massachusetts

HUGH L. HUFFMAN, JR., EXXON Research and Engineering Co., Linden, New Jersey

C. BRUCE KOONS, EXXON Production Research Co., Houston, Texas

EDWARD R. LONG, Environmental Research Laboratories, Marine Ecosystems Analysis, Puget Sound Project, Seattle, Washington

CLAYTON D. McAULIFFE, Chevron Oil Field Research Co., La Habra, California

SEELYE MARTIN, University of Washington, Seattle, Washington

WILLIAM T. ROUBAL, Northwest and Alaska Fisheries Center, Seattle, Washington

FRED T. WEISS, Shell Development Co., Houston, Texas

MARLEEN M. WEKELL, University of Washington, Seattle, Washington

CLAUDE E. ZoBELL, Scripps Institution of Oceanography, La Jolla, California

Contents of Volume II

BIOLOGICAL EFFECTS

Chapter 1

PETROLEUM: PROPERTIES AND ANALYSES IN BIOTIC AND ABIOTIC SYSTEMS

ROBERT C. CLARK, JR.
Environmental Conservation Division

DONALD W. BROWN
NOAA National Analytical Facility
Environmental Conservation Division

Northwest and Alaska Fisheries Center
National Marine Fisheries Service
National Oceanic and Atmospheric Administration
U.S. Department of Commerce
Seattle, Washington 98112

PETROLEUM: GENERAL DESCRIPTION

ORIGIN

Petroleum is a naturally occurring complex mixture of organic compounds. It was formed from the partial decomposition of animal and plant matter over geological time. During the early periods in the earth's history, the oceans had alternately covered and receded from land areas; at the same time animal and plant debris was carried to and deposited in the seas. The abundant animals and plants of the sea also contributed to the accumulation of sedimentary deposits. It is believed that these deposits constituted the raw material for the natural production of petroleum.

Petroleum may exist as a gas (natural gas), as a liquid (crude petroleum), as a solid (asphalt, tar, bitumen), or as any combination of these three states. It does not occur in underground lakes or pools, as once believed, but rather in porous and permeable rock. This rock is riddled with microscopic, connected cavities and channels which contain the petroleum much like a sponge. The fluid petroleum can move freely within these pores of the sedimentary rock. The petroleum bearing rocks are trapped, or blocked off, by denser nonpermeable strata; these areas are called structural or stratigraphic traps. The storage areas are known as oil reservoirs [1]. A description of what constitutes an oil field has been presented by Meyerhoff [2].

The physical properties and chemical composition of petroleum from different producing regions and even from different depths in the same well can vary markedly. Thus a precise definition of petroleum is not possible because no two samples are exactly alike. The fluid form may range in appearance and consistency from a colorless and light liquid like water to a black and heavy material like tar. Most crude oils are black by transmitted light but some are amber, brown, or red; they show a greenish fluorescence by reflected light.

Crude petroleum contains thousands of different chemical compounds; its chemical complexity probably resulted from a process of molecular scrambling which occurs when petroleum is formed in nature [3]. Of these chemical compounds, the hydrocarbons are most abundant. They range in molecular weight from 16 (methane) to considerably above 20,000. Apparently, the geochemical processes involved in the formation of petroleum tend to produce a wide variety of hydrocarbon compounds

[4]. Petroleum also contains traces of various compounds of nitrogen, sulfur, and oxygen as well as traces of organometallic compounds. Bestougeff [5] suggested that most crude oils contain the same structural groups of chemical compounds, but the differences between petroleums are a reflection of the relative amounts of the various compounds that are present.

RECOVERY

Drilling a well is only the first stage in a complicated and costly process designed to induce the oil to flow from a reservoir to the borehole through which it can be raised to the surface, sometimes by pent-up natural pressure, more often by some kind of pumping. What happens in the reservoir in terms of flow of oil is determined by the porosity and permeability of the rock and by the gravity and viscosity of the oil itself. Oil exists in a great variety of forms, ranging from a fluid approaching water in appearance and behavior to a heavy, sluggish substance that will hardly pour [1].

Once the natural pressure is expended, pumping or some other form of recovery is needed to force the oil from the reservoir rock to the borehole and up to the surface where it is collected. Crude oil as it comes from the ground is usually made up of both gases and liquids (petroleum and water, the latter often containing a high content of dissolved solids [6]). The water is removed by a separator prior to shipment of the crude oil to the refinery. "Natural gas" is generally regarded as that accumulation of gas which may or may not be associated with underground sources of oil. Natural gas is treated to remove non-hydrocarbons (carbon dioxide, hydrogen sulfide, water) and to separate natural gasoline, leaving a dry gas for movement through pipeline under pressure. The dry gas is principally methane with smaller amounts of ethane and propane [7]. Major methods of transportation of crude oil in the United States is by pipeline and tanker, although rail tank cars, tank trucks, and barges are also employed.

REFINING

Refining is essentially a manufacturing process in which crude petroleum is used as raw material to make numerous petroleum products and petrochemical starting materials. The petroleum refining industry is one of the largest manufacturing industries in the world. Aspects of refining as it relates to the description of crude oils and of refined products are discussed in subsequent sections of this chapter.

HISTORICAL USES

Crude petroleum, in the raw state, is of little use commercially. Occasionally, natural liquid petroleum may escape from a point or a fault in a structural trap and emerge at the surface of the earth as an oil seep. Oil-soaked, sandy shales having a kerosine-like odor were used by the local Indians along the west coast of the Olympic Peninsula in Washington State as medicinal "smell muds" [8]. In other localities, crude petroleum has been used to coat the bottom of sailing ships as a preservative measure. Today crude petroleum is used only as the raw material for the production of refined products such as fuels, lubricants, preservatives, construction materials, and medicinals, and, in addition, for the production of compounds that may be used as the starting material for the manufacture of petrochemicals.

PHYSICAL CHARACTERISTICS OF PETROLEUM

The specific gravity, viscosity, pour point, and fractional distillation temperatures are useful physical characteristics for describing crude petroleum since they are composite values based on all the components present. These values are related to the chemical nature, the molecular interactions, and the amounts of each component in petroleum. In general, boiling points of molecules increase with their molecular weight; specific gravity is influenced by molecular configuration; and viscosity is a function of both the molecular weight and configuration. The presence of waxes (high molecular weight hydrocarbons), for example, strongly influence the pour point temperature and the viscosity of petroleum [9].

Standard methods for the determination of the physical characteristics of petroleum have been established in the United States by the American Society for Testing and Materials (ASTM: Annual Book of ASTM Standards in 48 volumes, revised annually [10]) and in Great Britain by the Institute of Petroleum (IP Standards for Petroleum and its Products in two volumes, revised annually [11]).

SPECIFIC GRAVITY

The density of a substance is the ratio of the mass to a unit volume. In the industry, the density is usually measured in grams per milliliter at 16°C (60°F). Specific gravity of a given material is defined as the ratio of its density to the density of water at the same temperature. In the petroleum industry gravity is usually expressed in "Degrees API" in accordance with an arbitrary scale established by the American

Petroleum Institute. The relationship between specific gravity values and Degrees API Gravity at 16°C (60°F) is shown in the following formula:

$$\text{Specific gravity} = \frac{141.5}{131.5 + \text{Degrees API}}$$

Thus, a crude oil of 10° API Gravity is equal to a specific gravity of 1.0, the same as water.

Specific gravity and API Gravity can be determined by the use of hydrometers, calibrated in accordance with the scale used (ASTM Methods D-287 and D-1298 [10: part 23]). More accurate determinations are made with the use of a pycnometer (ASTM Methods 941, 1217, 1480, and 1481 [10: part 23]) or a specific gravity or density balance (IP Standard Method 59, Method C [11: part I; 12]).

Crude petroleums have specific gravities in the range of 0.79 to 1.00 (equivalent to API Gravities of 10 to 48) [13], although 44° API is the upper gravity limit reported for major crude oils [14]. Either specific gravity or API Gravity must be determined to make volume corrections at different temperatures [15]. Gravity may also be a rough index of whether an oil is naphthenic or paraffinic and is used as one of several tests for determining the quality and performance characteristics of various petroleum products.

VISCOSITY

The viscosity of a fluid relates to its internal friction or its resistance to flow. Absolute, or dynamic viscosity, is the force required to move a plane surface area of one square centimeter above another plane surface at the rate of one centimeter per second when the two surfaces are separated by a layer of fluid one centimeter in thickness. The unit of absolute viscosity is the poise.

The ratio of absolute viscosity of a fluid to its density is called the kinematic viscosity. The unit of kinematic viscosity is the stoke (St); because most fluids have low viscosity the centistoke (10^{-2} St = cSt) is frequently used. Although determination of absolute viscosity will give extremely accurate values, the method is difficult to carry out. Consequently, measurements of absolute viscosity are made only when accurate data are needed. Under practical operating conditions, the viscosity of petroleum is measured with a capillary viscometer. The kinematic viscosity (ν) is the measured flow time in seconds times the capillary tube viscometer instrument calibration constant in cSt per second at a specified temperature (ASTM Method D-445 [10: part 23] and IP Method 71 [11: part I; 12]). The absolute viscosity can then be calculated from the kinematic viscosity and the density (at the same temperature) by the equation:

$$\eta = \rho \cdot \nu$$

where η = absolute or dynamic viscosity in centipoise
ρ = density in grams per cubic centimeter
ν = kinematic viscosity in centistokes.

A liquid having a high viscosity value flows sluggishly like cold molasses; one with a low viscosity value flows freely like water.

The viscosity of petroleum is related to its chemical composition (molecular weight and configuration) and molecular interactions. The effect of molecular weight of the components on viscosity may not be the same as on specific gravity. For example, a Nigerian crude petroleum had a lower kinematic viscosity (5.16 cSt, at 38°C) than did an Iranian light crude oil (Agha Jari: 5.6 cSt), but had a higher specific gravity (0.867) than the latter (0.854) [16]. Viscosity is a function of temperature; as the temperature increases, the viscosity decreases and the oil flows more freely.

POUR POINT

The pour point is the lowest temperature at which an oil can be poured. The test is made under empirical, but standardized, conditions. Samples of petroleum are placed in clean, clear glass containers and are cooled slowly at a specific rate down to -51°C (-60°F). The samples are examined at each 3°C (5°F) drop in temperature by tilting the glass containers to a horizontal position and observing the sample for any movement of the fluid. The temperature is recorded at the interval where there is no visual movement of the fluid when the containers are held horizontally for five seconds. The pour point is arbitrarily designated as 3°C (5°F) above this recorded temperature (ASTM Method D-97 [10: part 23], IP Method 15 [11: part I; 12]).

The pour point approximates that temperature at which petroleum shows a kinematic viscosity of 300,000 cSt. The kinetic viscosity value of a petroleum at the pour point may or may not represent actual kinematic viscosity of the entire petroleum. At the pour point some components of the crude petroleum may crystallize out and form a stable gel mixture of liquid oil and waxy crystals. Under these conditions the kinematic viscosity may represent that of the liquid (oil) phase of the petroleum or that of the stable gel that was formed. In the latter case, the kinematic viscosity of the liquid oil phase might actually be considerably below 300,000 cSt.

Dean [16] listed the pour points of six representative world crude oils; the values ranged from -34°C to 7°C (-30°F to 45°F). A list of 93 major world export crude oil blends showed that the range can be from -43°C to 43°C (-45°F to 110°F) [14]. A low pour point is of particular significance when a fuel oil or lubricant is to be used in machines exposed to low temperatures. Also the pour point temperatures of crude petroleum may be an important aspect in case of accidental spills, particularly in arctic and subarctic regions. If the pour point of the petroleum is higher than the temperature of the environment, any spilled material would tend to become cohesive, simplifying cleanup of the spill material.

FRACTIONAL DISTILLATION TEMPERATURES

The hydrocarbon components of petroleum are separated by fractional distillation. The process of distillation is commonly used in the laboratory to determine the distribution of the components of crude petroleum and to test the quality of various petroleum products--particularly those which are vaporized in the course of their use, such as gasoline, kerosine, and solvents. By means of laboratory distillation tests, the distillation range of a given crude oil or product is determined, as well as the amount of material vaporized at various temperatures.

In a refinery, the distillates evaporating at certain temperatures are collected separately. A finer separation of components of the mixture may be achieved by further fractional distillation of the initial crude distillates. The fractional distillation of petroleum is a continuous process: a stream of heated petroleum is continuously charged to a distillation tower or column; the vaporized components escape from the top and the residue is removed at the bottom [15].

The initial distillation of crude petroleum in a refinery is carried out at atmospheric pressure and this phase of the process is called "stripping," "topping," or "reducing." The light ends are removed from the top of the distillation tower; the residue of this initial process is called the "topped crude" or "reduced crude." In the refinery the light gases are separated from the crude oil; methane and ethane may be used for refinery fuel and the propane and butane are compressed and sold as liquefied petroleum gases [7]. The topped crude contains the high-boiling components of petroleum. Thus, extremely high temperatures must be used to achieve separation of the remaining components by fractional distillation under atmospheric pressure. Such temperatures, however, may cause "cracking" (thermal decomposition or molecular rearrangement) of the chemical components. To reduce the amount of cracking, fractional distillation of the topped crude is carried out under vacuum. Distillation under reduced pressure (less than one atmosphere) lowers the boiling point of the components of the topped crude, thus lessening the risk of thermal degradation. The residue in the vacuum still is called the residuum [15].

The various fractional distillates of petroleum at atmospheric pressure consist of: (1) natural gas (less than 20°C); (2) straight-run gasoline (20°-200°C); (3) middle distillates--kerosine, heating oils, and jet, rocket, and gas turbine fuels (185°-345°C); (4) wide-cut gas oils--light and heavy lubricating oils, waxes, and feed stock for catalytic cracking to gasoline (345°-540°C); and (5) residual oil (above 540°C and vacuum distillation) [17]. The boiling ranges are shown in Figure 1. Fractional distillation procedures are described in detail in ASTM Method D-2892 [10: part 24] and IP Method 24 [11: part I; 12].

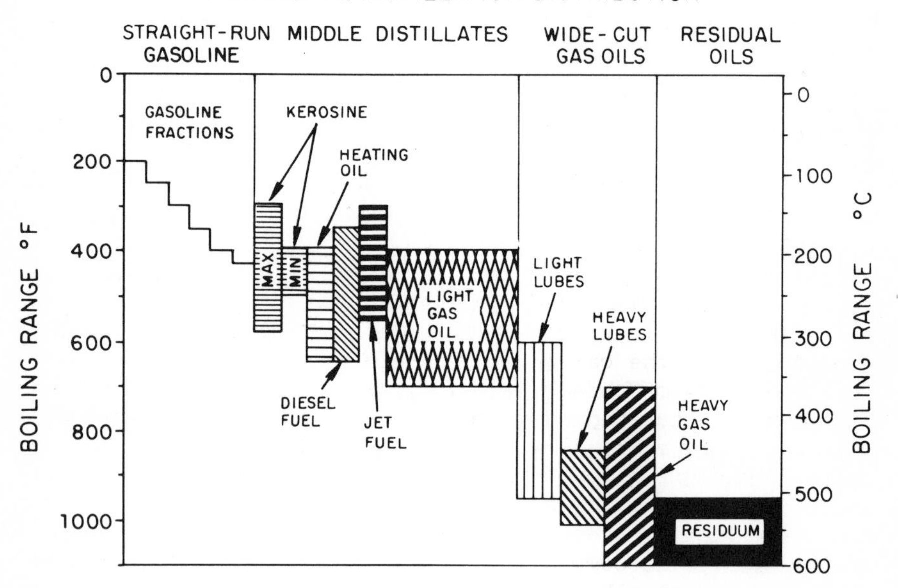

Fig. 1. Boiling point range of fractions of crude petroleum. (Adapted from Bureau of Naval Personnel [15]).

CHEMICAL PROPERTIES OF PETROLEUM AND PETROLEUM PRODUCTS

CHARACTERISTICS OF PETROLEUM

Hydrocarbon Components

Petroleum contains from 50 to 98% hydrocarbons with the majority of crude oils containing the higher concentrations of hydrocarbons. The variety of hydrocarbons may be divided into three general classes: aliphatic, alicyclic, and aromatic, each of which can be further divided into subclasses [17]. The many naturally occurring hydrocarbons of petroleum fall into each of these three general classes.

Aliphatic hydrocarbons are open-chain compounds, saturated or unsaturated. Open-chain means that the carbon atoms in the compounds are arranged in lines, either straight or branched. The atoms of a molecule are held together by strong attractive forces or bonds. A hydrocarbon compound with all single bonds, such as ethane, CH_3-CH_3, is saturated. An unsaturated hydrocarbon compound has one or more double bonds such as ethylene, CH_2=CH_2, or acetylene, CH≡CH. The unsaturated compounds tend to be less stable than the saturated and, thus, are more reactive chemically.

The saturated open-chain hydrocarbons are called paraffins or alkanes and have the empirical formula C_nH_{2n+2}. Paraffin is derived from the Latin *parum affinis* meaning slight affinity. The unsaturated hydrocarbons consist of several homologous series (a series of organic compounds that differ by a -CH_2- or some other group). These include: (1) those containing one double bond and having the formula C_nH_{2n} (olefins or alkenes); (2) those containing one triple bond and having the formula C_nH_{2n-2} (acetylenes or alkynes); (3) those having two double bonds and also having the formula C_nH_{2n-2} (diolefins or alkadienes); and (4) those having more than two double or triple bonds or both (alkatrienes, alkatetrienes, alkadynes, alkenynes, and alkadienynes).

The alicyclic hydrocarbons consist of compounds in which some or all of the carbon atoms are arranged in a ring. Individual members of this class may contain one or more rings. They may be saturated or unsaturated. These cyclic compounds fall into more subclasses than the aliphatic hydrocarbons because of the large number of the ring sizes possible. The saturated single ring (monocyclic) hydrocarbons in this class, C_nH_{2n}, are called cycloparaffins or cycloalkanes. The unsaturated monocyclic compounds include among others, the cycloolefins (or cycloalkenes) C_nH_{2n-2}, and the cyclodiolefins, C_nH_{2n-4}. Cyclic compounds with a ring containing a triple bond are highly unstable; they are rare in petroleum.

Olefin is often used in a general sense to include not only alkenes but also cycloalkenes and hydrocarbons containing more than one ethylenic double bond. The term olefins originated from the observation that ethylene, a gaseous unsaturated hydrocarbon (C_2H_4), reacted with chlorine to yield oily products. The oil-forming gaseous hydrocarbons were called olefiant gases; the designation was later contracted to olefins, denoting compounds related to ethylene [18].

The aromatic hydrocarbons contain at least one six-carbon benzene ring. Aromatic hydrocarbons owe their name to the fact that certain of these compounds have a pleasant aroma.

The three major components of petroleum consist of (1) paraffinic hydrocarbons, (2) saturated and unsaturated five- and six-carbon atom alicyclic hydrocarbons (naphthenes), and (3) a wide variety of aromatic compounds. The relative proportion of each type will vary with the source of the petroleum.

Paraffins

The paraffinic hydrocarbon compounds of petroleum contain from one to more than 78 carbon atoms. Paraffin hydrocarbons containing less than five carbon atoms are gases at room temperature and atmospheric pressure; those from 5 to 16 carbon atoms are liquid; and those with 17 carbon atoms and higher are semisolid or solid. These alkanes may be the straight-chain (normal or *n*-) or the branched-chain (e.g., *iso*) type.

Two or more compounds having the same elementary percentage composition and molecular weight but a different configuration--for example, *n*-butane, $CH_3(CH_2)_2CH_3$, and *iso*-butane, $CH_3CH(CH_3)_2$--are isomers. These saturated hydrocarbons are relatively stable at ordinary temperatures and are not readily affected by acid or alkali [15]. The melting point and boiling point of paraffins increase with an increase in molecular weight, but a *n*-paraffin has a higher boiling point than its branched-chain isomers. Paraffins also show lower density values than do other classes of hydrocarbons having the same number of carbon atoms.

Methane (CH_4) is predominant in the natural gas fraction of petroleum, but ethane (C_2H_6), propane (C_3H_8), and butane (C_4H_{10}) may also be present in substantial proportions. The paraffin waxes of petroleum contain chiefly *n*-paraffins from about 22 to 30 carbon atoms; petrolatum or petroleum jelly, contains *n*-paraffins from 30 to 70 carbon atoms. Various structural representations of some paraffin hydrocarbons found in petroleum are given in Figure 2.

Naphthenes

Naphthenes, the saturated and unsaturated alicyclic hydrocarbons, are found in most petroleums with the saturated forms predominating (Fig. 3). These compounds are fairly stable; they resist oxidation and are relatively insoluble in

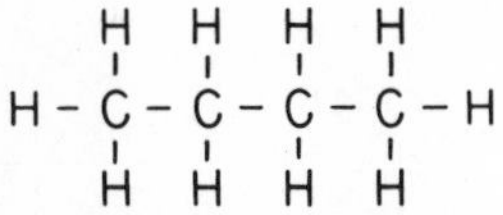

n-butane (n-C_4H_{10})

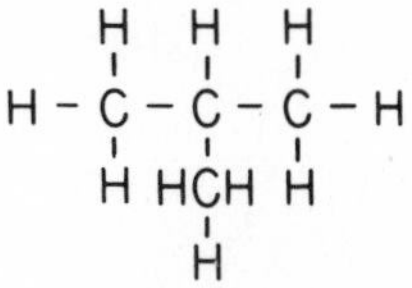

2-methyl propane,
iso-butane (iso-C_4H_{10})

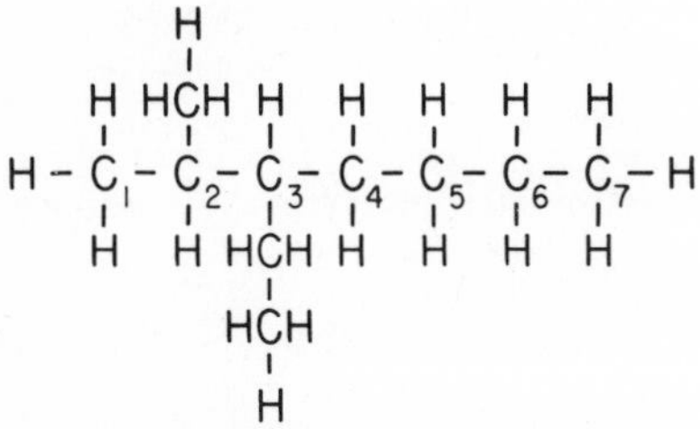

2-methyl-3-ethyl-heptane ($C_{10}H_{22}$)

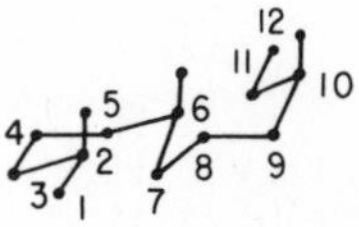

2,6,10-trimethyl dodecane,
farnesane ($C_{15}H_{32}$)

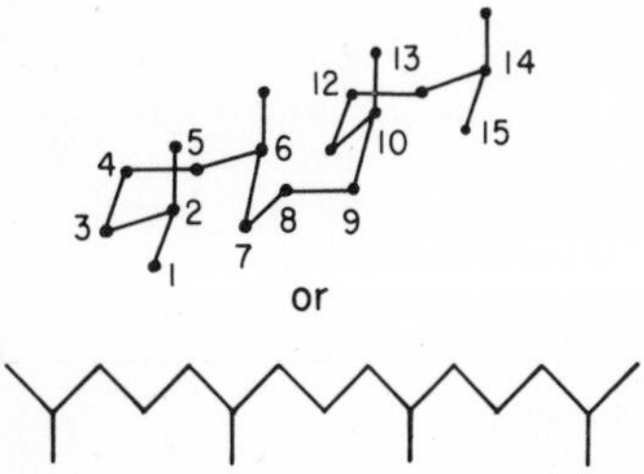

or

2,6,10,14-tetramethyl pentadecane,
pristane ($C_{19}H_{40}$)

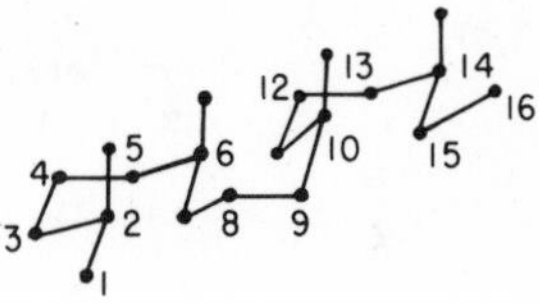

2,6,10,14-tetramethyl hexadecane,
phytane ($C_{20}H_{42}$)

n-heptadecane (n-$C_{17}H_{36}$)

Fig. 2. Paraffinic hydrocarbons found in petroleum.

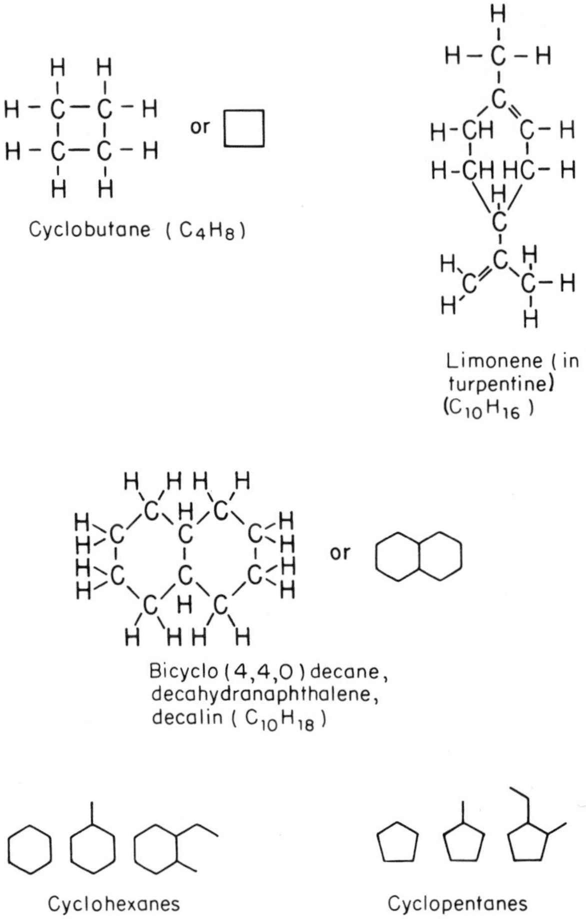

Cyclobutane (C_4H_8)

Limonene (in turpentine) ($C_{10}H_{16}$)

Bicyclo (4,4,0) decane, decahydranaphthalene, decalin ($C_{10}H_{18}$)

Cyclohexanes

Cyclopentanes

Fig. 3. Naphthenic hydrocarbons in petroleum.

strong sulfuric acid [15]. The boiling points of the naphthenes are 10° to 20°C higher than their corresponding open-chain hydrocarbons having the same number of carbon atoms. Also, naphthenes have higher specific gravities than the alkanes of the same boiling point [16].

Olefins

Olefins (Fig. 4) are usually absent in crude petroleum. These unsaturated compounds are relatively unstable and react readily with hydrogen or other elements such as oxygen, chlorine, and sulfur. Some olefins are found in refined petroleum products, but their presence in some fuels is undesirable

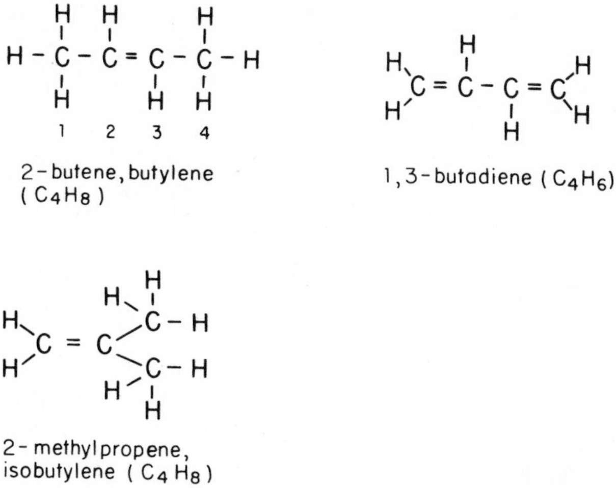

2-butene, butylene (C_4H_8)

1,3-butadiene (C_4H_6)

2-methylpropene, isobutylene (C_4H_8)

Fig. 4. Olefinic hydrocarbons found in petroleum.

because of their unstable nature. Gasolines contain olefins which are produced in the cracking processes in the refinery. Inhibitors and stabilizers are included in the gasoline to ensure proper storage stability [7].

Aromatic Compounds

Aromatic hydrocarbons are usually present in smaller amounts in petroleum than the paraffins and naphthenes. The simplest aromatic hydrocarbon is benzene (Fig. 5) consisting of a six-carbon atom ring. Naphthalene consists of two benzenoid rings fused together. Naphthalene should not be confused with the hydrocarbon subclass known as the naphthenes, which are the saturated and unsaturated alicyclic hydrocarbons. Any of the hydrogen atoms in an aromatic hydrocarbon, such as naphthalene, may be replaced by paraffinic, naphthenic, or olefinic compounds, or by other aromatic groups (e.g., biphenyl) (Fig. 5). Aromatic hydrocarbons boiling in the gasoline range of petroleum are chiefly alkylbenzenes such as toluene, the xylenes, and *p*-cumene. Higher boiling fractions (above 200°C) contain both fused-ring (e.g., naphthalene) and linked-ring (e.g., biphenyl) compounds, and other polynuclear aromatic components having three or more fused rings. Polynuclear aromatic hydrocarbons have two or more directly united (fused) benzenoid rings and have chemical properties intermediate between those of benzene and olefinic hydrocarbons [19].

Another group of compounds found in petroleum may contain up to ten aromatic and naphthenic rings as well as aliphatic groups. These components are called the naphthenoaromatics (or mixed structures, sometimes described as hybrid hydrocar-

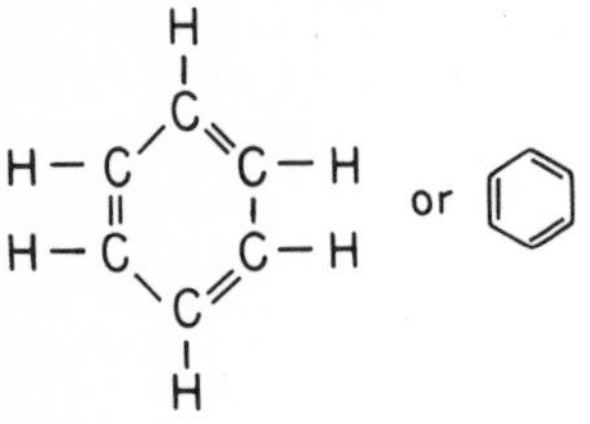

Benzene (C_6H_6)

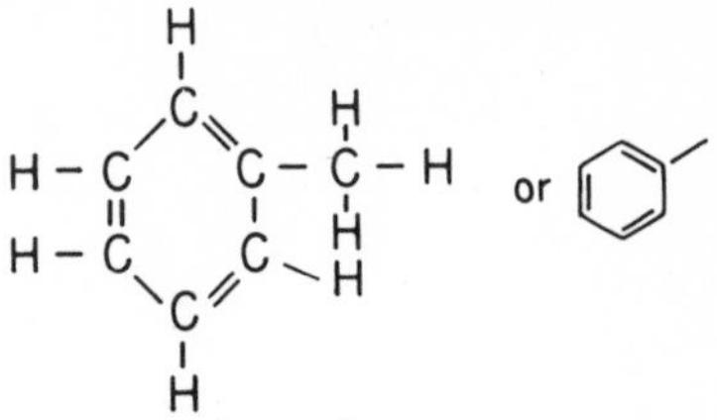

Toluene (C_7H_8)

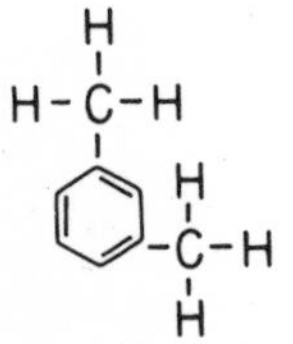

Xylene (C_8H_{10})
(meta-substituted)

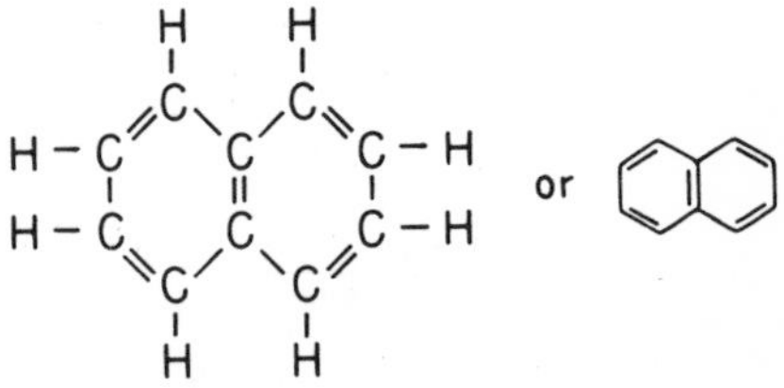

Naphthalene ($C_{10}H_8$)

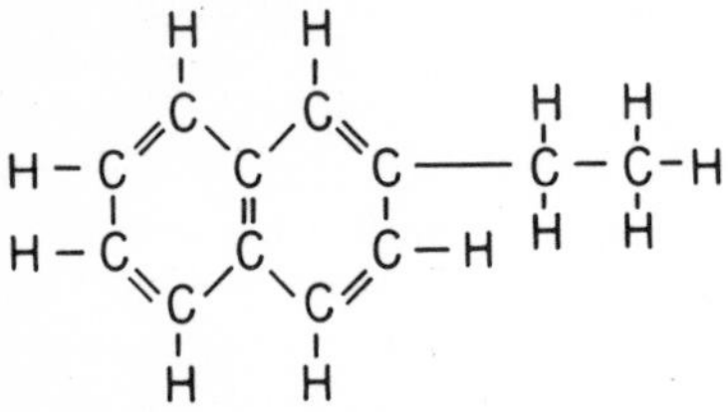

2-ethylnaphthalene ($C_{12}H_{12}$)

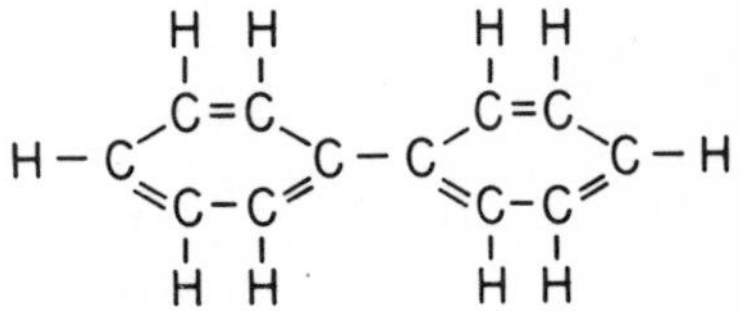

Biphenyl ($C_{12}H_{10}$)

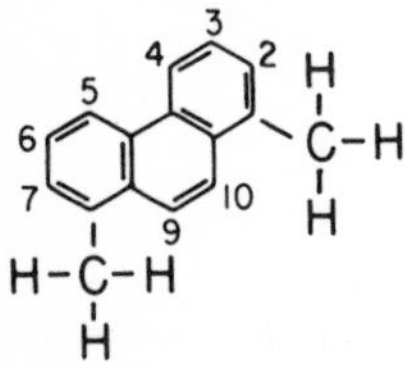

1,8-dimethylphenathrene

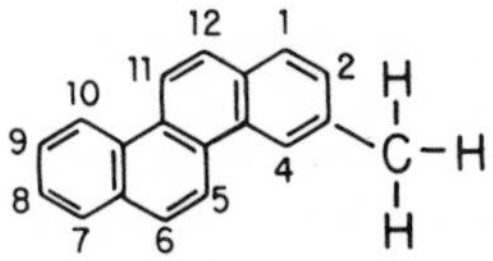

3-methylchrysene

Fig. 5 continued on next page

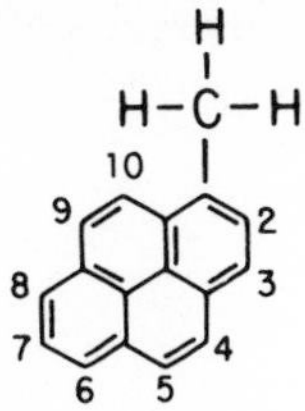

1-methylpyrene

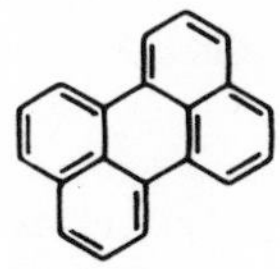

Perylene

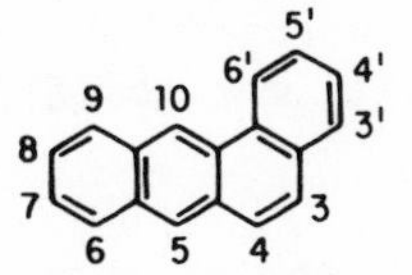

1,2-benzanthracene, benz [a] anthracene

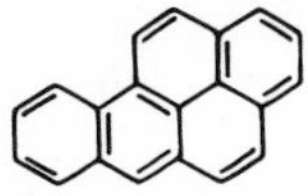

3,4-benzopyrene, benzo [a] pyrene

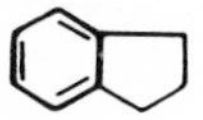

Indane (C_9H_{10})

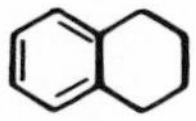

Tetralin ($C_{10}H_{12}$)

Fig. 5. Aromatic hydrocarbons found in petroleum.

bons). Indane, tetralin, and their homologs fall into this category. The naphthenoaromatics, together with the naphthenic hydrocarbon series, are the major constituents of higher boiling point petroleum fractions. Usually the different naphthenoaromatic components are classified according to the number of aromatic rings in their molecules. Their chemical structures and properties suggest that they may be related to resins, kerogen, and sterols. According to Bestougeff [5], it appears that practically all geochemical pathways leading to the formation of the different hydrocarbon classes found in crude petroleum involve, to some degree, the formation of naphthenoaromatic structures.

Non-Hydrocarbon Components

It is generally not appreciated to what extent crude petroleum may consist of non-hydrocarbons; a few crude oils may contain nearly 50% non-hydrocarbon components. The major non-hydrocarbons in crude petroleum are organic compounds containing nitrogen, sulfur, oxygen, and metals. Invariably these non-hydrocarbon components are concentrated during refining into the higher boiling fractions (above 350° to 400°C) [16].

Nitrogen Compounds

The total nitrogen content of petroleum can range from below detectable levels to about 0.9% by weight; most of these organic nitrogen compounds appear in the fraction boiling at 400°C and higher. Up to half of this content is in the form of basic, substituted pyridine and quinoline compounds, with the latter predominating. The non-basic nitrogen compounds include pyrroles, indoles, carbazoles, and benzcarbazoles. The nitrogen can also appear in combination with oxygen and certain metals in petroporphyrins (Fig. 6).

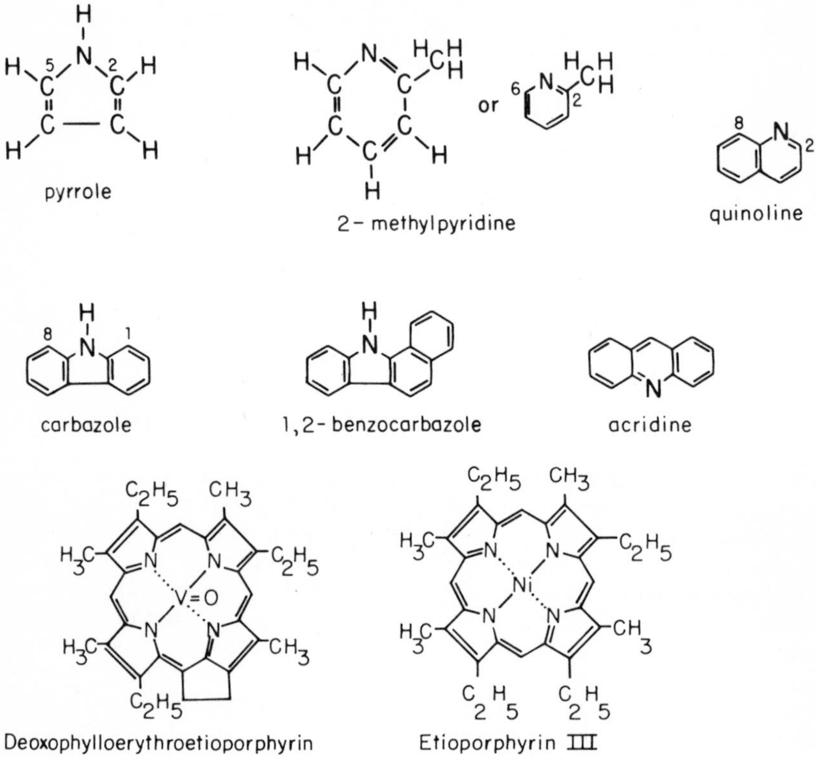

Fig. 6. Nitrogen compounds found in petroleum.

Sulfur Compounds

The total sulfur content of crude petroleum varies over a wide range, from traces to 5 to 6% by weight. For 93 major non-United States export crude oil blends, the sulfur content by weight ranges from 0.04 to 5.4% [14]. Sulfur may occur in crude oils as organic sulfide compounds, as hydrogen sulfide, as sulfur contained in aromatic structures, and as elemental sulfur. Petroleums with high hydrogen sulfide levels are termed "sour crudes." The presence of elemental sulfur may indicate that the crude petroleum had not been subjected to temperatures much higher than 100°C. However, most of the sulfur is present in organic compounds such as mercaptans (thiols), aliphatic sulfides, and cyclic sulfide compounds (Fig. 7). The mercaptans and sulfides exist as straight-chain compounds: for example, *n*-pentyl mercaptan and methyl ethyl sulfide, and as branched-chain compounds such as tertiary butyl mercaptan and methyl isopropyl sulfide. The cyclic sulfides consist of five- and six-membered ring compounds such as thiacyclopentanes and thiacyclohexanes [17].

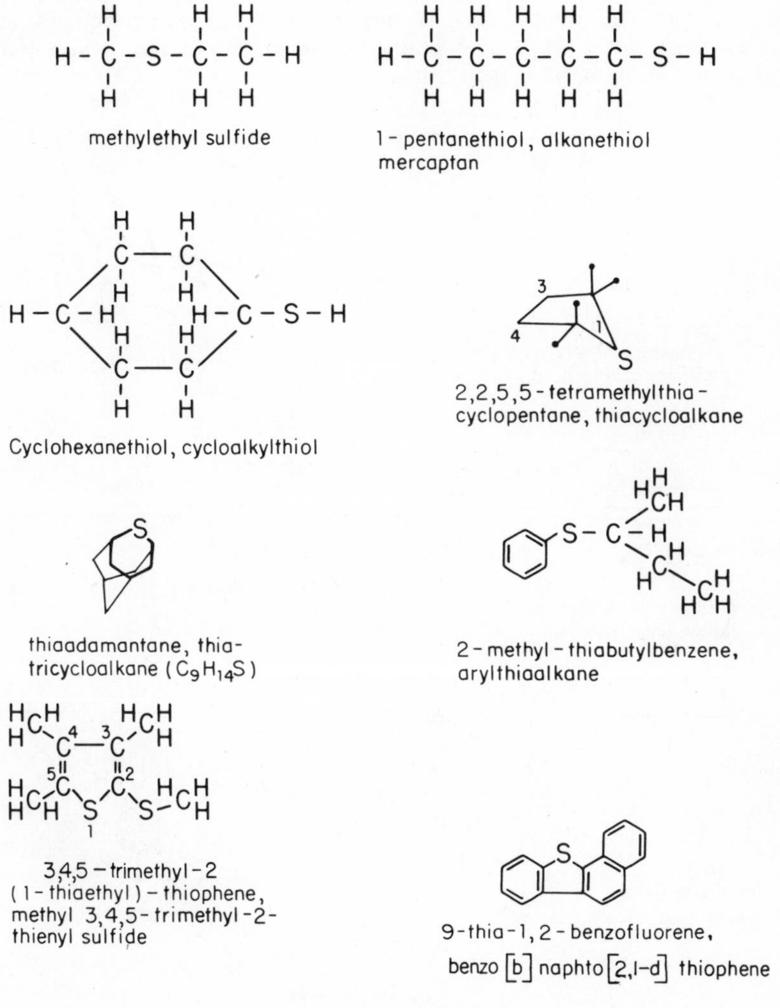

methylethyl sulfide

1-pentanethiol, alkanethiol mercaptan

Cyclohexanethiol, cycloalkylthiol

2,2,5,5-tetramethylthia-cyclopentane, thiacycloalkane

thiaadamantane, thia-tricycloalkane ($C_9H_{14}S$)

2-methyl-thiabutylbenzene, arylthiaalkane

3,4,5-trimethyl-2 (1-thiaethyl)-thiophene, methyl 3,4,5-trimethyl-2-thienyl sulfide

9-thia-1,2-benzofluorene, benzo [b] naphto [2,1-d] thiophene

Fig. 7. Sulfur compounds found in petroleum.

Oxygen Compounds

The oxygen content of petroleum is generally low but may reach as high as 2% in some crude oils. The oxygen content of the fractions of crude petroleum increases with boiling point and, as with other heteroatoms, the greater part of petroleum oxygen is found in distillation fractions boiling above 400°C [20]. The oxygen compounds consist principally of phenols and carboxylic acids. Ketones, esters, lactones, ethers, and anhydrides are known to be present. The carboxylic acids include the straight-chain and branched-chain acids containing up to 20 carbon atoms and also the derivatives of cyclopentane and cyclohexane. There is some indication that carboxylic acids containing mono- and dinuclear aromatic structures are also present [17,20]. The phenols comprise cresols and higher boiling alkyl phenols (Fig. 8).

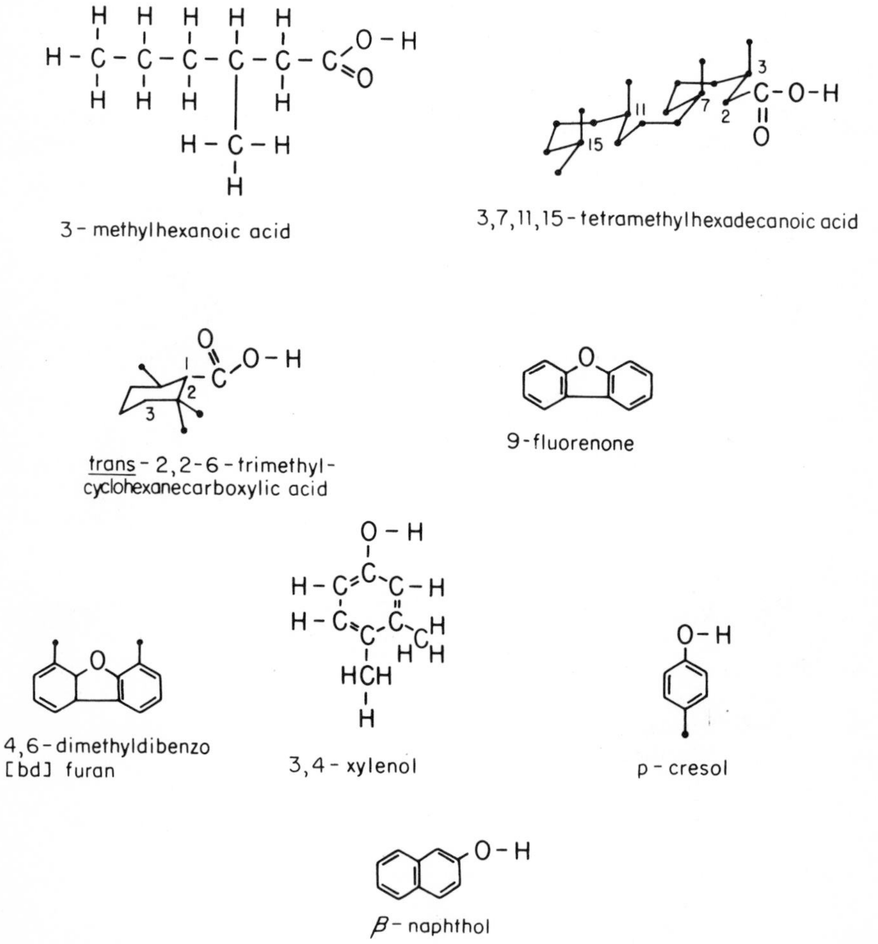

3-methylhexanoic acid

3,7,11,15-tetramethylhexadecanoic acid

trans-2,2-6-trimethylcyclohexanecarboxylic acid

9-fluorenone

4,6-dimethyldibenzo [bd] furan

3,4-xylenol

p-cresol

β-naphthol

Fig. 8. Oxygen compounds found in petroleum.

Metals and Organometallic Compounds

Over 50 different elements have been detected in crude petroleum; at least 40 of these are considered to be metals or have a metal-like character. Table 1 lists the elements found in crude oils.

Sodium and strontium may be present as aqueous solutions of salts finely dispersed in the petroleum; most of the other metals are present as oil-soluble salts of organic acids or as organometallic complexes. The chemical transition of naturally occurring pigments in extreme reducing conditions following geological burial, where metal exchange, decarboxylation, and hydrogenation reactions could take place, might yield the stable metalloporphyrins found in fossil fuels [20]. Nickel and vanadium, which are thought to be the most abundant porphyrin complexes, occur in the range of 0.03 to over 300 mg/l in crude petroleum. If it were possible to entirely recover the nickel and vanadium from crude oil, Smith et al. [22] have estimated that crude oil could provide up to 60% of the annual United States demand for nickel and nine times the annual demand for vanadium.

COMPOSITION OF CRUDE PETROLEUM

It is important to realize that crude petroleum is a complex mixture of many chemical compounds. There are a large number of different crude petroleums with widely differing chemical and physical properties; for example, there are heavy crude oils having a specific gravity near 1.0, a sulfur content of 5%, and very little material boiling below 270°C. On the other hand, there are light crude oils having virtually no elements other than carbon and hydrogen, a specific gravity of 0.8 or less, and practically all of the components distilling below 270°C [16].

Even crude petroleums from different wells in the same field or from different producing depths within the same well may show variation in their hydrocarbon components. No concise system has been developed for adequately classifying different petroleums. The chemist defines crude petroleum in terms of the chemical composition, that is, by the quantity of the various classes and types of hydrocarbons and other components. The refinery manager might define crude oil by the amount of the various types of commercial products that are obtained by fractional distillation and refining (Fig. 1 and Table 2).

Crude petroleums may be classified into three types in accordance with their general composition: (1) paraffin-base, (2) mixed-base, and (3) asphalt-base. The paraffin-base crude oils, such as those from Pennsylvania and Michigan, contain large amounts of paraffin wax and practically no asphalt. The

TABLE 1

Elements found in crude petroleum.

Element	Concentration range in crude petroleum (μg/l)	Element	Concentration range in crude petroleum (μg/l)
Hydrogen	As hydrocarbons	Gallium	11 to 810
Helium	Present	Germanium	$\leqq 100$
Beryllium	Trace	Arsenic	2 to 2×10^3
Boron	$(7.2 \text{ to } 104) \times 10^3$	Selenium	9 to 1.4×10^3
Carbon	As hydrocarbons	Bromine	72 to 2.2×10^3
Nitrogen		Rubidium	10 to 720
as N_2	2.8×10^3	Strontium	0.35 to 250
as NO_3^-	9.5×10^5	Yttrium	Trace
as organic	1.0×10^5 to 9.0×10^6	Zirconium	2.4 to 9.8
Oxygen	$\leqq 2.0 \times 10$	Molybdenum	1 to 7.3×10^3
Sodium	0.5 to 2.8×10^4	Silver	Trace
Magnesium	34 to 1.2×10^3	Cadmium	0.2 to 29
Aluminum	24 to 2.1×10^3	Indium	$\leqq 0.1$
Silicon	3.0×10^3 to 1.0×10^4	Tin	5×10^{-3} to 190
Phosphorus	$\leqq 9.8 \times 10^4$	Antimony	6 to 300
Sulfur	1.0×10^4 to 5.5×10^7	Iodine	10 to 9.0×10^3
Chlorine	1.1×10^3 to 1.0×10^6	Cesium	4 to 68
Argon	Present	Barium	0.2 to 308
Potassium	390 to 7.7×10^3	Lanthanum	0.03 to 39
Calcium	$\leqq 1.9 \times 10^5$	Cerium	Trace
Scandium	0.27 to 200	Neodymium	Trace
Titanium	0.4 to 230	Samarium	$\leqq 0.78$
Vanadium	24 to 1.3×10^6	Europium	0.6 to 23.2
Chromium	1.4 to 690	Dysprosium	Trace
Manganese	0.14 to 3.8×10^3	Rhenium	$\leqq 200$
Iron	20 to 1.2×10^5	Gold	2.4×10^{-2} to 3.0
Cobalt	1.3 to 1.4×10^4	Mercury	23 to 3.0×10^4
Nickel	27 to 3.4×10^5	Lead	0.5 to 430
Copper	3.5 to 6.3×10^3	Bismuth	Trace
Zinc	3.5 to 1.6×10^5	Uranium	1×10^{-2} to 434

Adapted from Clark [21].

TABLE 2

Hydrocarbon classes, types, and structures found in petroleum refined products

Class	Type	Molecular structure	
		Arrangement of carbon atoms	Characteristic bonding
Aliphatic	Paraffin (alkanes)	Open chain, straight or branched	Saturated
	Olefins (alkenes)	Open chain, straight or branched	Unsaturated
Alicyclic	Cycloparaffins	Closed chain (cyclic or ring)	Saturated
	Cyclo-olefins	Closed chain (cyclic or ring)	Unsaturated
Aromatic	(various types)[a]	Closed chain (benzenoid ring)	Unsaturated[b]

Information adopted from Bureau of Naval Personnel [15].
Not all classes or types of hydrocarbons are found in any one crude oil or specific refined product.

a Includes types such as benzene, alkylbenzenes, cycloalkylbenzenes, fused-ring polynuclear hydrocarbons (naphthalene), and polynuclear aromatic hydrocarbons.

b The benzene ring contains three conjugated double bonds, however these bonds do not show the reactive character associated with alkenes or cycloalkenes.

mixed-base crude oils, such as those of the Mid-Continent United States and the Middle East, contain both paraffin wax and asphalt. The asphalt-base or naphthenic crude petroleums, such as those of the United States Gulf Coast, California, and Venezuela, contain asphalt but almost no paraffin wax. Asphalt is the residue from the fractional distillation of crude oil and consists of a complex mixture of high-boiling (>540°C) aromatic, naphthenic, and non-hydrocarbon compounds. It is a strong cement, readily adhesive, highly waterproof, and durable and is highly resistant to most alkalies, acids, and salts [15].

Data on the chemical composition of a crude petroleum from one oil producing region does not necessarily define the chemical nature of all crude oils from that region. Nevertheless, such chemical analyses serve as a basis for comparing the representative types of crude petroleum produced in different regions. The following is a general description of these major crude oils from three different producing regions of the world.

The Oil and Gas Journal [14] and the International Petroleum Encyclopedia [13] have published a guide to world crude oil export streams listing the major physical characteristics and chemical properties. This listing is of non-United States export streams which may be blends of crude oils from several individual oil fields that are shipped from major production centers; many, if not all, of these crude oil streams are moved by tankers over the major ocean shipping lanes (Fig. 1 of Chapter 2).

Prudhoe Bay Crude Oil

The crude petroleum producing area of Prudhoe Bay, Alaska, is located along the north coast of the state facing the Arctic Ocean. It covers only about 120 km^2 (46 mi^2), but is thought to contain the largest known petroleum reservoir in North America. The estimates of recoverable reserves are: from 10 to 20 billion bbl (420 to 840 billion gal or 1,590 to 3,180 billion l) of crude oil and from 680 to 790 billion m^3 (24 to 28 trillion ft^3) of natural gas. The petroleum is contained in the Sadlerochit sand formation which is from 60 to 90 m (200 to 300 ft) thick and is located at depths from 1,500 to 2,700 m (5,000 to 9,000 ft).

General characteristics of Prudhoe Bay crude oil, determined by a Bureau of Mines routine analysis, show it to be an intermediate gravity, high-sulfur, high-nitrogen oil that contains a high percentage of aromatics. It is brownish-black with a high carbon-residue and asphalt content and a pour point of -9°C (15°F) [23].

Extensive onshore explorations for petroleum have taken place in the Arctic region in Alaska on the North Slope near Prudhoe Bay, in Canada along the Mackenzie River Delta and on the Arctic Islands, and in the USSR near the Kara Sea and near the Sea of Okhotsk. Extensive offshore development and production are taking place in the North Sea off Scotland and Norway and in Cook Inlet in south-central Alaska. Large-scale production of petroleum has not been realized yet in either the onshore or offshore areas of the Arctic. As the technology of oil drilling advances, exploration in the ice-covered areas of the Arctic undoubtedly will increase [13].

The physical characteristics and chemical properties of a typical Prudhoe Bay crude petroleum are given in Table 3.

South Louisiana Crude Oil

In order to compare the Prudhoe Bay crude oil characteristics and properties with other oils, a South Louisiana crude petroleum is used as an example of a typical major Gulf of Mexico crude oil (Table 4); variations in crude oil characteristics and properties can be expected between this crude oil and those from other production formations or nearby Gulf Coast oil fields. This crude oil has been transported extensively by tankers along the coastal waters of the eastern

TABLE 3

Typical Prudhoe Bay crude oil analysis

Characteristic or component	Unit	Crude oil	Natural gas	Naphtha		Middle Distillate	Wide-cut gas oils	Residuum
				Gasoline	Kerosine			
Boiling point range	°C	---	<20	20-190	190-205	205-343	343-565	565+
Specific gravity	(15°C)	0.8883	---	0.7531	0.818	0.8581	0.9279	1.0231
API gravity	°API	27.8	---	56.9	41.5	33.4	21.0	6.8
Pour point	°C	-10	---	---	<-60	-23	35	52
Viscosity, Saybolt (38°C)[a]	sec	73.5	---	---	---	36.1	85-200	>200
Kinematic (38°C)	cSt	14.0	---	---	---	3.05	>30	---
Yield: Crude oil	vol%	100	3.08	18.0	2.1	24.6	35.0	17.6
Paraffins	vol%	27.3	100	47.3	41.9	8.9	9.3	
Naphthenes	vol%	36.8	0	36.8	38.1	14.4	22.8	
Aromatics	vol%	25.3	0	15.9	20.0	76.7[b]	67.9[b]	
Others[c]	vol%	10.6	0	0	0			
Composition:								
Sulfur	wt%	0.94	---	0.011	0.04	0.34	1.05	2.30
Mercaptan sulfur	ppm	20	---	5	---	---	---	---
Nitrogen	wt%	0.23	---	0.02	0.02	0.04	0.16	0.68
Oxygen	wt%	0.01	---	---	---	---	---	---
Vanadium	ppm	18	0	0	0	0	<1	93
Nickel	ppm	10	0	0	0	0	<1	46
Iron	ppm	4	0	0	0	0	<1	25

Data from Thompson et al. [23] *and Coleman et al.* [24].

a Saybolt viscosity: the time in seconds for 60 ml of a sample to flow through a calibrated Universal orifice under specified conditions, according to ASTM method D-88 [10:part 23].

b Includes naptheno-aromatic compounds and nonhydrocarbons.

c Polar compounds, non-volatile aromatic hydrocarbons, and column holdup in fractions boiling about 205°C.

TABLE 4

Physical characteristics and chemical properties of several crude oils

Characteristic or component	Crude oil		
	Prudhoe Bay[a]	South Louisiana[b]	Kuwait[b]
API gravity (20°C)(°API)	27.8	34.5	31.4
Sulfur (wt%)	0.94	0.25	2.44
Nitrogen (wt%)	0.23	0.69	0.14
Nickel (ppm)	10	2.2	7.7
Vanadium (ppm)	20	1.9	28
Naphtha fraction[c] (wt%)	23.2	18.6	22.7
Paraffins	12.5	8.8	16.2
Naphthenes	7.4	7.7	4.1
Aromatics	3.2	2.1	2.4
Benzenes	0.3[d]	0.2	0.1
Toluene	0.6	0.4	0.4
C_8 aromatics	0.5	0.7	0.8
C_9 aromatics	0.06	0.5	0.6
C_{10} aromatics	-	0.2	0.3
C_{11} aromatics	-	0.1	0.1
Indans	-	-	0.1
High-boiling fraction[e] (wt%)	76.8	81.4	77.3
Saturates	14.4[f]	56.3	34.0
n-paraffins	5.8[g]	5.2	4.7
C_{11}	0.12	0.06	0.12
C_{12}	0.25	0.24	0.28
C_{13}	0.42	0.41	0.38
C_{14}	0.50	0.56	0.44
C_{15}	0.44	0.54	0.43
C_{16}	0.50	0.58	0.45
C_{17}	0.51	0.59	0.41
C_{18}	0.47	0.40	0.35
C_{19}	0.43	0.38	0.33
C_{20}	0.37	0.28	0.25
C_{21}	0.32	0.20	0.20
C_{22}	0.24	0.15	0.17
C_{23}	0.21	0.16	0.15
C_{24}	0.20	0.13	0.12
C_{25}	0.17	0.12	0.10
C_{26}	0.15	0.09	0.09
C_{27}	0.10	0.06	0.06
C_{28}	0.09	0.05	0.06
C_{29}	0.08	0.05	0.05
C_{30}	0.08	0.04	0.07
C_{31}	0.08	0.04	0.06
C_{32} plus	0.07	0	0.06
Iso-paraffins	-	14.0	13.2
1-ring cycloparaffins	9.9	12.4	6.2
2-ring cycloparaffins	7.7	9.4	4.5
3-ring cycloparaffins	5.5	6.8	3.3
4-ring cycloparaffins	5.4	4.8	1.8
5-ring cycloparaffins	-	3.2	0.4
6-ring cycloparaffins	-	1.1	-

continued on next page

TABLE 4 *continued*

Characteristic or component	Crude oil		
	Prudhoe Bay[a]	South Louisiana[b]	Kuwait[b]
Aromatics (wt%)	25.0	16.5	21.9
Benzenes	7.0	3.9	4.8
Indans and tetralins	-	2.4	2.2
Dinaphtheno benzenes	-	2.9	2.0
Naphthalenes	9.9	1.3	0.7
Acenaphthenes	-	1.4	0.9
Phenanthenes	3.1	0.9	0.3
Acenaphthalenes	-	2.8	1.5
Pyrenes	1.5	-	-
Chrysenes	-	-	0.2
Benzothiophenes	1.7	0.5	5.4
Dibenzothiophenes	1.3	0.4	3.3
Indanothiophenes	-	-	0.6
Polar materials[h] (wt%)	2.9	8.4	17.9
Insolubles[i] (wt%)	1.2	0.2	3.5

These analyses represent values for one typical crude oil from each of the geographical regions; variations in composition can be expected for oils produced from different formations or fields within each region.

a *Adapted from Thompson et al. [23] and Coleman et al. [24].*

b *From Pancirov [25].*

c Fraction boiling from 20° to 205°C.

d Reported for fraction boiling from 20° to 150°C.

e Fraction boiling above 205°C.

f Reported for fraction boiling above 220°C.

g Prudhoe Bay crude oil weathered two weeks to duplicate fractional distillation equivalent to approximately 205°C *n*-
percents from gas chromatography over the range C_{11} to C_{32} plus for the Prudhoe Bay crude oil sample only. Unpublished data (R.C. Clark, Jr.)

h Polar material: Clay-gel separation according to ASTM method D-2007 [10; part 24] using pentane on unweathered sample.

i Insolubles: Pentane-insoluble materials according to ASTM method D-893 [10; part 23].

United States. It is part of a suite of four reference oils (two crude oils and two refined products) provided by the American Petroleum Institute in the early 1970's for biological effects studies [26].

Kuwait Crude Oil

The Kuwait crude petroleum was selected as an example of an oil transported by tankers along major ocean shipping lanes, especially from the Middle East to western Europe and to Japan. This petroleum is the second crude oil in the API suite of reference oils (Table 4).

COMPOSITION OF REFINED PETROLEUM PRODUCTS

In refining crude petroleum into the desired products, four processes are performed based on the physical characteristics and chemical composition of the crude oil: (1) direct separation by distillation of the oil to recover the desired products; (2) breaking the remaining large chemical compounds into smaller chemical compounds by cracking; (3) building up the desired chemical compounds by chemical reactions, such as polymerization, reforming, alkylation, and isomerization; and (4) blending these components to produce refined products with the desired characteristics. Refined products such as gasoline, kerosine, diesel oil, and others are mixtures of chemical compounds that are prepared by combining certain fractional distillates and various products manufactured in the refining process. A list of the fractions of crude petroleum and the products prepared from them are shown in Figure 9 and Table 5.

The terminology for refined petroleum products among countries differs considerably; thus, it is often difficult to properly identify the refined products and to make comparisons of their composition. Furthermore, petroleum products are formulated for specific commercial uses. These formulations will reflect the original chemical nature of the crude oil, the variation in refinery plant operations, and the blending practices. Gasoline, for example, is a special type of fuel formulated for use in automobiles; its chemical components, however, will vary from sample to sample, a reflection of the geochemical and refinery processes involved in its production.

A typical Venezuelan crude oil, defined in terms of product yield and hydrocarbon type, contains by weight [27]:

Fractions:

Gasoline	(C_5-C_{10})	10%
Kerosine	$(C_{10}-C_{12})$	5%
Light distillate oils	$(C_{12}-C_{20})$	20%
Heavy distillate oils	$(C_{20}-C_{40})$	30%
Residuum	$(>C_{40})$	35%

Compound Types:

Paraffins	10%
Naphthenes	45%
Aromatics	25%
Non-hydrocarbons	20%

TABLE 5

Refinery fractions by hydrocarbon types from crude petroleum

Distillation fraction	Approximate boiling range (°C)	Hydrocarbon types	Range of carbon atoms	Typical refined products
Natural gas	<20	Paraffins	1 - 6	Natural gas
Gasoline & naphtha	20 - 200	Paraffins Aromatics Naphthenes	4 - 12	Gasoline
Middle distillate	185 - 345	Paraffins Aromatics Naphthenes	10 - 20	Kerosine Jet fuel Heating oils Diesel oils
Wide-cut gas oil	345 - 540	Paraffins Aromatics Naphthenes	18 - 45	Catalytic Cracking Feed stock Lube oils, wax
Residuum	>540	Complex aromatic and naphthenic compounds	>40	Residual oils Asphalt Coke

Adapted from Bureau of Naval Personnel [15], Encyclopedia of Science and Technology [17], and National Academy of Sciences [27].

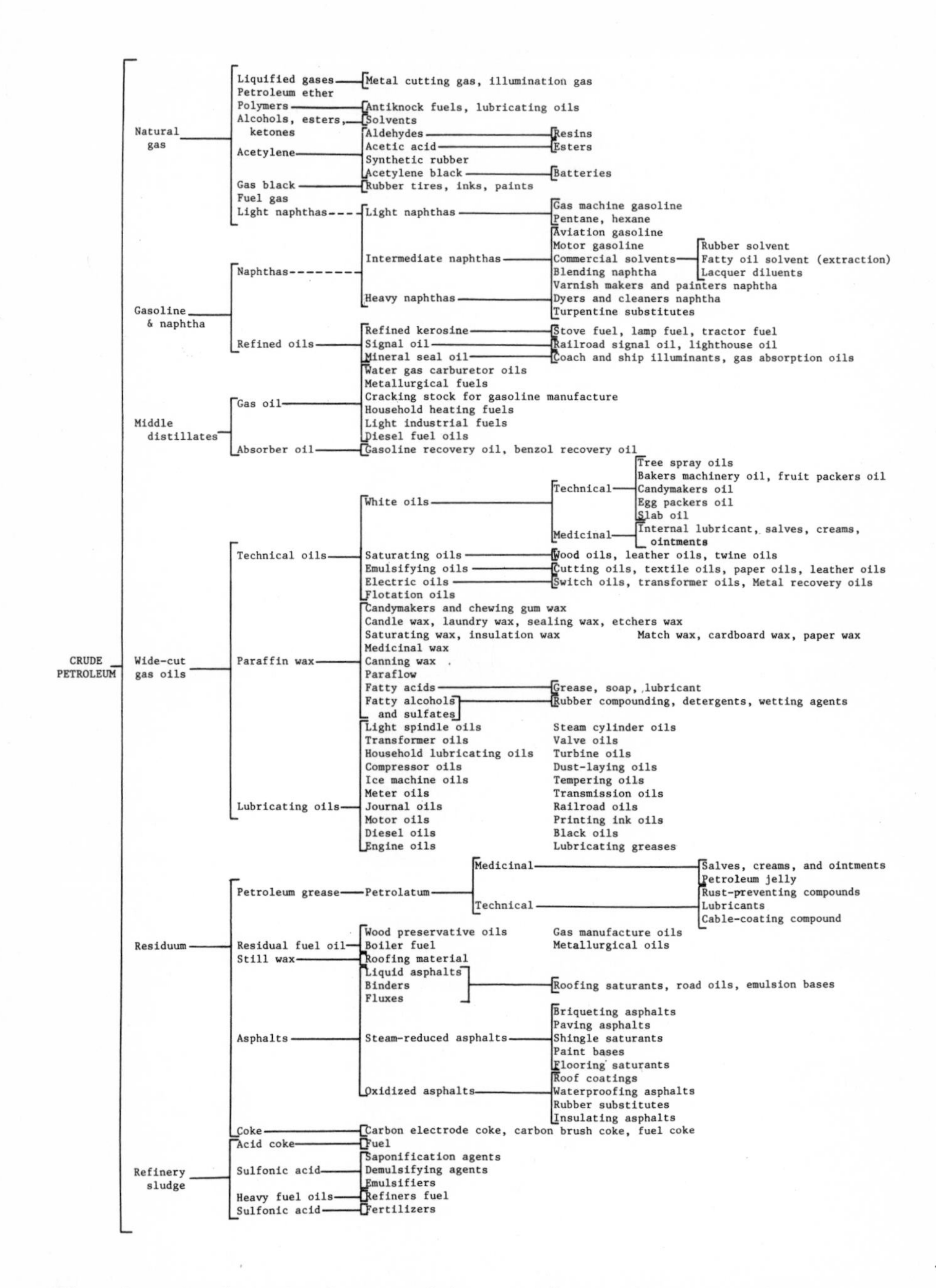

Fig. 9. Crude petroleum and its products (From P. Albert Washer, Texas A & M College Extension Division. (Adapted from McGraw-Hill Encyclopedia of Science and Technology [17]. Used with permission of McGraw-Hill Book Company.)

No. 2 Fuel Oil

No. 2 fuel oils, or diesel oils, are produced from the middle distillate fraction of crude petroleum. These oils are usually blends of virgin stock (distillate fractions directly from the initial crude oil distillations) and catalytically- or thermally-cracked products from refinery processing virgin stocks. Some of these fuel oils are prepared entirely of virgin (or straight-run) stocks. In North America, a 1973 refinery would generally produce about 30 bbl (1 bbl = 42 U.S. gal) of No. 2 fuel oil from every 100 bbl of crude oil processed [13]. A typical 1976 refinery would produce up to 50% gasoline, 15 to 20% No. 2 fuel oil, and the remainder as jet fuels, lubricating oil stocks, petro-chemical feeds, and residual fuel oils [7].

The specific gravities of No. 2 fuel oils range from 0.82 to 0.85, the viscosity is about 40 cP, and the pour point is about -20°C. A typical No. 2 fuel oil would contain about 30% paraffins, 45% naphthenes, and 25% aromatics. These hydrocarbons would contain from 12 to 25 carbon atoms, with a predominance of 15 to 17 carbon atoms [27].

Data on the physical properties and chemical composition of an unusually high-aromatic content No. 2 fuel oil is presented in Table 6. The sample is one of the four petroleum materials made available by the American Petroleum Institute as reference oils. This No. 2 fuel oil is a blend of virgin stock and cracked products from the intermediate distillate of petroleum [26].

Other fuel oil products, such as kerosine and gasoline, are prepared from virgin stocks (light and intermediate distillates) and cracked products. A typical kerosine has a specific gravity of about 0.80 and a viscosity of 1 to 2 cP. It contains about 35% paraffins, 50% naphthenes, and 15% aromatics; these hydrocarbons contain from 10 to 12 carbon atoms [27]. A typical gasoline has a specific gravity of about 0.70 and a viscosity of about 1 cP. Gasoline prepared from virgin stock contains about 50% paraffins, 40% naphthenes, and 10% aromatics; these hydrocarbons contain from 5 to 10 carbon atoms. Blended gasolines, prepared from virgin stocks and cracked products, contain about 20 to 30% aromatics [27].

Bunker C Fuel Oil

Bunker C fuel oil (No. 6 fuel oil) is the highest boiling fraction of the heavy distillates from petroleum. It is used primarily for firing steam boilers in the generation of electrical power, for large-scale heating, and for powering marine vessels. Bunker C fuel oils represent about 5 to 8% of the original crude petroleum, but the yield will often depend upon the source of petroleum, refinery design and operations,

TABLE 6

Physical characteristics and chemical properties of two refined products

Characteristic or component	No. 2 fuel oil[a]	Bunker C fuel oil
API gravity (20°C) *(°API)*	31.6	7.3
Sulfur *(wt%)*	0.32	1.46
Nitrogen *(wt%)*	0.024	0.94
Nickel *(ppm)*	0.5	89
Vanadium *(ppm)*	1.5	73
Saturates *(wt%)*	61.8	21.1
n-paraffins	8.07	1.73
$C_{10} + C_{11}$	1.26	0
C_{12}	0.84	0
C_{13}	0.96	0.07
C_{14}	1.03	0.11
C_{15}	1.13	0.12
C_{16}	1.05	0.14
C_{17}	0.65	0.15
C_{18}	0.55	0.12
C_{19}	0.33	0.14
C_{20}	0.18	0.12
C_{21}	0.09	0.11
C_{22}	0	0.10
C_{23}	0	0.09
C_{24}	0	0.08
C_{25}	0	0.07
C_{26}	0	0.05
C_{27}	0	0.04
C_{28}	0	0.05
C_{29}	0	0.04
C_{30}	0	0.04
C_{31}	0	0.04
C_{32} plus	0	0.05
Iso-paraffins	22.3	5.0
1-ring cycloparaffins	17.5	3.9
2-ring cycloparaffins	9.4	3.4
3-ring cycloparaffins	4.5	2.9
4-ring cycloparaffins	0	2.7
5-ring cycloparaffins	0	1.9
6-ring cycloparaffins	0	0.4
Aromatics *(wt%)*	38.2	34.2
Benzenes	10.3	1.9
Indans and tetralins	7.3	2.1
Dinaphtheno benzenes	4.6	2.0
Naphthalene	0.2[b]	
Methylnaphthalenes	2.1[b]	2.6
Dimethylnaphthalenes	3.2[b]	
Other naphthalenes	0.4	
Acenaphthenes	3.8	3.1
Acenaphthalenes	5.4	7.0
Phenanthrenes	0	11.6
Pyrenes	0	1.7
Chrysenes	0	0
Benzothiophenes	0.9	1.5
Dibenzothiophenes	0	0.7
Polar materials[c] *(wt%)*	0	30.3
Insolubles (pentane)[c] *(wt%)*	0	14.4

These analyses represent typical values for two different refined products; variations in composition can be expected for similar materials from different crude oil stocks and different refineries. *From Pancirov [25].*

[a] This is a high aromatic material; a typical No. 2 fuel oil would have an aromatic content closer to 20-25%. *From Vaughan [26].*

[b] *From Vaughan [26].*

[c] See footnotes *h* and *i* for Table 4.

and product requirements.

Analysis of a typical Bunker C fuel oil is given in Table 6. This oil was refined in the United States from a Venezuelan crude oil; it is one of the four petroleum reference materials provided by the American Petroleum Institute. A typical Bunker C fuel oil has a specific gravity of about 1.00, a viscosity of about 1,000 cP (at 38°C), and a pour point of about 21°C. It contains about 15% paraffins, 45% naphthenes, 25% aromatics, and 15% non-hydrocarbon compounds; the hydrocarbons contain 30 and greater carbon atoms [27].

METHODS OF ANALYSIS FOR PETROLEUM HYDROCARBONS

The effects of petroleum on the marine environment and organisms may involve complex physical, chemical, and biological reactions. A study of these effects requires experts trained in a wide variety of disciplines including chemistry, biology, ecology, geology, and engineering. To carry out such studies, comprehensive data are needed on the nature and on the amount and distribution of petroleum and petroleum hydrocarbons in marine systems. The analytical procedures for the identification and quantification of petroleum hydrocarbons in marine systems involve four basic steps:

1. Collection and preservation of samples.
2. Extraction (including saponification of the fats) of the organic matter from the cellular matrix of organisms, from the inorganic matrix of sediments, and from the dissolved and particulate fractions in seawater.
3. Separation of the petroleum hydrocarbons from the lipoid material by chromatography.
4. Identification and quantification of the petroleum hydrocarbons.

The analytical procedures used to determine hydrocarbons in marine systems must be designed on the basis of the nature and type of sample, the nature and type of hydrocarbon components to be studied, the equipment available, and the type of information required. Flow diagrams that depict some analytical systems that have been used are given in Figure 10 for saturated hydrocarbons in aquatic organisms [28], in Figure 11 for hydrocarbon compound types in marine organisms [29], in Figure 12 for hydrocarbon compound types in water [27], in Figure 13 for saturated hydrocarbons in sediment [30], and in Figure 14 for aromatic hydrocarbons in sediment [31]. A discussion of the reasons and principles for each step in the analytical system for determining hydrocarbons in marine samples was given by Farrington and Meyers [32] and by Farrington et al. [33].

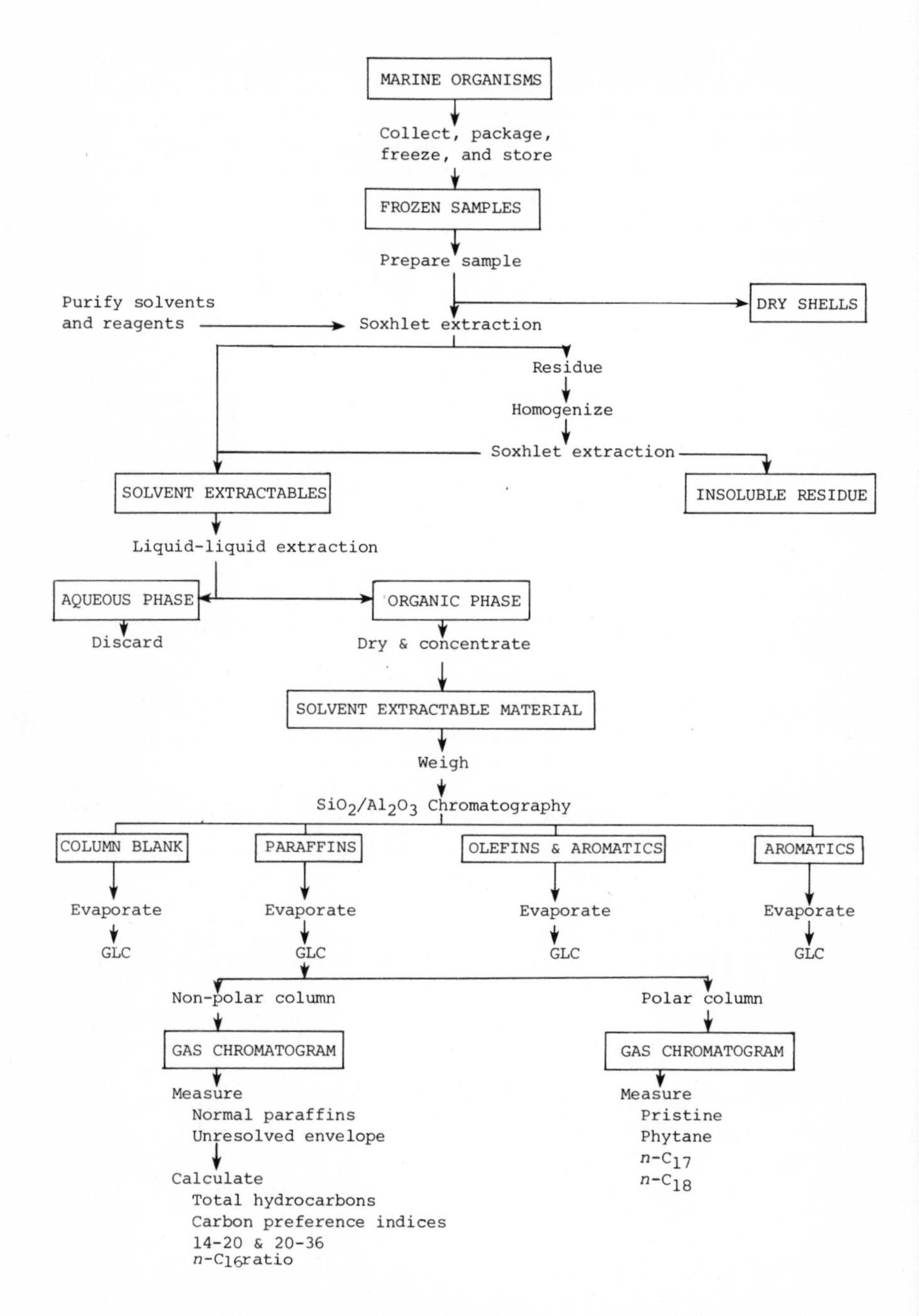

Fig 10. Flow diagram for analysis of saturated hydrocarbons in marine orgamisms. (Adapted from Clark & Finley [28]; refer to original paper for details and definitions of parameters.)

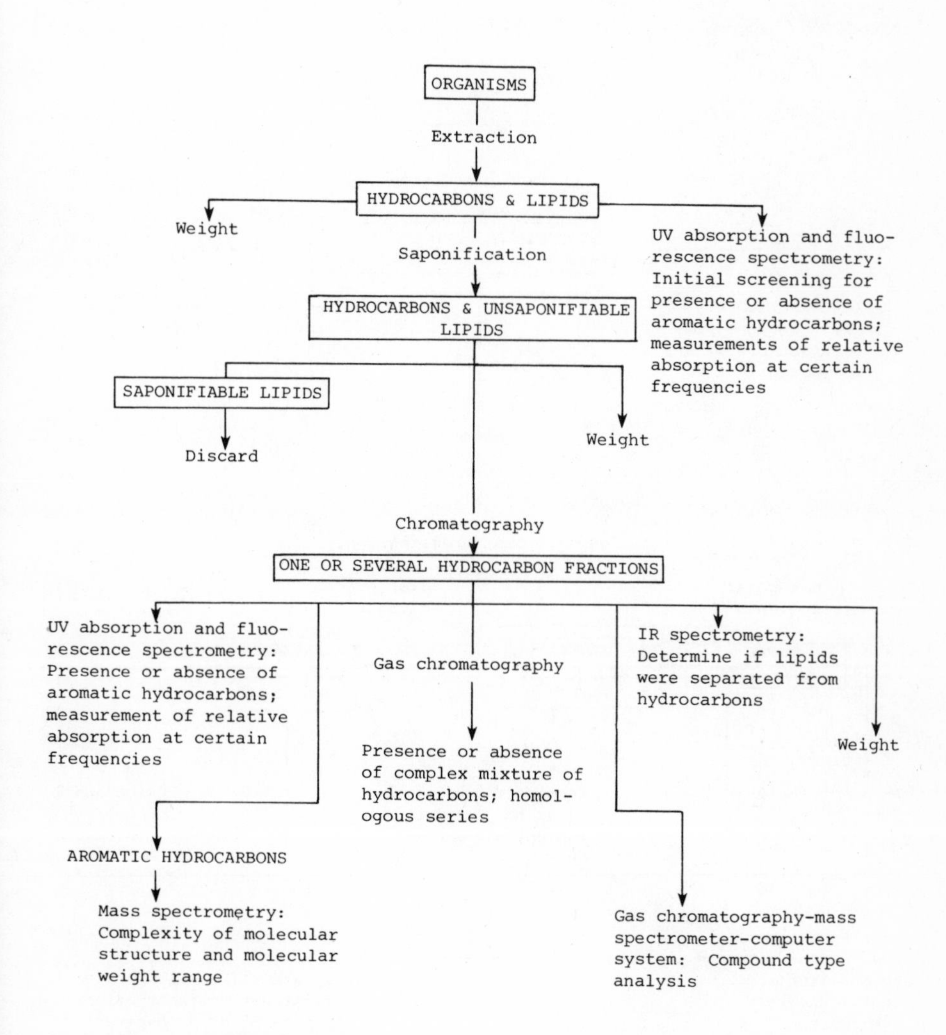

Fig. 11. Flow diagram for analytical techniques to detect and estimate petroleum contamination in marine organisms. (Adapted from Farrington [29].)

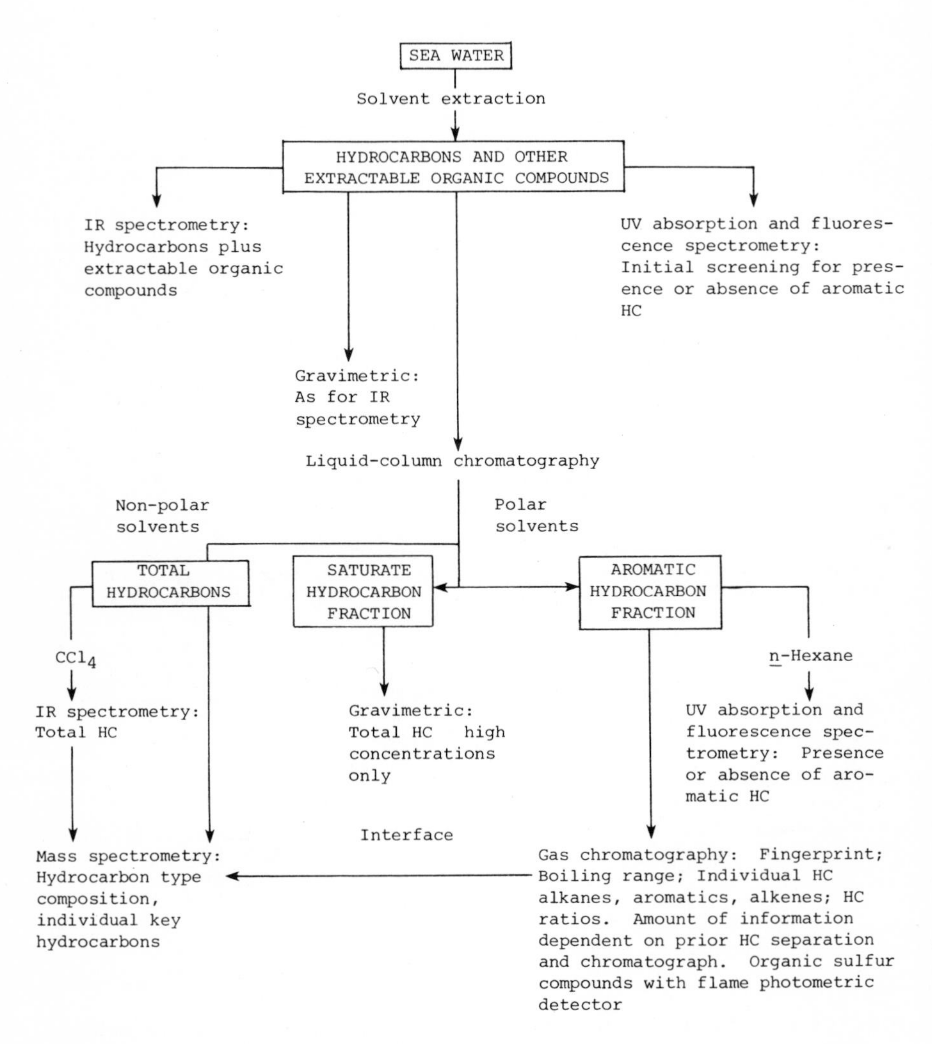

Fig. 12. Flow diagram for analytical techniques to measure C_{11}+ hydrocarbons in seawater. (Adapted from National Academy of Sciences [27].)

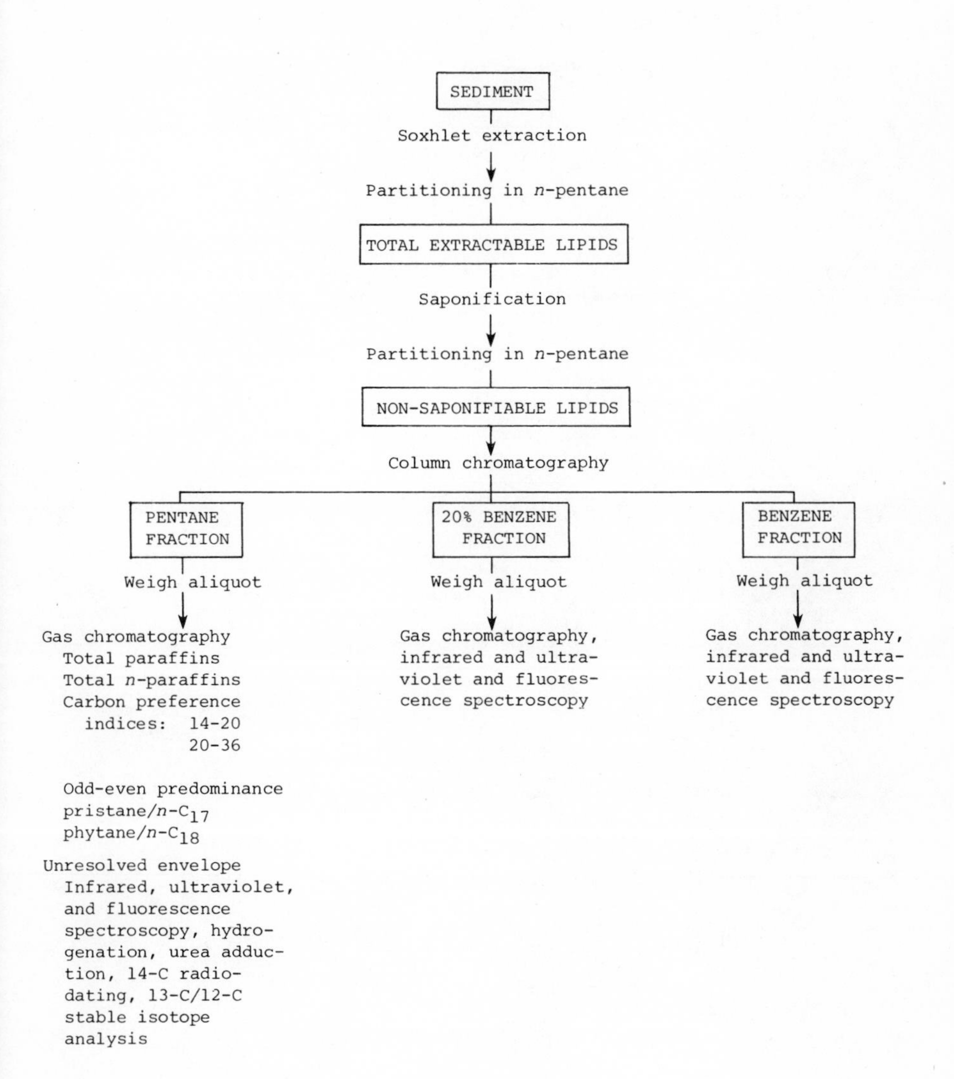

Fig. 13. Flow diagram for the analysis of hydrocarbons in sediments. (Not all procedures outlined apply to all samples. Adapted from Wakeham [30]; refer to original paper for details and definition of parameters.)

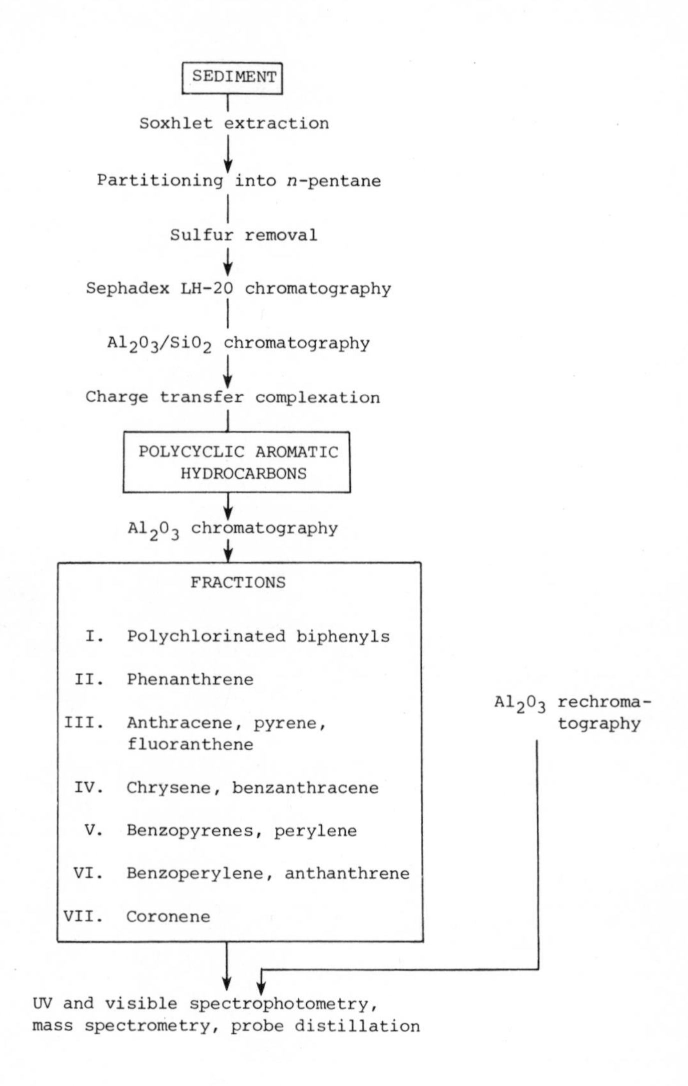

Fig. 14. Flow diagram for the analysis of polycyclic aromatic hydrocarbons in sediments. (Adapted from Giger & Blumer [31]. Used with permission from Analytical Chemistry. Copyright by The American Chemical Society.)

Because of the ubiquitous nature of hydrocarbons and the many unavoidable sources of contamination, each analytical method must provide for the establishment of background values for hydrocarbons inherent to the system. These so-called blank values represent the hydrocarbon levels contributed by the analytical system over and above that in the sample under investigation. Basically, these values reflect the hydrocarbon contamination contributed by the equipment, apparatus, chemicals, and atmosphere, and the artifacts resulting from the limitations of the particular instrumentation. The analysis of blanks must be carried out exactly in the same manner and along with the samples. Usually, one blank for every seven to twelve samples is sufficient [28,33].

Very few interlaboratory studies have been carried out to establish the precision and accuracy of various analytical procedures for determining hydrocarbons in the marine environment. Interlaboratory calibration studies on hydrocarbons in the marine system were made as part of the International Decade of Ocean Exploration Baseline Program for Hydrocarbon Analysis [34,35]. The National Bureau of Standards is currently in the process of determining what standardized reference materials would prove useful for investigations of petroleum contamination in the marine environment. Additional collaborative studies on analytical procedures are needed to establish a comparative basis for evaluating data on the distribution and levels of hydrocarbons in the marine environment.

SAMPLE COLLECTION AND PRESERVATION

Hydrocarbon compounds are common residents of the environment and of the laboratory. Thus, any sampling program must be designed not only to obtain representative samples but also to avoid the many contaminants that may be introduced at any stage of collecting, preserving, and storage of samples. Even the slightest contamination of certain samples may introduce errors sufficient to completely nullify the results. For example, measurement of certain hydrocarbon components in seawater are made to a sensitivity of one nanogram (10^{-9} g); contamination by a single fingerprint accidentally left on the inside of a piece of glassware could add several micrograms (10^{-6} g) of hydrocarbons, introducing a substantial error in the results [28,36].

Laboratory Aspects

Containers

Glass bottles with screw-on lids are suitable for storing samples. The wax-coated liner of the lid should be replaced with aluminum foil or better yet, with solvent-extracted

Teflon. Aluminum or stainless steel containers may also be used. Other types of containers, including those made of plastic, rubber, fiberglass, wood, paper, and unglazed ceramics, must be considered contaminated and unsuitable [10: Method D-3324, in part 31].

All containers should be cleaned with detergent and hot water and then rinsed with clean solvents, first with acetone then a second solvent such as pentane, hexane, or methylene chloride. Care must be exercised to avoid contamination by materials which are common to a laboratory and which may come in contact with the containers. These are lubricants (silicones, plasticizers--phthalate and sebacate esters), reagent-grade chemicals, solvents, rubber gloves, and even polluted air [37].

Storage

Generally, all samples (water, sediment, organisms) should be frozen immediately after they are collected if extraction and analysis cannot be started immediately. If freezing facilities are not readily available, or it is impractical to freeze the material (such as large water samples in glass containers), the samples should be chilled, but, where possible, should be frozen at the earliest opportunity. Chilling and freezing will help reduce any losses of hydrocarbons that might occur due to volatilization and will minimize any changes that might develop through enzymatic modification of the sample components. Solid samples (organisms) can be wrapped well in aluminum foil or packaged in suitable containers to prevent contamination. Storage temperatures should be -20°C or lower. Additional procedures are given in ASTM Method D-3324 [10: part 31].

Field Aspects

Intertidal Samples

Intertidal organisms such as clams, oysters, mussels, snails, shore crabs, starfish, and seaweed are conveniently collected at low tide, labeled, and sealed in heavy aluminum foil. Subtidal organisms collected by divers can be enclosed in containers underwater so that they are not contaminated when the samples are passed through the hydrocarbon-rich film at the sea surface.

Oceanic Samples

The air-sea interface is a major source of contamination common to all oceanic sampling. This microlayer boundary is relatively rich in hydrocarbons, fatty acids, and pesticides [38,39]. Surface-active compounds and compounds whose molecules contain both hydrophobic and hydrophilic components tend to concentrate in the surface boundary [40]. For example, the surface layer of ocean water collected in the Gulf of Mexico off Florida and Louisiana contained on the average about 35 μg of alkanes per square meter of surface; the total saturated hydrocarbon content was an order of magnitude greater [41]. Thus, any sample of water, sediment, or organisms drawn through the surface layer of the sea would be subject to contamination.

Research Vessel Operations. Practically every day-to-day activity associated with operating a research vessel at sea is a potential source of contamination. Specifically, serious problems can occur from: (1) pumping of the bilge; (2) discharge of the sanitary system; (3) dumping of garbage, debris, or engine room wastes; (4) cleaning of boilers, especially on steam vessels, by blowing the tubes; (6) chipping of paint by deck personnel; (7) falling of soot from improperly adjusted diesel engines; and (8) contacting petroleum-lubricated hydrographic winch cables [42]. It is essential to obtain a suite of samples of possible contaminants from the vessel at the same time as the environmental samples are collected: lubricating oils and greases, fuel oil, paint being used or chipped, and exhaust stack fallout. Collection of samples aboard ship must of necessity require close cooperation between scientists, vessel officers, and crew as well as rigid coordination of the activities of all three groups [33]. Each sampling plan must be specifically developed in light of vessel operating conditions, environmental aspects, and the types of samples required.

Collecting Small Organisms with Plankton Nets. Clayton and Pavlou [43] designed a plankton net that could be closed during launching and recovery, thus avoiding contamination from the surface layer of the sea. A plankton net may be rinsed with suitable organic solvents or with hydrocarbon-free water prior to launching. Harvey and Teal [44] found that several short tows with a plankton net may be more effective than one long tow. During the towing operation plankton nets will strain out particles of tar suspended in the water column. Nylon mesh netting has a tendency to adsorb or desorb hydrocarbons in the sea as it is moved through areas of high or low concentrations, respectively. Also, plankton entrapped in the nylon net may absorb hydrocarbons from the collected tar particles and even from the netting. All these factors must be taken into consideration in the assessment of the samples and the interpretation of analytical data. Once the net is retrieved, the cod-end cup is allowed to drain thoroughly, and the cake of plankton is transferred to a sample bottle by means of a metal spatula that had been thoroughly rinsed with suitable organic solvents [38].

Collecting Large Organisms. Trawls or other fish-sampling gear and their contents should not be allowed to come in contact with the surface of the research vessel. The sorting and classifying of specimens should begin immediately after the net is retrieved. Sorting tray and non-plastic tools should be rinsed with suitable hydrocarbon-free organic solvents, such as pharmaceutical-grade 95% ethanol (prepared from ethylene hydration) [42]. Fish and other large organisms may be rinsed with solvent to remove surface contamination if such a solvent will not interfere with subsequent analyses. Benthic organisms should be rinsed with freshly-dipped seawater. The organisms can then be packaged in suitable containers or in aluminum foil and chilled for later freezing or placed directly in a freezer at -20°C or lower.

Collecting Sediment Samples. Many types of surface sediment sampling devices [45] are available to the investigator; the choice will depend on a combination of criteria unique to each investigation. Bulk sediment samples from prerinsed grab samplers should be placed in prerinsed glass or metal containers and frozen as soon as possible after collection. Core samples in rinsed core liners or barrels should be transferred as quickly as possible to a freezer (-20°C or lower) in which the sample can be kept upright until fully frozen. After the core samples are frozen, the end of the core tube should be wiped free of sediment and then covered with aluminum foil. The samples should be kept in frozen storage until needed for analysis. In preparation for analysis, the core is first extruded from the cylinder while still frozen whenever possible. Cross-section cuts are then made of selected portions of the core. The outer one-half centimeter rind is trimmed off to yield samples with a minimum of contamination [38].

Collecting Water Samples. Large-volume surface water samples have been collected by stainless steel bucket [46] and 4-liter glass bottles [47]; deep-water samples have been collected with 12-liter Niskin bottles [48], 1.25-liter Knudsen bottles [49], and the 20-liter Blumer bottles [33,50]. Adsorption of fluorescent compounds from seawater onto the plastic and metal parts of samplers going through the surface layer in an open configuration (normal cocked position on deployment) has been reported [51]. On the basis of laboratory and field tests, Gordon and Keizer [51] concluded that the collection of subsurface seawater samples for hydrocarbon analysis obtained by lowering sampling bottles (such as Nansen, Knudsen, VanDorn, NIO or Niskin) through the surface film and water column in the open state is unsatisfactory. In order to obtain an uncontaminated water sample, the sampling device must be lowered closed and opened only at the collection depth.

Samples to be analyzed for low molecular weight hydrocarbons (from 1 to 10 carbon atoms) require special handling to avoid volatilization of these components. Water samples have been collected from all depths with devices such as VanDorn water bottles. The rubber seals on these sampling devices apparently are satisfactory when water is to be analyzed for only low-molecular weight hydrocarbons. The retrieved water should be transferred through clean Teflon tubing with a minimum of agitation into a clean glass bottle, and the bottle should be hermetically sealed with a cap lined with Teflon [27]. Mercuric chloride or sodium azide may be added to the container as a bactericide prior to sealing although the efficiency of these compounds as bactericides may be in question [52]. Sulfuric acid has been added to water samples as an algicide [53]. For accurate results, the samples should be analyzed immediately [54,55], otherwise, the samples should be stored under refrigeration.

Samples to be analyzed for high molecular weight hydrocarbons (above C_{10}) are more subject to contamination than samples to be analyzed for low-molecular weight hydrocarbons. This is true because the higher molecular weight fossil hydrocarbon contaminants are less volatile and are common constituents of extraneous materials (lubricants, fuels, plastics, waste effluents) associated with the operation of a vessel. The samplers should be deployed and retrieved so as to minimize contamination from the surface layer. Sample size should be from 0.5 to about 20 liters depending upon the requirements of the analytical methods used. In one method of extraction, hydrocarbon-free carbon tetrachloride, 3.5% sodium chloride, and enough hydrochloric acid to make the pH less than one are used. The sample bottle should be stored upright in the dark, preferably under refrigeration. The carbon tetrachloride serves as a bactericidal agent as well as extractant and the sample storage bottle can be used for the subsequent extraction step as well [27].

A system for the continuous collection and analysis of hydrocarbons in surface water was developed by Brooks et al. [56] for light hydrocarbons (less than C_{10}). The integrated system for the continuous analysis of low-molecular weight hydrocarbons in seawater called the "Sniffer" is used by the petroleum industry for the routine sampling, measuring, and mapping of variations of dissolved hydrocarbons in potential oil and gas seep areas. The system has four parts: towing and pumping, instrumentation, analytical, and data processing. One analyzer provides a continuous chromatographic analysis (>5 ppb by volume) at 90-second intervals for methane, ethylene, ethane, propane, *iso*-butane, and butane. A second analyzer records the total hydrocarbons dissolved in the water [57].

EXTRACTION OF THE ORGANIC MATTER

Various techniques are used for extracting petroleum hydrocarbons from biological material, seawater, and sediment. Table 7 summarizes the techniques used for extracting petroleum hydrocarbons from biological material; Table 8 contains information on extraction techniques for seawater. Extraction procedures for sediments are frequently similar to those for biological material; however, cold solvent extraction [63] and cold solvent extraction combined with ultrasonication [58,60] have been tried. Comparison studies on the extraction of sediments were carried out by Farrington and Tripp [69] and Rohrback and Reed [58] and on the extraction of biological materials by Farrington and Medeiros [96]. These studies indicate that certain solvent systems and techniques are more efficient than others and that the extraction efficiency depended on the sample matrix.

Biological Material

The petroleum hydrocarbons present in marine plants and animals are concentrated in the lipoid substance of the organisms. The lipoid material and accompanying hydrocarbons are extracted from biological material usually by one of two basic methods: (1) extraction with organic solvents in a refluxing apparatus, or (2) a combination of alkaline digestion followed by extraction into an organic solvent. Other methods that have been tried involve steam distillation [61,62] and head-space gas stripping [97]. Farrington and Medeiros [96] found that Soxhlet extraction of a clam homogenate using a benzene:methanol mixture was slightly more efficient than digestion in alcoholic alkali; both were superior to extraction with anhydrous sodium sulfate and pentane in a VirTis Homogenizer.

The advantages of the digestion procedure is that it disrupts the cellular matrix, extracts the lipoid material, and saponifies the lipoid material all in one step, thereby reducing sample handling. Saponification is desirable to allow the removal of the esters of fatty acids (waxes and glycerides) that can interfere with subsequent isolation of alkene and aromatic hydrocarbon fractions. Complete saponification is confirmed by the absence of the carbonyl absorption band for esters in the infrared spectra of the material (extractable lipoid) remaining after saponification [27].

A disadvantage of the digestion procedure is the possible molecular rearrangement of biogenic molecules, such as transesterification and the loss of some metastable metabolites. For instance, this procedure may be incompatible with the simultaneous determination of hydrocarbons and chlorinated hydrocarbons in the DDT family because of chemical alteration of certain DDT metabolites [96]. The digestion procedure has

TABLE 7

Techniques for the extraction of petroleum components from biological material

Extraction techniques	Application	Sample requirements	Equipment requirements	Limitations	Total time	Operator time	References
Soxhlet separation with MeOH, benzene combinations or other solvents of similar polarity. Xylene may be substituted for benzene for reasons of lab health safety.	Frozen samples. Lyophilized samples. Wet or fresh samples.	As little as 1 g wet or fresh weight.	Soxhlet apparatus. Separatory funnel. Rotary evaporator for concentrating extract.	Recovery incomplete for hydrocarbons boiling below n-C_{15}.	24-48 hr	1-2 hr	28, 58
Solvent extraction using sonication to improve efficiency. CCl_4 and solvents given above.	Frozen samples. Lyophilized samples. Wet or fresh samples.	As little as 1 g wet or fresh weight.	Sonicator probe or sonicator bath. Separatory funnel. Rotary evaporator.	Also question of artifact formation.	½ hr	½ hr	59, 60
Solvent extraction coupled with steam distillation. Solvent was ethyl ether.	Frozen samples. Wet or fresh samples.	1-10 g wet or fresh weight.	Extraction flasks. Steam distillation apparatus.	Limited upper boiling range.	1-2 hr	1 hr	61, 62
Homogenization-solvent extraction. Anhydrous Na_2SO_4 may be used with wet samples to promote cell disruption and remove water.	Frozen samples. Lyophilized samples. Wet or fresh samples.	As little as 1 g wet or fresh weight.	Waring blender. Virtis homogenizer. Tissue grinders or similar equipment. IMPORTANT to avoid lubrication of bearings and motor shafts.	Recovery incomplete for hydrocarbons boiling below n-C_{10}.	15 min to 2 hr	15 min to 2 hr	63-65
Digestion in KOH with or without alcohol.	Frozen samples. Lyophilized samples. Wet or fresh samples.	1-10 g wet or fresh weight.	Heating baths. Centrifuge.	Trans-esterification possible.	30 min to 2 hr	30 min	66, 67

Adapted from Anderson et al. [68].

TABLE 8

Techniques for the extraction and identification of petroleum components in water

Technique	Component determined	Sample size	Advantages	Disadvantages	Equipment (Approximate cost in dollars)	Analysis time: Operator (elapsed)	References
C_1-C_{10} Hydrocarbons							
Gas equilibration	Individual hydrocarbons and HC type.	50-250 ml	Parts per trillion sensitivity; separate hydrocarbons from non-hydrocarbons. No sample preparation.	Analysis time relatively long.	Gas chromatograph (15,000).	10-30 min (0.5-2 hr)	69-71
Gas stripping	Individual HC & HC type.	1-2 l	Measure background levels of C_1, C_2, C_2=, C_3, C_3=, i-, n-C_4 in open ocean water.	Nonhydrocarbons can interfere. Analysis time relatively long.	Gas chromatograph (15,000).	10-30 min (0.5-1 hr)	54
Vacuum degassing	Individual HC & HC type.	4-20 l	As above, can be used to continuously measure HC in water.	Normally used to measure C_1-C_4 in sea-water. Equipment expensive.	Complete system-pumps, gas chromatograph (300,000; 2,500 per day rental).	3-20 min (3-30 min)	72, 73
C_{11} plus Hydrocarbons							
Gravimetric	Nonvolatile extractables.	1-4 l	Simple, minimum equipment.	Nondiagnostic, conc. between 0.3-1,000 mg/liter.	Glassware, balance (1,000).	20 min (40 min)	74
UV absorption spectrometry	Conjugated polyalkanes aromatics.	1 l	Useful for conc. >10 μg/liter.	Not very diagnostic, less sensitive than fluorescense spectrometry. No information on saturated HC.	UV absorption spectrometer (3,000-5,000).	20 min (20 min)	75

continued on next page

TABLE 8 *continued*

Technique	Component determined	Sample size	Advantages	Disadvantages	Equipment (Approximate cost in dollars)	Analysis time: Operator (elapsed)	References
UV fluorescence spectrometry	Unsaturated compounds, aromatics.	1 l	Useful for conc. >10 μg/liter; measure HC in open ocean water.	Not very diagnostic, no information on saturated HC. Fluorescence may be quenched	Fluorescence spectrometer (10,000).	20 min (20 min)	75-79
Infrared spectrometry	Methyl, methylene carbonyl, aromatic. Total HC	1-4 l	Information on functional groups. Identify contaminants such as silicons, plasticizers.	Concentrations >3μg/liter; >0.1 mg. Not very diagnostic.	Low or high resolution infrared spectrometers (3,500-35,000).	5-10 min (20-40 min after separation from water).	46, 81-85
Gas chromatography (low resolution)	HC profiles and boiling range of sample, C_{11}-C_{50}.	1-20 l	Quick examination, reasonably diagnostic.	Little information from highly weathered or biodegraded oils.	Gas chromatograph (10,000-15,000).	10 min (2 hr)	86, 87
Gas chromatography (high resolution), special detectors	More detailed HC profiles. Sulfur profiles, individual HC ratios, C_{11}-C_{40}.	1-20 l	Better diagnostic power. Sulfur compounds assist in identification.	Little information from highly weathered or biodegraded oils.	Gas chromatograph (15,000-20,000).	10 min (2 hr)	88-92
Mass spectrometry	HC types.	1-10 l	Provides complete HC type information.	Complex and expensive equipment. Requires computer interface.	Low resolution, mass spectrometers (60,000).	10 min (2 hr)	93, 94
Gas chromatography mass spectrometry	Specific HC C_4-C_{30}.	1-10 l	Identify and measure individual HC.	Very complex and expensive equipment.	Add gas chromatography cost to above.	2-4 hr (2-4 hr)	67, 95

Hydrocarbons are extracted from water and then separated from nonhydrocarbons by column or thin-layer chromatography.
From National Academy of Sciences [27].

been used principally on the soft tissue of shellfish [26,65-67], whereas the Soxhlet extraction procedure has been used for a wide variety of hard and soft biological and geological matrices [28,30,31,98-100]. Warner [67] used an aqueous alkaline digestion method for soft-bodied organisms exposed to parts-per-million levels of petroleum under bioassay conditions. The emulsion problem, which sometimes occurs in the presence of alcohol during the partitioning of the non-saponifiable lipids into the non-polar solvent (pentane, hexane, petroleum ether), can be minimized or avoided.

Ultrasonic extraction methods [59,60,101] are rapid but tend to produce artifacts in certain solvent systems. Wet halogenated solvents may decompose when exposed to ultrasonic radiation, and chlorine gas [101] and phosgene may be produced instantaneously.

Sediment

Hydrocarbons can be extracted from sediments by some of the same methods used for biological materials. They include gas equilibration techniques [27], head-space gas stripping [97], organic solvent extractions, alkali digestions [102, 103], and direct heating of a sample in the inlet system of a mass spectrometer [104]. High molecular weight hydrocarbons may be extracted from sediments with solvent in a Soxhlet apparatus [58,100,105] or by alkaline digestion followed by extraction of the non-saponifiables into an organic solvent [96,102,103]. Farrington and Tripp [96] found that Soxhlet extraction and alkali digestion in benzene and methanol gave essentially the same results for alkane, cycloalkane, and 2- and 3-ring aromatic hydrocarbons. Rohrbach and Reed [58] compared seven pre-extraction techniques, five extraction methods, and a series of pure solvent and solvent mixtures and found that acid-treated, water-washed, freeze-dried sediment extracted with toluene and methanol (3:7) for 100 hr in a Soxhlet apparatus appeared to be the best all-round procedure.

Elemental sulfur must be removed from the extracts of the sediment to prevent interference with the chromatographic separations and analyses. This can be accomplished by passing the extract through a column containing activated copper [106].

Seawater

Hydrocarbons may exist in seawater in the dissolved state; and, they may be absorbed or adsorbed by living or non-living particulate matter or be part of the microscopic oil droplets suspended in the water column. For analysis, the seawater may be extracted as received or may be filtered depending upon the type of information required. Analysis of the seawater sample as received would reflect the amount of dissolved hydrocarbons as well as those associated with the suspended particulate matter. The suspended matter can be separated by filtration and separate analyses can be made on the filtrate and the particulate matter contained on the filter. Filtering the seawater, however, may introduce errors by altering the distribution of the hydrocarbons. For example, the filter material may adsorb dissolved hydrocarbons from the seawater; also, a rapid rate of filtering could rupture the cells of marine organisms releasing any hydrocarbons that they might contain and introduce lipoid material that might interfere with subsequent analysis.

Hydrocarbons can be extracted from seawater by various means depending upon the molecular weight range of the compounds under consideration. Low-molecular weight hydrocarbons (less than C_{10}) can be extracted by gas equilibrium, gas stripping, or by vacuum degassing techniques. McAuliffe [70, 71] described gas equilibrium techniques for estimating alkanes, alkenes, cycloalkanes, and aromatics in seawater solutions. In the gas equilibrium extraction process, successive gas chromatographic analyses are made after repeated equilibration of a hydrocarbon-free gas with an aqueous sample containing the dissolved hydrocarbons. The method is sensitive to 1-3 parts per trillion (10^{-12}) for the aliphatics and to 8-10 parts per trillion for aromatic hydrocarbons having less than 11 carbon atoms. Because the method specifically identifies hydrocarbons, it can be used to determine the amount of hydrocarbons in a mixture containing non-hydrocarbon compounds (alcohols, organic acids, aldehydes, and others) without additional sample preparation.

Gas stripping techniques involve purging the aqueous phase with a hydrocarbon-free gas by bubbling and trapping the hydrocarbons on some type of substance where the hydrocarbons are retained for subsequent analyses. By this extraction method, the volatile hydrocarbons from seawater can be concentrated into a small volume at levels consistent with the sensitivity of the analytical instrument. This procedure can be used to determine background levels of low-molecular weight hydrocarbons (C_1 to C_4) in ocean water [54,55]. Unfortunately, non-hydrocarbon components can interfere with the analysis. Wasik and Boyd [107] have used an unusual electrolytic stripping technique in which the seawater sample served as the electrolyte; micro-size bubbles of pure hydrogen gas were produced, effectively stripping the volatile hydrocarbons from the water. Nitrogen and helium can also be used for gas stripping but they must be free of hydrocarbons. The organic components from a nitrogen stream were trapped on a short Tenax column for subsequent gas chromatographic analysis [61]. Grob [92] proposed a recirculating air-stripping technique to remove low levels (0.1 part-per-trillion) of organic substances from fresh water samples.

A vacuum degassing method for extracting low-molecular weight hydrocarbons from seawater is used in certain commercially-available gas chromatographic partitioning systems [27]. Such systems have been used for continuous monitoring of ocean water from a vessel while underway. The sensitivity of the method is in the subpart-per-trillion ($<10^{-12}$) level for C_2 to C_4 hydrocarbons.

High molecular weight hydrocarbons (C_{11} and above) are extracted from seawater with organic solvents although gas stripping methods onto active carbon or Tenax traps show promise. Both hydrocarbons and non-hydrocarbon organic compounds can be extracted from seawater using methylene chloride, CH_2Cl_2 [79,108], chloroform, $CHCl_3$ [109], carbon tetrachloride, CCl_4 [46,81], pentane, n-C_5H_{12} [110], petroleum ether [53], or Freon, $C_2Cl_3F_3$ [84] by shaking the seawater in the presence of the solvent in a large separatory funnel.

SEPARATION

The lipoid material from the marine samples are now contained in the solvent extracts. The purpose of the separation stage is to isolate the hydrocarbons from the extraneous non-hydrocarbon material for identification and quantification of individual compounds. The procedure is the same for all extracts regardless of the nature of the original sample. Table 9 summarizes the principal techniques used for separating petroleum components from solvent extracts. An extensive bibliography on the subject is contained in the report of Farrington et al. [33].

Column chromatography using silica gel (SiO_2) may be used for the isolation of saturated hydrocarbons from the total extract for hydrocarbons of petroleum pollutants having a boiling point less than 350°C. Silica gel provides good resolution of saturated and aromatic hydrocarbons. A short alumina (Al_2O_3) layer on top of the silica gel column may be used to retain the high molecular weight polar compounds. The alumina and silica gel column may be partially deactivated by water (using about 5% by weight) to prevent the formation of hydrocarbon artifacts, such as the production of phytadienes from phytol [112]. Column chromatography is used for the separation of saturated and olefinic hydrocarbons from plant and animal extracts and for the separation of fuel oil components taken up by sediment and marine organisms [101].

High pressure liquid chromatography is gaining wider acceptance with the development of improved column packing substances and detectors [108]. Miles et al. [111] used high pressure liquid chromatography to analyze aromatic

TABLE 9

Techniques for separation of petroleum components from solvent extracts

Separation	Components separated	Application	Sample requirements (lower limits)	Equipment requirements	Limitations	Time required	Reported analysis	References
Column chromatography	Hydrocarbons from lipids. Classes of HC: Alkanes, alkenes, and aromatics. From NSO heterocyclics.	Sediment and biological material extracts.	>100 μg	Standard column chromatography equipment.	Molecular weight range n-C_5 to n-C_{50}. Requires repeated applications to completely separate certain classes of compounds. Precautions needed to avoid formation of artifacts.	3/4-2 hr	Extensively used.	25,31, 67,110
Thin-layer chromatography	Hydrocarbons from lipids. Classes of HC: Alkanes, alkenes, and aromatics. From NSO heterocyclics.	Sediment and biological material extracts.	<100 μg	Standard TLC equipment.	Also problem of large surface area exposed to contamination from laboratory air.	3/4-2 hr	Less extensively used.	63, 64
Paper chromatography	Hydrocarbons from lipids. Classes of HC: Alkanes, alkenes, and aromatics. From NSO heterocyclics.	Sediment and biological material extracts.	<100 μg	Standard paper chromatography equipment.	TLC preferred and used more extensively. TLC can accommodate larger amounts.	3/4-2 hr	Few reports of recent applications.	
High pressure liquid chromatography	Ring number separations of aromatic HC. Separation of classes of HC.	Sediment, biota, and water extracts.	1 mg	High pressure liquid chromatography	Experience to date is too little to discuss limitations.	1 hr	Few reports. Appears to be promising.	30,108, 111
Gel permeation chromatography	Separation according to molecular weight.	Primarily for heavier molecular weight components.	1 mg	GPC equipment.	Nonideal behavior, i.e., absorption of compounds.	1-4 hr	Sediment analyses.	31

Adapted from Anderson et al. [68].

hydrocarbons in marine organisms. Zsolnay [113] used high performance liquid chromatography to determine aromatic and total hydrocarbon content of aqueous solutions.

Giger and Blumer [31] separated various polynuclear aromatic hydrocarbons from extracts of sediments according to ring type with a combination of techniques involving Sephadex LH-20 gel permeation chromatography followed by a charge transfer complexation with trinitrofluorenone and alumina column chromatography. Larson and Weston [114] used Sephadex LH-20 gel permeation chromatography to analyze chloroform extractable petroleum hydrocarbons from crude oil-in-water mixtures.

Hydrocarbon compounds can be separated in accordance to their molecular configuration and size by use of molecular sieves. The sieves are available commercially with effective pore diameters of 3,4,5,9, and 10 angstroms (A = 10^{-10} m). A 5 A pore diameter molecular sieve will retain straight-chain hydrocarbons while excluding branched hydrocarbons and compounds containing rings of four carbons or more. A 9 A pore diameter sieve will remove most normal and *iso*-paraffins and olefins from the aromatic hydrocarbons. Schenck and Eisma [115] applied this characteristic separation feature in the direct quantitative gas chromatographic determination of *n*-alkanes in crude oil.

Straight-chain paraffins can be separated from their branched-chain analogs by preparing their urea or thiourea adduction clathrates [116]. Aromatic hydrocarbons can also be complexed with certain nitro compounds [31]. Morris [117] experimented with the formation of silver ion complexes with unsaturated compounds. Weiss [118] presented a detailed discussion of methods for the separation and determination of the various classes of organic compounds.

ANALYSIS

The principal methods for identifying and determining the amount of hydrocarbon components in extracts are summarized in Table 10. Some methods only provide an estimate of how much material is present within the total extract (gravimetric) or in a portion of the extract having a specific photometric response (ultraviolet, infrared, and fluorescence spectrometry). Gas chromatography provides a means for separating various components in the extract as well as for estimating their content; mass spectrometry yields data to identify a particular compound. Recent advances in the identification and analysis of trace organic contaminants in predominantly non-marine waters has been compiled by Keith [122].

TABLE 10

Quantitative techniques for the determination and characterization of petroleum components in extracted fractions

Techniques	Components determined	Applications	Extract weight	Limitations	Advantages	Equipment	Time required	References
Gravimetric	Nonvolatile extractables.	High concentrations.	10^{-4} to 1g	Loss of volatiles - all residues (unless separation).	Simple, minimum amount equipment required.	Glassware, balance.	20 min	65, 119
UV absorption	Conjugated polyolefins, aromatics.	Highest contribution from polynuclear aromatics.	10^{-3} g	Less sensitive than fluorescence. Depends on calibration. Provides no information on saturated HC. Unsuitable for most ocean waters because of low concentration.	Quick, simple, inexpensive.	Separatory funnels, reagents, UV absorption spectrophotometer.	20 min	75, 120
Fluorescence spectrophotometry	Unsaturated compounds, aromatics, etc., depending on λ-excitation, λ-emission, extraction, and volatity.	Useful for large numbers of samples.	10^{-6} g in CCl	Provides no information on saturated HC. Possible quenching fluorescence. Provides only marginal information on compound type. Depends on calibration; need pollutant.	Applicable to low concentrations. Simple, quick. Probable use at sea; relatively unaffected by weathering. High sensitivity.	Separatory funnels, reagents, fluorescence spectrophotometer.	20 min	51, 77, 121
Infrared spectrometry	C-H_2 stretching frequency (2930 cm^{-1}).		10^{-3} g	Requires separations. Diagnostic power poor.	Provides information on functional groups; possible on age also. Quick, simple, inexpensive.	Low resolution infrared spectrophotometer	40 min	27
	Total HC.		10^{-3} g in CCl_4	Not too convenient for large numbers of samples; loss of volatiles.	Identify contaminants that generally mask silicones, plasticizers.	High resolution infrared spectrophotometer with scale expansion.	20 min	46, 82, 85

continued on next page

TABLE 10 *continued*

Techniques	Components determined	Applications	Extract weight	Limitations	Advantages	Equipment	Time required	References
Gas chromatography	Individual C_5-C_{40}.	Sensitive to parts per trillion.	10^{-4} to 10^{-6} g	Analysis time is relatively long.	Quantitative.	Gas chromatograph, temperature program with flame ionization detector.	3 hr	28, 29,
	Depends on column and detector.	HC profiles, sulfur profiles, *n*-paraffin/isoprenoid ratios.		Preseparation into compound classes.	Provides information on compound types; boiling range; provides information on origin.		1-2 hr	92, 103
Mass spectrometry	Hydrocarbon type C_{10}-C_{40}.		10^{-5} g	Very complex; preseparation; requires computer output.	Provides complete type composition.	Low resolution MS of of 60-600 mass range. Preferable to interface w/digitizer and/or computer.	2 hr	93
Gas chromatograph/Mass spectrometry	Preselected C_4-C_{30}.		10^{-5} g	Very complex; preseparation; requires computer output.	Identify and measure key individual HC.	Gas chromatography/Mass spectrometry preferable to interface w/digitizer and/or computer.	4 hr	95

Adapted from Anderson et al. [68].

Gravimetric Method

In the gravimetric method an aliquot of the solvent fraction is evaporated to dryness and the residue weighed on a sensitive balance (ASTM Method D-2778 [10: part 31]). An electronic balance will weigh a sample as little as 10^{-6} g [119]. This method gives the total amount of non-volatile extractable material (such as paraffinic, olefinic and aromatic, and high-molecular weight aromatic hydrocarbons) in each fraction after column chromatography. The method is subject to errors because of possible contamination by dust and because the highly volatile solvents used may not dissolve all the compounds quantitatively. Also, some of the lower molecular weight hydrocarbons (C_{12} to C_{15}) are sufficiently volatile that some may be lost during the solvent evaporation procedures. However, the method is rapid, does not require expensive analytical equipment, and is not destructive. Farrington and Medeiros [65] discussed the application of electronic microbalance measurements. For mass balance considerations in a complete analytical sequence, it is desirable to have gravimetric data at various stages in the procedure, such as starting wet weight of sample, dry extracted weight of any biological material remaining, total extractable lipids, and total fraction-eluant weights.

Ultraviolet Absorption Spectrophotometry

Ultraviolet absorption spectrophotometry may be used to determine conjugated polyolefins and aromatic hydrocarbons. In addition, it is useful as a qualitative indication of petroleum hydrocarbon components in extracted fractions, for the rapid screening of extracts to identify those that contain aromatic hydrocarbons, and to determine the upper concentration limits of aromatic hydrocarbons.

Ultraviolet absorption spectra of the solvent extracts are made over the wavelength range of 210 to 350 nanometers (nm, 10^{-9} m). The absorbance of the test extracts at 256 nm is compared with the absorbance at the same wavelength for a sample of petroleum or for a pure aromatic hydrocarbon. Any substance that is extracted and separated by a particular solvent system and absorbs ultraviolet light at 256 nm is arbitrarily interpreted as being from petroleum sources [27]. Ultraviolet absorption spectrophotometry does not give much information on all the compounds in an extract; the spectra provide no indication of the nature or the molecular weight of the absorbing molecules in a mixture.

Ultraviolet spectra of refined petroleum products often show absorption maxima at 315 to 325 nm, 250 to 270 nm, and 225 to 235 nm. Several sharp maxima, caused by the presence of polynuclear aromatic hydrocarbons, are at times present and may be helpful in the identification of spilled oil by comparison with oil from the suspected source. Quantitative data on the ultraviolet spectra of several Canadian fuel oils are presented in Table 11 [77].

TABLE 11

Quantification of several Canadian fuel oils by ultraviolet spectroscopy (1 cm cell)

Fuel oil	Number of samples	Absorbance/mg per ml		
		320 nm	250-270 nm	220 nm
Stove	4	0.069	2.06	23.1
Arctic diesel	1	0.054	1.59	23.3
Diesel	3	0.209	3.93	42.6
Marine diesel	1	0.615	11.40	59.6
Furnace	9	0.229	4.53	47.1
Bunker C	5	-	24.7	38.0

From Zitko and Carson [77].

Fluorescence Spectrometry

Fluorescence spectrometry involves both the ultraviolet (200 to 400 nm) and the visible spectra (350 to 600 nm). In conventional fluorometry, the fluorescence is recorded as a function of the light intensity at the emission wavelength while the excitation wavelength is held constant or *vice versa*. Ultraviolet fluorescence spectrometry, including synchronous or offset scanning, has been used to identify petroleum aromatic hydrocarbons in aquatic samples [33,75,123,124]. Fluorescence spectrometry is as rapid as ultraviolet absorption spectrometry and can provide more details about the components in a mixture. It is recommended [125] for the initial screening of seawater extracts for dissolved hydrocarbons.

Fluorescence spectrometry has been used to analyze for aromatic hydrocarbons in the marine environment. Zitko and Carson [77], Thurston and Knight [76], and Coakley [126] have summarized the use of ultraviolet fluorescence spectrometry to identify oils by differences in their aromatic composition. An example of the fluorescence spectrometry applied to the analysis of four API reference oils has been prepared by Wakeham [124] and is reproduced in Figure 15A.

Fluorescence contour diagrams (excitation vs. emission) for fluorescent aromatic compounds extracted from sediments have been prepared with computer assistance by Hornig [78] and Hargrave and Phillips [127]. Fluorescence spectrometry

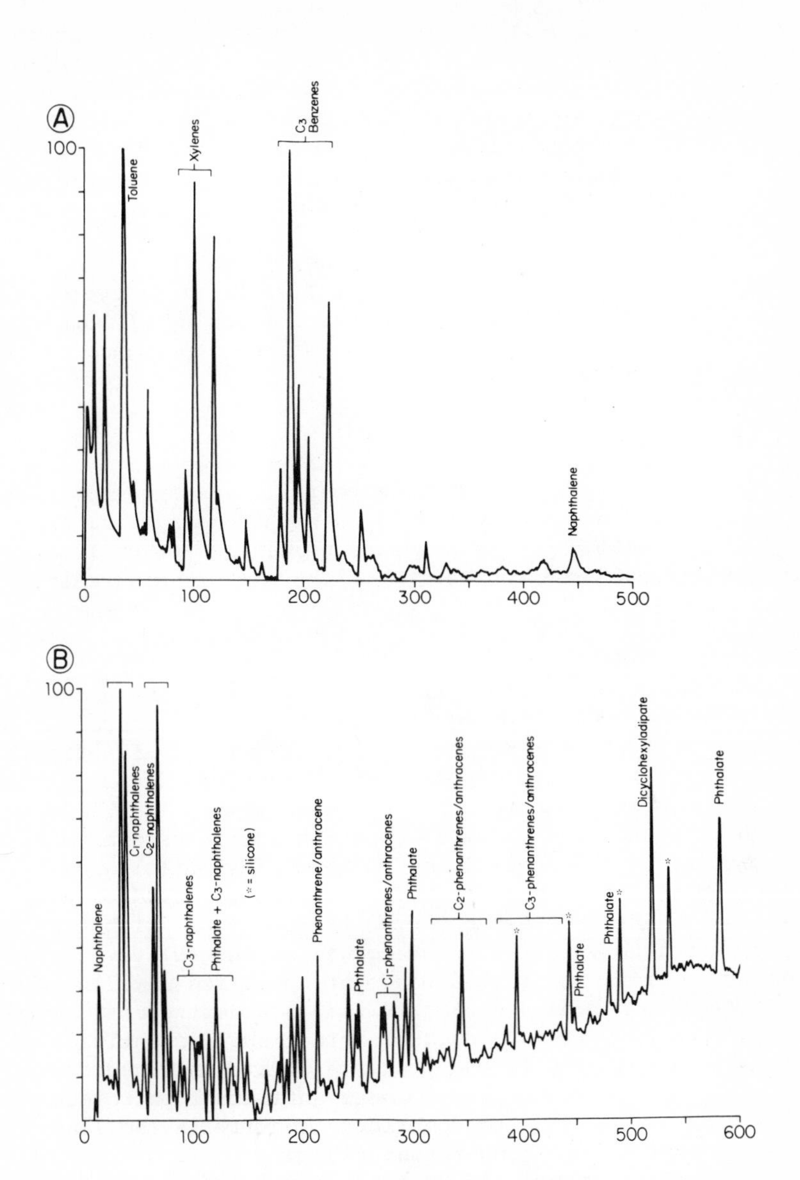

Fig. 15 (A) Conventional fluorescence spectroscopy emission (excitation wavelength, 310 nm) spectra of four API reference oils. Concentrations are 40 µg/10 ml. (B) Synchronous (25 nm offset scan) fluorescence spectra of the same four API reference oils with approximate ring number designations. (Adapted from Wakeham [124]. Used with permission from Environmental Science and Technology. Copyright by The American Chemical Society.)

requires the use of a standardization mixture (containing the same components as the test sample) for calibration. Thus, the interpretation of the results are difficult in environmental monitoring surveys where the nature of the contaminants is not known. However, this technique is rapid and sensitive for checking known oil pollution incidents or for bioassay monitoring in the laboratory where the aromatic hydrocarbon content pattern can be established.

In synchronous excitation fluorometry the difference between the emission wavelength and the excitation wavelength remains constant during the simultaneous scanning of both wavelengths [92]. Considerably more information regarding the nature of the hydrocarbons can be obtained by that procedure over conventional spectrofluorometry; for example, it is possible to calibrate a spectrum according to aromatic ring number. Synchronous spectra of four API reference oils are presented in Figure 15B [124]. Neither fluorescence spectrometry nor ultraviolet absorption spectrophotometry will provide any information on the presence of alkanes or cycloparaffins or indicate the complexity of the mixture of aromatic hydrocarbons [33].

Infrared Spectrometry

Infrared spectrometry, especially over the wavelength region from 2.5 to 15 micrometers, can be used to characterize individual molecules or molecular structure (e.g., functional groups) but is of little use in identifying individual hydrocarbons in a mixture containing more than a very limited number of components. Infrared spectrometry is used primarily to characterize crude petroleum, residual fuel oils, and asphalts [83]. Identification of oil types is based on the comparison of the recorded spectrum of the sample with that of a known petroleum material or by comparing the relative intensities of specific absorption bands at specified wavelengths [84] (ASTM Method D-3414 [10: part 31]). Nevertheless, infrared spectrometry cannot be used to detect small quantities of petroleum hydrocarbons in the presence of biogenic hydrocarbons, except in those instances where sufficient petroleum hydrocarbons are present to contribute heavily in the long wavelength part of the fingerprint region.

Brown and coworkers [46,82,128] used infrared spectrometry to estimate total hydrocarbons in oceanic waters. In this method the dissolved hydrocarbons from seawater are extracted into carbon tetrachloride and the absorbance of the extract is measured at a wavelength of 3.42 μm (about 2,930 cm^{-1}: carbon-hydrogen stretching frequency) in a long-path length (5 or 10 cm) cell (sodium chloride windows) using a high resolution spectrophotometer. The absorbance of the test sample is compared with that of the calibrated known sample.

Infrared spectrometry is particularly useful for checking the efficiency of various separation techniques and for the characterization of the structure of isolated specific hydrocarbon compounds.

Application of such non-destructive techniques to petroleum samples provides a measure of the functional groups in the various molecules in the mixture. On the whole, however, these techniques produce better and more useful data, both on petroleum and on the lipoid extracts of organisms or sediments, if the techniques are applied to sub-fractions obtained by fractionation procedures, such as column or thin-layer chromatography. Thus, infrared analysis after a thin-layer chromatographic separation provides much more information on the separated saturated and olefinic hydrocarbon fractions than the infrared analysis of the total bulk sample. Similarly, infrared or ultraviolet spectrometry applied to a polycyclic aromatic hydrocarbon fraction collected from the effluent of a gas chromatographic column may identify ring number and degree of, as well as type of, substitition; such information would be difficult or impossible to obtain on a bulk sample [101].

Gas Chromatography

Gas-liquid chromatography has been extensively developed for use in the petroleum industry and for the study of organic pollutants in the environment. In this separation technique, a gaseous moving phase of organic compounds is passed over a stationary liquid phase (held on an inert structure) in a long, thin chromatographic column. The sample is flash-evaporated and carried through the column by an inert gas. Selective separation of the organic components is achieved and is related to the boiling points and polarity of the individual compounds and the nature of the liquid adsorbant base. Petroleum hydrocarbons are tentatively identified by comparing their retention times with those obtained from authentic samples of known hydrocarbons or by determining the Kováts retention indices of peaks on the gas chromatograms [32,129].

The high discrimination afforded by gas chromatography in correlating petroleum and refined petroleum products has been described by Zafiriou and coworkers [89]. The gas chromatogram can be used as a pattern or "fingerprint" to identify the source of the petroleum product; at high resolution substantial information can be obtained on the levels of individual alkenes and alkanes [27]. Ehrhardt and Blumer [86] described the detailed analysis of petroleum and refined products; they were able to distinguish between hydrocarbons from petroleum sources and hydrocarbons from biogenic sources.

It has been found that when crude petroleum and fuel oils are analyzed by gas-liquid chromatography, the chromatograms often show some unresolved material beneath the resolved peaks (called the "envelope" or unresolved complex mixture). Figure 16 shows three gas chromatograms taken on 15-meter long SCOT columns. The top pattern displays the hydrocarbon distribution in a clam extract made from samples collected at Washburn Island, Falmouth, Massachusetts [65]; the middle pattern is of the same sample containing 10 ppm No. 2 fuel oil; and the bottom pattern is of the No. 2 fuel oil alone. These patterns show the differences in aliphatic hydrocarbons found in a biotic sample (clam, upper pattern) and an abiotic sample (fuel oil, bottom pattern), as well as the occurrence and position of the unresolved complex mixture or unresolved "envelope" in a contaminated extract (middle pattern) [33].

Some investigators have determined the amount of this unresolved envelope area and used it as a measure of the amount of petroleum hydrocarbons present in a sample containing both petroleum and biogenic hydrocarbons [102,130-133]. This practice can be subject to considerable error, however, if applied without a complete understanding of the principles involved, because (1) the area of the unresolved envelope or complex mixture is not always included in the total estimate of hydrocarbons presented by all authors and (2) the amount of unresolved envelope may be a function of the resolution of the particular gas-liquid chromatography system employed [27]. In another approach, some workers have used the ratio of a resolved hydrocarbon peak height to the height of the unresolved envelope beneath that peak [28,89,98,99,110,134]. This method may be subject to error because it is based on the assumption that weathering or biological incorporation has not altered the ratio of resolved hydrocarbons to unresolved hydrocarbons within the envelope [101]. Considerable information on the use of gas-liquid chromatography for the estimation of petroleum contamination of marine organisms has been reported by the National Academy of Sciences [27], Farrington et al. [33], Clark [36], and Blumer et al. [101].

Mass Spectrometry

Mass spectrometry is used to identify individual hydrocarbon structures and groups of hydrocarbons having similar fragmentation ions in extracts from biological materials, seawater, sediments, and petroleum materials. The mass spectrum gives a record of the mass distribution of ionized atoms, molecules, and fragments of molecules. Brown [93] reviewed the use of mass spectrometry in the petroleum industry.

Molecular ions and charged fragments produced by bombardment of the parent molecules by a beam of medium-energy electrons are separated and recorded as mass spectra. These spectra are fragmentation patterns which provide molecular fingerprints of extraordinary detail on a few nanograms of

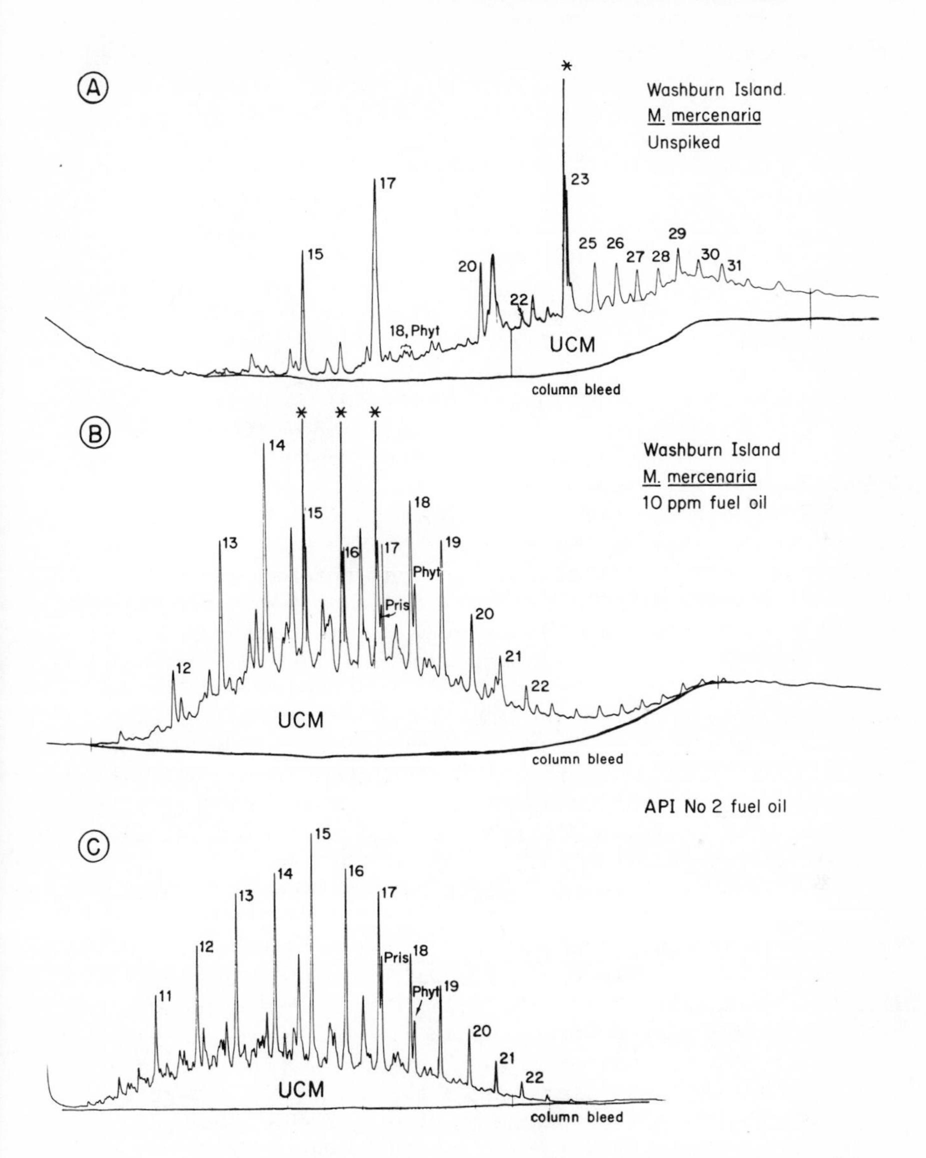

Fig. 16. Gas chromatograms of alkanes, cycloalkanes, and alkenes isolated from clams (Mercenaria mercenaria) *and clams spiked with the API reference No. 2 fuel oil. Fifteen meter SCOT OV columns, temperature programmed from 70°-100°C to 270°C at 4-6°/min and held at 270°C until* n-C_{32} *eluted. Numbers 11-31 refer to number of carbon atoms in* n-*alkane eluting at the position indicated. Pris: pristane; Phyt: phytane; UCM: unresolved complex mixture or envelope; and *: automatic attenuation of peak. (Adapted from Farrington et al. [33]. Used with permission of John Wiley & Sons, Inc.)*

sample or less. With high-resolution mass spectrometers the exact mass of the molecular ion can provide a precise molecular weight (and often, elemental composition) to three or more decimal places. Identification of an unknown compound can be made by matching the fragmentation pattern from the purified unknown sample with the fragmentation pattern of a pure authentic compound [20]. Simoneit et al. [53] applied real-time high-resolution mass spectrometry to total mixtures of petroleum ether extractions of estuarine waters. Their analyses yielded specific data in terms of elemental composition and high accuracy for all ion intensities in the spectrum. Accurate mass spectrometric determinations allow estimation of the heteroatom content (especially for oxygen) as well as the degree of unsaturation. Polar compounds and compounds of low volatility, which might not pass through a gas chromatographic column, can be determined.

High-resolution gas-liquid chromatography has been coupled to the mass spectrometer so that a mixture can now be analyzed by mass spectrometry. The gas chromatograph separates the mixture into individual compounds which are then rapidly scanned for their individual fragmentation patterns; by this method both gas chromatographic retention times and mass fragment patterns are obtained. This method has proved useful for demonstrating the occurrence of low levels of petroleum aromatic hydrocarbons in marine organisms and sediments [135,136]. The hydrocarbons isolated from shellfish and sediments exhibited molecular fragments of a series of alkylated aromatics having a distribution and complexity similar to those found in petroleum [32].

Gas chromatography-mass spectrometry systems interfaced with a computer make possible the powerful technique of selective reconstruction of mass spectrograms of molecular and fragment ions characteristic of specific compounds [33,46,67, 137,138]. Simoneit et al. [53] used computer-coupled gas chromatography-mass spectrometry to analyze petroleum ether extractable organic matter from filtered San Francisco Bay water. Extracts containing 2.5 to 102 μg/l provided sufficient material to allow identification of major compounds and the tentative classification of others by functional component. The use of capillary gas chromatographic columns with their high resolving capacity combined with computerized mass spectrometry is one of the most promising methods for identifying and quantifying petroleum hydrocarbons in the marine environment [94,95]. A typical reconstructed gas chromatogram of a fuel oil aromatic fraction performed on a capillary column coupled to a quadrupole mass spectrometer is presented in Figure 17. It depicts the resolving capability of the capillary gas chromatographic technique and the ability to

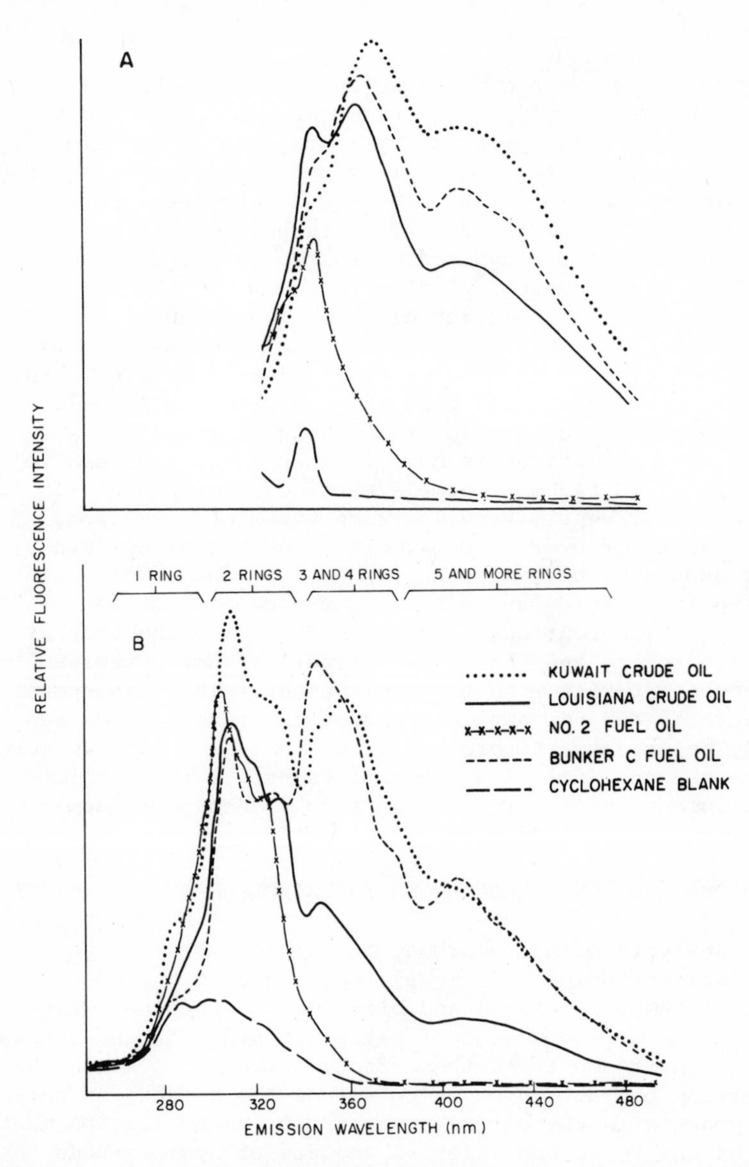

Fig. 17. Reconstructed gas chromatograms of water-soluble extracts of a Prudhoe Bay crude oil run on a SE-30 capillary column in a computerized gas chromatograph/mass spectrometer. (A) Low-molecular weight aromatic hydrocarbon fraction. (B) Wide-range aromatic hydrocarbon fraction. The silicone peaks represent septum contamination, and phthalates occur in the solvent blanks.

identify these peaks by a computer-controlled mass spectrometer [139].

Integration of Methods

Highly selective and specific analytical techniques, especially high-resolution gas chromatography alone and in combination with mass spectrometry and to a lesser degree, gel permeation chromatography, are capable of separating many of the complex mixtures in petroleum and petroleum-contaminated environmental samplesA variety of techniques and methods for analyzing petroleum hydrocarbons in biotic and abiotic systems have been developed and published. Yet the state of the art is still far from providing us with a single "cookbook" procedure for separating, analyzing, and quantifying all the petroleum hydrocarbon components at levels greater than 1-10 ppb in the presence of biogenic hydrocarbons for all marine samples. In an attempt to minimize the analytical problems, most investigators have limited their approach to certain specified types of sample matrices, certain levels of contamination or sensitivity, certain types of pollution, certain classes of hydrocarbons, or to the use of certain types of apparatus or equipment. These approaches provide cumulative information about the occurrence of hydrocarbons in the marine environment but frequently frustrate the acquisition of a complete and integrated picture of how petroleum hydrocarbons interact with the environment. The development of new methods and new applications of existing methods can hopefully be expected to proceed at an increased pace as more environmental and analytical scientists attack the problems involved in analyzing marine samples to detect petroleum contamination.

HYDROCARBONS AND NON-HYDROCARBONS IN PETROLEUM AND ORGANISMS

The analysis of hydrocarbons is only one step in petroleum pollution studies. Of equal importance is an understanding of the geochemical and biochemical processes which control the input and routes of petroleum hydrocarbons through the marine environment and the effects these hydrocarbons have on the marine biota. In order to assess the effects of petroleum components in the marine environment, an evaluation must be made of the types and relative amounts of hydrocarbons that normally occur in or that are biosynthesized by marine organisms. In addition, data must be obtained on the nature and distribution of hydrocarbons in petroleum and on the effects of these components, singly and in combination, on marine organisms.

Blumer [140] prepared a detailed discussion of the characteristic differences between hydrocarbons synthesized by organisms and those present in petroleum. Farrington et al. [141] and Anderson, Clark, and Stegeman [68] enlarged and updated Blumer's approach. In general, biogenic hydrocarbons cover a rather narrow molecular weight range; in addition, one hydrocarbon may predominate over all others in certain marine organisms. Biosynthesized hydrocarbons include, predominantly, the saturated and unsaturated aliphatic hydrocarbons, as well as branched-chain hydrocarbons, especially the isoprenoids. Naphthenic and aromatic hydrocarbons occur at very low levels in marine organisms.

Crude petroleum, on the other hand, contains high levels of many types of saturated aliphatic hydrocarbons covering a wide range of molecular weights. Olefins are seldom found in crude petroleum but are produced in the cracking processes and are present in gasolines. Aromatic and naphthenic compounds are abundant in petroleum.

Geochemical processes such as submarine and coastal land oil seeps, weathering of soils and ancient sediments followed by aeolian or fluvial transport, forest fires, and, to a minor extent, early diagenesis of organic matter in the marine environment contribute hydrocarbons to the sea [33]. Furthermore, the chemical analysis of marine samples is complicated by the fact that certain hydrocarbons found in petroleum are also synthesized by most, if not all, living marine organisms or occur in their food source [142]. Because of their stability, these hydrocarbons may spread from their source into water masses, where they may become adsorbed onto the surfaces of particulate matter and re-enter the food web [68].

Baker and Murphy [143] compiled a list of 504 organic compounds found in marine organisms as reported in the scientific literature through 1973. They listed sesquiterpenes, prostaglandins, halogenated derivatives, furans, carotenoids, sterols, phenols, quinones, nitrogen compounds, and hydrocarbons, although the only paraffinic hydrocarbon listed was pristane.

The following discussion is a summary of experimental observations and conclusions on the parameters or characteristics that have been considered and used to differentiate between biogenic and petroleum hydrocarbons in the environment. For convenience the topic is broken down according to hydrocarbon type and volatility; some of the effects on organisms are described although this aspect will be covered in detail in Volume II. Emphasis is placed on marine organisms where data are available, however, pertinent information on microbial and terrestrial biota examples may be used for perspective.

HYDROCARBONS

Volatile Normal Paraffins

The volatile paraffinic hydrocarbons will be defined as those containing ten carbon atoms and less (n-C_{10} boils at 174°C). Crude petroleum can contain up to 30% of volatile paraffins [27]. The natural gas fraction contains the highly volatile paraffin components (C_1 to C_5); the gasoline fraction contains the less-volatile liquid paraffin components. The *n*-paraffin content generally decreases with increasing carbon number from C_6 to C_{10}. Methane gas is a major hydrocarbon produced by bacteria in highly-reducing, organic-rich sediments (marshes, rice paddies, anoxic basins), although most of the volatile normal paraffinic hydrocarbons have been detected somewhere in the marine environment [142].

Until quite recently, low-boiling *n*-alkanes have been considered relatively harmless to the marine environment. It has now been suggested that some alkanes, which are slightly soluble in seawater, can at low concentrations produce anaesthesia and narcosis and, at higher concentrations, can cause cell damage and death in lower animals [144].

Volatile Branched Paraffins

Petroleum contains significant levels of isobutane, isopentane, and higher homologs of this series; their levels in petroleum vary inversely with their molecular weight. Other branched paraffins, especially 2-, 3-, and 4-methyl-substituted paraffins to C_{10}, occur at high levels in petroleum [145]. Mair and coworkers [146] isolated 2,6-dimethyloctane and 2-methyl-3-ethylheptane from the gasoline fraction of Ponca City (Oklahoma) crude petroleum. These two C_{10} branched hydrocarbons were present in relatively large amounts (0.50% and 0.64%, respectively, by volume of the crude oil) and could easily be derived from the monoterpenoids (e.g., biogenic precursor like limonene, Fig. 3). Published data on the levels of volatile branched biogenic hydrocarbons in this molecular weight range in marine organisms were not available.

Non-Volatile Paraffins

Normal paraffin hydrocarbons in crude petroleum occur from C_1 to beyond C_{78}. Over the carbon chain length range from C_{10} to above C_{40} the ratio of abundance between odd-numbered and even-numbered carbon chains of this homologous series in petroleum is usually equal to unity and the adjacent members of the series are present at similar concentration levels [20]. Several homologous branched-chain paraffin series are also present in petroleum, including a series of C_{12} to C_{22} isoprenoid paraffins (e.g., farnesane, pristane, and phytane); the ratio of levels between even- to odd-carbon chain compounds also approaches unity. In the distillation processes during refining this odd to even ratio is usually retained in the final products which contain similar component distributions. The paraffin content of the lubricating oil fraction from petroleum is low because the higher-melting paraffin waxes are intentionally removed in the refining process.

Intermediate between volatile (C_{10} and below) and the non-volatile paraffin hydrocarbons (above C_{16} or C_{17}) are a group of compounds that boil between 175° and 275°C. These moderately volatile paraffins (above C_{10} to about C_{15}) can sometimes be lost from samples during extraction and separation unless suitable precautions are taken [36,66,67]. In order to identify an oil spilled at sea, samples of known oils are artificially weathered by fractional distillation at 280°C (ASTM Method D-3326 [10: part 31]), and the resulting compositional patterns are compared with that of the unknown oil. The weathering achieved artificially is generally equivalent to that produced by environmental exposure of oil on seawater for several days [88,147].

The abundance of methyl-substituted branched paraffins seems to be a characteristic of petroleum and is considered by Meinschein [148] as evidence to support the theory that biogenic terpenoids are the major sources of some petroleum hydrocarbons. Farnesane, pristane, and phytane, also, have been isolated from petroleum distillates.

Marine organisms synthesize straight-chain and some branched-chain paraffin hydrocarbons; the odd-numbered carbon chains predominate although the even-numbered carbon chains are also produced [142]. In certain organisms, one or two odd-numbered carbon chain *n*-paraffins predominate over all other non-volatile hydrocarbons (Fig. 18) [140]. In marsh grasses and pelagic macroalgae, C_{21} through C_{31} compounds predominate with the odd-numbered paraffins in the majority; and in phytoplankton, C_{15}, C_{17}, C_{19}, and C_{21} compounds predominate [100,149]. Marine sponges and corals have little, if any, odd-numbered carbon chain compounds between C_{25} and C_{34} [150]. Some marine bacteria contain equal amounts of even-numbered and odd-numbered *n*-alkanes between C_{25} and C_{32} [151].

Branched-chain alkanes, including pristane, have been found in aquatic organisms. In some plankton and fish, pristane is the most abundant alkane present [142]; however, another branched alkane, phytane, has not been detected in copepods from the Gulf of Maine, in basking shark [152], or in benthic or pelagic algae [100,149]. An example of how organisms can produce hydrocarbons from precursor compounds obtained from their food is the conversion of pristane from phytol [153]. The presence of high levels of pristane in

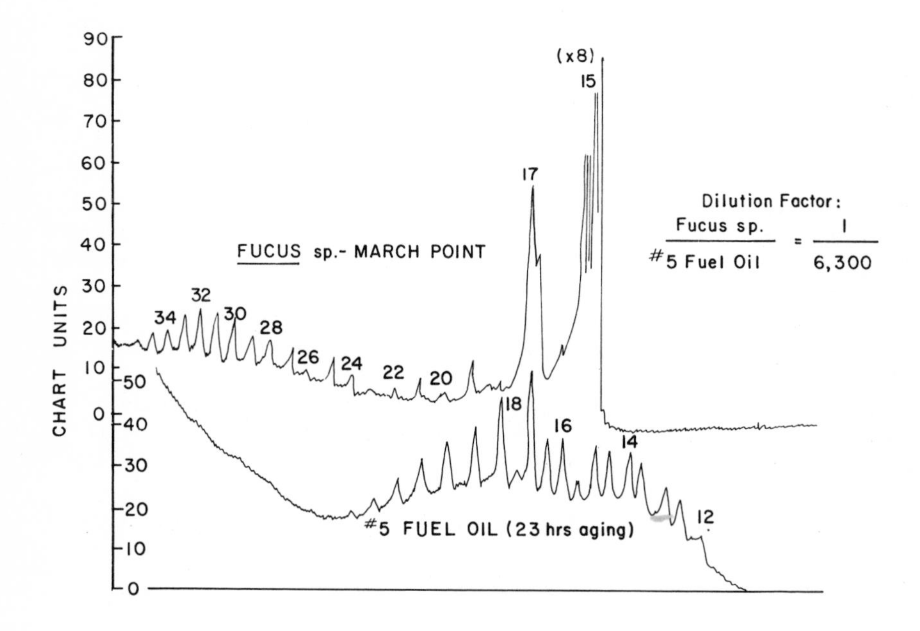

Fig. 18. Gas chromatograms run on a packed silicone rubber column showing the aliphatic hydrocarbon pattern of a biotic (an alga: Fucus sp.*) and an abiotic (No. 5 fuel oil) sample. Sample size adjusted to show minimum attenuation for graphic presentation in this figure: the petroleum pattern represents a solution of 1 to 6,300 times that of the alga pattern. Note the increased resolution or sharpness of separations between peaks in the chromatograms from SCOT (Fig. 16) and capillary (Fig. 17) columns. (From Clark and Finley [28].)*

certain marine organisms and in petroleum is considered as evidence to support the theory that biogenic terpenoids are a major source of some petroleum hydrocarbons. The higher-boiling saturated hydrocarbons occur naturally in many aquatic organisms and probably are not directly toxic, although some petroleum contaminants may interfere with nutrition and possibly with the chemical cues which are necessary for feeding and communication between many aquatic macroorganisms (chemoreception) [101,154].

Naphthenes

Petroleum contains a complex mixture of naphthenes and cycloparaffins; those containing five or six carbon atoms in the ring are most abundant. The alkyl, particularly methyl, substituted cyclopentanes and cyclohexanes are more common than the unsubstituted compounds. The abundance of the methyl substituents relative to larger alkyl substituents in cycloparaffin rings seems to be characteristic of petroleum and may be related to biogenic terpenoid precursors [148]. Hills and Whitehead [155] identified a group of optically-active triterpanes in a high-boiling petroleum distillate. Other naphthenes in petroleum may have been derived from such biogenic compounds as carotenes and plant phenols.

Hydrocarbons containing one to three alicyclic rings are prominent in certain herbs and other land plants. Most of these hydrocarbons appear to be terpenoidal [140]. An alkyl cyclopropane has been tentatively identified in marine algae [149].

Olefins

Olefins are rarely present in crude petroleum and then only in trace amounts [5]; however, they are formed in some refining processes and are present in certain cracked products, such as gasoline [140]. Olefins may occur at levels up to 30% in gasoline and up to 1% in jet fuels [27].

Alkenes may account for a large percentage of the hydrocarbons in certain aquatic macroorganisms. Isoprenoid C_{19} and C_{20} mono-, di-, and triolefins are present in copepods and some fishes [156]. Straight-chain olefins containing up to six double bonds have been found at relatively large concentrations in certain marine organisms [149, 157-160]. Squalene is a major constituent of the liver oils of cod and the basking shark. Carotenes (polyolefins) are found in considerable quantities in many aquatic macroorganisms [140]. The biological role of alkenes is poorly understood; it has been suggested that they may be involved, for example, in the chemoreception functions of marine organisms [161].

Aromatics

Petroleum contains a complex mixture of aromatic hydrocarbons including benzene and its derivatives, naphthalenes, naphthenoaromatics, and polynuclear aromatic hydrocarbons. The naphthenoaromatic molecules contain a combination of benzene ring and cycloalkane units [27]. Mair and Barnwell [162,163] suggested that the alkylbenzene compounds of petroleum were of terpenoid origin. They isolated 21 aromatic compounds in the carbon atom range from C_{14} to C_{18}; these compounds included 2,3,4,7-tetramethylnaphthalene, phenanthrene and its 1-, 2-, 3-, and 9-methyl and 1,8-dimethyl derivatives, and 1,2-cyclopentenophenanthrene and its 3'-methyl derivative. Pyrene and its 4-methyl derivative and a series of alkylated fluorenes were also identified. These hydrocarbons are similar to those commonly produced by the dehydrogenation of steroids and other natural products [20].

Chrysene and its 3-methyl derivative have been reported to be present in petroleum, and other studies have indicated the presence of a series of anthracenes and phenanthrenes, 1-methylpyrene, benzo[a]fluorene, and triphenylene, in Kuwait crude oil. Perylene, 2,9-trimethylpicene, and several benzanthracenes have also been found in petroleum. The trimethyl tetrahydropicene may be a dehydrogenation product of the triterpenoids [20]. The various aromatic classes occurring in distillates from Prudhoe Bay crude oil boiling between 370° and 535°C have been characterized by Coleman and coworkers [24].

Pancirov and Brown [164] found eleven to sixteen individual polynuclear aromatic hydrocarbons in samples of a South Louisiana and a Kuwait crude oil as well as in a No. 2 fuel oil (high aromatic content) and a Bunker C fuel oil (all API reference oils). These compounds included the principal 3- to 5-ring aromatics found in petroleum and one 6-ring (benzo-[ghi]perylene). The tricyclics (phenanthrene and methylphenanthrenes) were generally present in the highest concentrations and occurred in the concentration range from 26 to 7,677 ppm by weight.

A review by Gerarde and Gerarde [165] contains references to data on the occurrence of biosynthesized aromatic hydrocarbons in land plants and the spices prepared from them. Aromatic hydrocarbons have not been isolated from plankton remote from petroleum input sources according to Blumer et al. [156]. However, several German investigators [166-168] have claimed that algae and higher plants biosynthesize polynuclear aromatics but others have disputed this claim [169]. Many species of marine organisms, including bacteria (See Chapter 3), are able to metabolize aromatic hydrocarbons and excrete the oxidized products [27,170].

ZoBell [171] reviewed the sources and biological degradation of polynuclear aromatic hydrocarbons in the marine environment. In 1970, Andelman and Snodgrass [172] discussed the occurrence and significance of polynuclear aromatic hydrocarbons in the aquatic environment with particular emphasis on health aspects associated with such compounds. However, their statement that no polynuclear aromatic hydrocarbons have been isolated from uncracked crude oil is incorrect; a number of polynuclear aromatic hydrocarbons have been isolated from crude oil, a few of which are carcinogenic [158]. Harrison [19] in 1975 reviewed the sources of polynuclear aromatic hydrocarbons in the aquatic environment. Blumer [173] presented a discussion of the geochemical occurrence of polynuclear aromatic hydrocarbons in the environment with reference to their stability and possible modes of formation. Martin and Blumer [174] compiled a listing of 1,055 references on the occurrence and analysis of polycyclic hydrocarbons.

The aromatic hydrocarbons constitute less than one percent of the total hydrocarbons in marine organisms; however, even this figure is open to question because most of the biological samples examined to date had been collected from areas of suspected pollution [27] or analytical techniques may have been employed in which contamination could have been a problem. The ubiquity of some polynuclear aromatic hydrocarbons in the marine environment suggests the possibility of indigenous formation in plants and microorganisms. This is based on the fact that the wide variety of polynuclear hydrocarbons found in the marine environment may not have been formed solely by a pyrolytic (burning) process which is thought to be a major pollution pathway for the introduction of polynuclear aromatic hydrocarbons [175].

The low-molecular weight aromatic hydrocarbons constitute the inherent toxic components of oil pollutants; benzene, toluene, and the xylenes in particular are acute poisons for man as well as for marine organisms (See Chapter 3 of Volume II). These volatile aromatic hydrocarbons are more water-soluble than equivalent molecular weight paraffins and are lethal to aquatic organisms either by direct contact or in dilute solutions [68]. Under certain conditions, naphthalene and phenanthrene can be even more toxic to fish than benzene or the alkylated benzenes [176]. Certain of the non-volatile aromatics are suspected of being cumulative toxicants. Some, such as the alkylated 4- and 5-ring aromatics as well as benzo[a]pyrene, are potent inducers of tumors in test animals [177].

NON-HYDROCARBON COMPONENTS

The level of non-hydrocarbon components in crude petroleum is generally only a few percent, although it can reach as high as 50%. These compounds are of particular interest, because many of them are toxic to marine organisms and are relatively soluble in water. The toxic non-hydrocarbon compounds in petroleum include cresols, xylenols, naphthols, carboxylic acids, quinoline and substituted quinolines, hydroxybenzoquinolines, and substituted pyridines.

Farrington et al. [140] stated they were unaware of any published information on the analyses for non-hydrocarbon components which have been used for estimating petroleum contamination in aquatic organisms except for certain reports on ultraviolet fluorescence studies [117,121] that included fluorescent non-hydrocarbon compounds as well as fluorescent aromatic hydrocarbons. No reports of studies on the degradation of the non-hydrocarbon compounds from petroleum in marine organisms were found in the published literature [68].

BASIC DIFFERENCES BETWEEN ABIOTIC AND BIOTIC SYSTEMS

Only a few systematic studies have been reported on the background, or biogenic, content and distribution patterns of hydrocarbons and non-hydrocarbon compounds in marine organisms (See Chapter 2). In most studies, analytical techniques were used that would identify only one or two specific classes of hydrocarbons, so that other classes of hydrocarbons, even if present, would not have been identified [33]. It is possible, therefore, that other classes of hydrocarbons are more prevalent in nature than the limited number of published comprehensive analyses would suggest.

Marine organisms can contain many types of hydrocarbons. Nevertheless, analyses of individual species of organisms have revealed only a few members of each homologous hydrocarbon series. In some marine organisms, the biogenic hydrocarbons consist of only a few members of a single class, of which one member may be predominant. Exceptions may be the microorganisms (bacteria, yeasts, and molds) which may contain hydrocarbon classes covering a wide range of molecular weights and a variety of molecular structures [33,151].

The following information taken from Farrington, Teal, and Parker [33] summarizes some of the principal differences between biotic and abiotic (petroleum) hydrocarbons. Not all differences apply to all marine macroorganisms, nor to all crude petroleums or refined petroleum products.

(1) Petroleum contains a much more complex mixture of hydrocarbons with a much greater range of molecular structures and molecular weights than has been reported for biogenic

hydrocarbons found in marine organisms.

(2) Petroleum contains many homologous series within each of the hydrocarbon classes with the members within each series usually present in nearly equal concentrations.

(3) Petroleum contains a complex mixture of cycloparaffins and aromatic hydrocarbon compounds; marine organisms appear to contain only a few compounds from each of these groups, based on published results.

(4) Petroleum contains numerous naphthenoaromatic hydrocarbons, none of which have been found in marine organisms.

(5) Petroleum contains many non-hydrocarbon compounds which appear to be absent in marine organisms based on present analytical results.

(6) Crude petroleum is generally devoid of olefins. The compounds are sometimes formed in the cracking process and may appear in some refined products. Marine organisms contain a wide variety of olefins.

(7) Petroleum can contain isoprenoid alkanes over a range from C_{12} to C_{22}. The isoprenoid alkanes in marine organisms are generally limited to those containing 19 and 20 carbon atoms.

(8) Petroleum has no marked odd-number carbon chain length predominance over the range from C_{14} to C_{22}. The aliphatic hydrocarbons of most marine macroorganisms are predominantly of odd-numbered carbon chain length over this range.

(9) Organisms possesss specific biosynthetic processes or pathways which favor the production of hydrocarbons in preferred molecular size ranges. Thus, calanoid copepods contain, almost exclusively, a branched-chain hydrocarbon containing 19 carbon atoms. Many algae contain high concentrations of *n*-pentadecane and *n*-heptadecane and virtually no other normal paraffins [108].

(10) Pentacyclic triterpane structures may be a useful petroleum pollution indicator in recent marine sediments. A series of "molecular pollution markers" ranging from C_{27} to C_{35} has been isolated from sediments off the west coast of France and has been identified as the (17αH, 21β)-hopane series. Dastillung and Albrecht [178] described the use of these polycyclic biological markers.

The differentiation between hydrocarbons introduced naturally from geochemical sources, such as oil seeps or weathering of petrolic rocks, and those introduced by man (pollution) may be extremely difficult in environmental samples exposed to both sources of contamination.

An approach for discriminating between biotic and abiotic origin of hydrocarbons is the use of carbon-14 age dating techniques, provided sufficient hydrocarbon material (>0.5 g) can be isolated. An age older than Recent (10,000 years) indicates a predominance of petroleum or abiotic hydrocarbons. This assumes that recently biosynthesized biotic hydrocarbons do not have a fossil carbon source and that the contribution of hydrocarbons from weathering of ancient sediments is negligible [32].

Teal and Farrington [179] compared the hydrocarbon content and distribution pattern in benthic animals with that of the estuarine and marsh habitats of these animals. When exposed to oil spilled in the environment, some marine animals will absorb and incorporate the hydrocarbons in the same compositional pattern present in the oil; other marine animals will take up the hydrocarbons but will metabolize petroleum hydrocarbons until the hydrocarbon patterns appear similar to those in the organisms during pre-spill conditions. Field studies have demonstrated the usefulness of sensitive hydrocarbon analyses to discriminate between biotic (baseline) and abiotic (petroleum) hydrocarbons in marine organisms [98,99, 108,180].

PROSPECTUS

Petroleum is a complex mixture containing primarily a staggering array of parallel homologous series of hydrocarbons, as well as relatively small amounts of nitrogen-, sulfur-, and oxygen-containing organic compounds, and certain trace metals, either in elemental form or as salts or organometallic compounds. The naturally occurring petroleum was formed from the partial degradation of plant and animal matter over geological time, as influenced by environmental conditions. Its components, therefore, may bear little resemblance to their biochemical ancestors. The wide range of chemical properties and physical characteristics exhibited by the world's crude oils arise from the various combinations of the thousands of constituents that make up the petroleum. Yet, in spite of the large number of complicated mixtures possible, the various crude oils do show general trends in distribution of their various components and in physical characteristics. Refining of crude petroleum, the processes used to convert petroleum to useful products, further alters its chemical make-up and physical characteristics. The environmental research analyst, therefore, is confronted with several almost insuperable problems that must be resolved in order to assess the impact of petroleum in the marine environment.

First, there is the problem of identification and quantification of the thousands of chemical components of petroleum and petroleum products and the classification of the chemical and physical properties of these components singly and in combination. Secondly, there is the problem of identifying and classifying the chemical and physical changes that occur or are produced by petroleum components which can undergo differential changes and dispersion throughout the marine environment following oil spills or chronic discharges. And finally, there is the problem of identifying and assessing the levels in the environment of petroleum-type hydrocarbons produced naturally by living marine organisms.

The ability to study the low levels of hydrocarbons (less than one ppm) in the environment depends upon the availability of highly sophisticated analytical equipment and ingenious techniques. Suitable equipment has been developed and has become available only quite recently. These instruments are highly sensitive and selective to the various organic components present in petroleum. Considerable effort is being carried out with these instruments in the adaptation of analytical methodology for identifying and quantifying the petroleum constituents in marine biota, water, and sediment. But the use of sophisticated analytical systems creates additional problems, particularly when the techniques are transferred from the laboratory to field conditions or environmental samples. Because the analytical procedures involve the assessment of target components to the parts per million or even parts per billion level, the possibility of error through contamination is serious. Problem areas, under field conditions particularly, involve sampling and storage of biological tissue, bulk water, and moist sediments in an environment containing ubiquitous fossil hydrocarbons. These hydrocarbons arise from the sea surface film, aerosols, fuels, lubricants, preservatives, exhausts from vessels, and others, and from bacterial action on petroleum and other products. In order to provide useful data, the level of hydrocarbon contamination introduced into the samples must be no greater than 10 ppb or less than 10% of that in the test samples, whichever is smaller.

Considerable research should be carried out on the development of rapid, simple, and low-cost, but reliable, systems for extracting and separating hydrocarbons from biological and geological matrices. Suitable techniques are presently available, but they are complicated and expensive. Until economical systems are developed, the amount of research will necessarily be limited.

Marine organisms not only assimilate and recycle petroleum hydrocarbons from the environment, but also synthesize similar compounds. We need more information on the nature and role of the biogenic hydrocarbons in order to accurately discriminate between hydrocarbons from abiotic sources and those from biotic systems. Such research involves the fields of marine ecology, marine geochemistry, marine biochemistry, and analytical chemistry. Such knowledge may best be acquired by interdisciplinary efforts.

Absolute, or baseline, levels of petroleum hydrocarbons and other man-contributed hydrocarbon contaminants in the various compartments of the marine environment (water, sediment, atmosphere, biota) should be established soon. Such information is needed for two principal reasons. First, baseline, or benchmark, levels provide reference points in determining future changes brought about by natural or man-made influences. Secondly, since some petroleum hydrocarbons and non-hydrocarbon organic compounds exert short- and long-term effects on biological systems, benchmark data will provide the basis for designing laboratory bioassay studies to assess the effect of such hydrocarbons on marine ecosystems. The principal targets for continued monitoring programs will be those components which display lethal or sublethal effects on the biota based on realistic laboratory and field effects studies.

The success of both research and baseline studies will depend upon the development of standardized techniques for the classification and quantification of petroleum hydrocarbons. Although it has been difficult to achieve, intralaboratory and interlaboratory standardization of the analytical systems should be established worldwide. At the least, some specifically-detailed and scientifically-sound criteria should be established for interlaboratory comparison of procedures and for the interpretation of the results. The nature of the problem and the state of the art have progressed to the point where standardization of techniques and reliable interpretation of results are required to fully assess the global effect of petroleum contamination on the marine environment.

While we do not expect that the analytical problems for characterizing abiotic petroleum hydrocarbons in the presence of biotic hydrocarbons to be significantly different in arctic and subarctic environments compared to more temperate marine systems, we are hampered by the lack of research results to confirm them. Crude oils from these high latitude regions appear to be similar in chemical and physical nature to those from more southerly regions. Suitable research is most noticeably lacking in the field of fate and effects of petroleum on arctic and subarctic marine environments.

As additional accurate and pertinent data are collected, new questions will undoubtedly arise, quite likely faster than we are able to answer the questions which we originally set out to investigate. Max Blumer [181] said that the rapid expansion of our knowledge about organic geochemistry frequently occurs after a lag period following the discovery of a major geochemical principle or property. This period of slow growth is not due to a lack of activity in the field but rather to a limitation in the resolving power of the methods or analytical instruments then in use. With the application of a new generation of methods and instrumentation, insight into the complexity (but also the order) of environmental and geochemical principles blossoms; this has happened, for example, in the investigation of environmental polycyclic aromatic hydrocarbons. We suggest that we are now entering the initial, exciting upsurge following a relatively quiet period of much preparatory activity but of few major breakthroughs. The period started in the late Sixties with the increased awareness of the potential impacts of petroleum pollution on the marine environment and, hopefully, if our predictions are correct, we should experience a large increase in the immediate future in the basic understanding of the routes, rates, and reservoirs of hydrocarbons in the marine environment.

REFERENCES

1. Bishop, P.W. (1969). Petroleum. Smithsonian Institution Press, Washington, D.C. Smithson. Publ. 4751, 31 p.
2. Meyerhoff, A.A. (1976). Economic impact and geopolitical implications of giant petroleum fields. Am. Sci. 64:536-41.
3. Davis, J.B. (1967). Petroleum Microbiology. Elsevier Publishing Co., Amsterdam, 604 p.
4. Eglinton, G. (1969). Organic geochemistry. The organic chemist's approach. In: Organic Geochemistry: Methods and Results (G. Eglinton and M.T.J. Murphy, eds.), p. 20-73. Springer-Verlag, New York.
5. Bestougeff, M.A. (1967). Petroleum hydrocarbons. In: Fundamental Aspects of Petroleum Geochemistry (B. Nagy and U. Colombo, eds.), p. 77-108. Elsevier Publishing Co., Amsterdam.
6. Koons, C.B., C.D. McAuliffe, and F.T. Weiss (1976). Environmental aspects of produced waters from oil and gas extraction operations in offshore and coastal waters. Paper OTC 2447. Offshore Technology Conference, 3-6 May 1976. Offshore Technology Conference, Dallas, Tex., 11 p.
7. Weiss, F.T. (1976). Personal communication. Shell Development Center, Houston, Tex.
8. Conkling, Inc. (1966). The Potential for the Petroleum Industry in the Pacific Northwest. Vol. II, Part 11C. Pacific Northwest Economic Base Study for Power Markets. Bonneville Power Administration, U.S. Dep. of the Interior, Portland, Oregon, 196 p.
9. Nelson-Smith, A. (1973). Oil Pollution and Marine Ecology. Plenum Press, New York, 260 p.

10. American Society for Testing and Materials (1976). Annual Book of ASTM Standards (48 parts, published annually), Philadelphia.
11. Institute of Petroleum (1974). IP Standards for Petroleum and its Products (2 parts, published annually), Applied Science Publishers, Barking, Essex, England.
12. Institute of Petroleum Oil Pollution Analysis Committee (1974). Marine Pollution by Oil. Applied Science Publishers, Barking, Essex, England, 198 p.
13. International Petroleum Encyclopedia. (1974) (1975) (1976) Petroleum Publishing Co., Tulsa, Okla., 468 p., 480 p., 456 p.
14. Aalund, L.R. (1976). Wide variety of world crudes gives refiners range of charge stocks. Oil Gas J. 74(13):87-122; 74(15):72-8; 74(17): 112-26; 74(19):85-94; 74(21): 80-7; 74(23):139-48; 74(25):137-52; 74(27):98-108.
15. Bureau of Naval Personnel (1965). Fundamentals of Petroleum. NAVPERS 10883-A. U.S. Navy, Washington, D.C., 200 p.
16. Dean, R.A. (1968). The chemistry of crude oils in relation to their spillage on the sea. In: The Biological Effects of Oil Pollution on Littoral Communities (J.D. Carthy and D.R. Arthur, eds.), Suppl. to Field Studies, Vol. 2, p. 1-6. Obtainable from E.W. Classey, Ltd., Hampton, Middx., England.
17. Encyclopedia of Science and Technology (1971). Petroleum, Petroleum Processing, and Petroleum Products, Vol. 10, p. 64-6, 74-82. McGraw-Hill, New York.
18. Meinschein, W.G. (1969). Hydrocarbons - saturated, unsaturated and aromatic. In: Organic Geochemistry: Methods and Results (G. Eglinton and M.T.J. Murphy, eds.), p. 330-56. Springer-Verlag, New York.
19. Harrison, R.M., R. Perry, and R.A. Wellings (1975). Polynuclear aromatic hydrocarbons in raw, potable and waste waters. Water Res. 9:331-46.
20. Speers, G.C. and E.V. Whitehead (1969). Crude petroleum. In: Organic Geochemistry: Methods and Results (G. Eglinton and M.T.J. Murphy, eds.), p. 638-75. Springer-Verlag, New York.
21. Clark, R.C., Jr. (1976). Sources and levels of trace metals in the marine environment, Unpublished report. Northwest and Alaska Fisheries Center, NMFS, NOAA, U.S. Dep. of Commerce, Seattle, Washington, 47 p.
22. Smith, I.C., T.L. Ferguson, and B.L. Carson (1975). Metals in new and used petroleum products and by-products: Quantities and consequences. In: The Role of Trace Metals in Petroleum (T.F. Yen, ed.), p. 123-48. Ann Arbor Science Publishers, Ann Arbor, Michigan.

23. Thompson, C.J., H.J. Coleman, J.E. Dooley, and D.E. Hirsch (1971). Bumines analysis shows characteristics of Prudhoe Bay crude. Oil Gas J. 69(43):112-20.
24. Coleman, H.J., J.E. Dooley, D.E. Hirsch, and C.J. Thompson (1973). Compositional studies of a high-boiling 370-535°C distillate from Prudhoe Bay, Alaska, crude oil. Anal. Chem. 45:1724-37.
Alyeska Pipeline Service Co. (1971). Supplement to description of marine transportation system - Valdez to West Coast ports. Letter submitted to U.S. Dep. of the Interior, 24 Sept. 1971, 4 p., 6 attach.
25. Pancirov, R.J. (1974). Compositional data on API reference oils used in biological studies: A #2 fuel oil, a Bunker C, Kuwait crude oil and South Louisiana crude oil. Rept. AID.1BA.74. ESSO Research and Engineering Co., Linden, New Jersey, 6 p.
26. Vaughan, B.E. (1973). Effects of oil and chemically dispersed oil on selected marine biota - A laboratory study. Am. Petrol. Inst. Publ. 4191, 105 p.
27. National Academy of Sciences (1975). Petroleum in the Marine Environment. Washington, D.C., 107 p.
28. Clark, R.C., Jr. and J.S. Finley (1973). Techniques for analysis of paraffin hydrocarbons and for interpretation of data to assess oil spill effects in marine organisms. In: Proceedings of 1973 Joint Conference on Prevention and Control of Oil Spills, p. 161-72. American Petroleum Institute, Washington, D.C.
29. Farrington, J.W. (1973). Analytical techniques for the determination of petroleum contamination in marine organisms. Unpublished manuscript. Woods Hole Oceanogr. Inst. Tech. Rep. 73-57, 27 p.
30. Wakeham, S.G. (1976). The geochemistry of hydrocarbons in Lake Washington. Ph.D. Thesis, University of Washington, Seattle, 192 p.
31. Giger, W. and M. Blumer (1974). Polycyclic aromatic hydrocarbons in the environment: Isolation and characterization by chromatography, visible, ultraviolet and mass spectrometry. Anal. Chem. 46:1663-71.
32. Farrington, J.W. and P.A. Meyers (1975). Hydrocarbons in the marine environment. In: Environmental Chemistry (G. Eglinton, ed.), Vol. I, p. 109-36. The Chemical Society, London.
33. Farrington, J.W., J.M. Teal, and P.L. Parker (1976). Petroleum hydrocarbons. In: Strategies for Marine Pollution Monitoring (E.D. Goldberg, ed.), p. 3-34. Wiley-Interscience, New York.

34. Farrington, J.W., J.M. Teal, J.G. Quinn, P.L. Parker, J.K. Winters, T.L. Wade, and K.A. Burns (1974). Analyses of hydrocarbons in marine organisms: Results of IDOE intercalibration exercises. In: Marine Pollution Monitoring (Petroleum). Natl. Bur. Stand. Spec. Publ. 409, p. 163-6.
35. Medeiros, G.C. and J.W. Farrington (1974). IDOE-5 intercalibration sample: Results of analysis after sixteen months storage. In: Marine Pollution Monitoring (Petroleum). Natl. Bur. Stand. Spec. Publ. 409, p. 167-9.
36. Clark, R.C., Jr. (1974). Methods for establishing levels of petroleum contamination in organisms and sediment as related to marine pollution monitoring. In: Marine Pollution Monitoring (Petroleum). Natl. Bur. Stand. Spec. Publ. 409, p. 189-94.
37. Blumer, M. (1965). Contamination of a laboratory building by air filters. Contam. Control 4(5):13, 15.
38. Duce, R.A., J.G. Quinn, C.E. Olney, S.R. Piotrowicz, B.J. Ray, and T.L. Wade (1972). Enrichment of heavy metals and organic compounds in the surface microlayer of Narragansett Bay, Rhode Island. Science 176:161-3.
39. Wade, T.L. and J.G. Quinn (1975). Hydrocarbons in the Sargasso Sea surface microlayer. Mar. Pollut. Bull. 6:54-7.
40. MacIntyre, F. (1974). The top millimeter of the ocean. Sci. Am. 230(5):62-77.
41. Ledet, E.J. and J.L. Laseter (1974). Alkanes at the air-sea interface from offshore Louisiana and Florida. Science 186:261-3.
42. Grice, G.D., G.R. Harvey, V.T. Bowen, and R.H. Backus (1972). The collection and preservation of open ocean marine organisms for pollution analysis. Bull. Environ. Contam. Toxicol. 7:125-32.
43. Clayton, J.R., Jr. and S.P. Pavlou (1976). A zooplankton net for excluding surface film components. Unpublished report. University of Washington, Seattle, 15 p.
44. Harvey, G.R. and J.M. Teal (1973). PCB and hydrocarbon contamination of plankton by nets. Bull. Environ. Contam. Toxicol. 9:287-90.
45. Barnes, H. (1959). Oceanography and Marine Biology, A Book of Techniques. George Allen and Unwin, London, 218 p.
46. Brown, R.A., T.D. Searl, J.J. Elliott, B.G. Phillips, D.E. Brandon, and P.H. Monaghan (1973). Distribution of heavy hydrocarbons in some Atlantic Ocean waters. In: Proceedings of 1973 Joint Conference on Prevention and Control of Oil Spills, p. 505-19. American Petroleum Institute, Washington, D.C.

47. Gump, B.H., H.S. Hertz, W.E. May, S.N. Chesler, S.M. Dyszel, and D.P. Enagonio (1975). Drop sampler for obtaining fresh and sea water samples for organic compound analysis. Anal. Chem. 47:1223-4.
48. Keizer, P.D. and D.C. Gordon, Jr. (1973). Detection of trace amounts of oil in sea water by fluorescence spectroscopy. J. Fish. Res. Board Can. 30:1039-46.
49. Gordon, D.C., Jr., P.D. Keizer, and J. Dale (1974). Estimates using fluorescence spectroscopy of the present state of petroleum hydrocarbon contamination in the water column of the Northwest Atlantic Ocean. Mar. Chem. 2:251-61.
50. Clark, R.C., Jr., M. Blumer, and S.O. Raymond (1967). A large water sampler, rupture-disc triggered, for studies of dissolved organic compounds. Deep-Sea Res. 14:125-8.
51. Gordon, D.C., Jr. and P.D. Keizer (1974). Estimation of petroleum hydrocarbons in seawater by fluorescence spectroscopy: Improved sampling and analytical methods. Fish. Res. Board Can. Tech. Rep. 481, 28 p.
52. Colwell, R.R. (1975). Personal communication. Dep. of Microbiology, University of Maryland, College Park.
53. Simoneit, B.R., D.H. Smith, G. Eglinton, and A.L. Burlingame (1973). Application of real-time mass spectrometric techniques to environmental organic geochemistry. II. Organic matter in San Francisco Bay area water. Arch. Environ. Contam. Toxicol. 1:193-208.
54. Swinnerton, J.W. and V.J. Linnenbom (1967). Determinaton of the C_1 to C_4 hydrocarbons in sea water by gas chromatography. J. Gas Chromatogr. 5:570-3.
55. Swinnerton, J.W. and R.A. Lamontagne (1974). Oceanic distribution of low-molecular-weight hydrocarbons. Baseline measurements. Environ. Sci. Technol. 8:657-63.
56. Brooks, J.M., A.D. Fredericks, W.M. Sackett, and J.W. Swinnerton (1973). Baseline concentrations of light hydrocarbons in Gulf of Mexico. Environ. Sci. Technol. 7:639-42.
57. Prough, R.G. (1976). Sniffing for hydrocarbons in the sea. Ocean Ind. 11(6):110-2.
58. Rohrback, B.G. and W.E. Reed (1976). Evaluation of extraction techniques for hydrocarbons in marine sediments. University of California, Los Angeles, Inst. Geophys. Planet. Phys. Publ. 1537, 41 p.
59. McIver, R.D. (1962). Ultrasonics - A rapid method for removing soluble organic matter from sediments. Geochim. Cosmochim. Acta 26:343-5.
60. Golden, C. and E. Sawicki (1975). Ultrasonic extraction of total particulate aromatic hydrocarbons (TpAH) from airborne particles at room temperature. Intern. J. Environ. Anal. Chem. 4:9-23.

61. Ackman, R.G. and D. Noble (1973). Steam distillation: A simple technique for recovery of petroleum hydrocarbons from tainted fish. J. Fish. Res. Board Can. 30:711-4.
62. Vale, G.L., G.S. Sidhu, W.A. Montgomery, and A.R. Johnson (1970). Studies on a kerosene-like taint in mullet (*Mugil cephalus*). I. General nature of the taint. J. Sci. Food Agric. 21:429-32.
63. Hunter, L., H.E. Guard, and L.H. DiSalvo (1974). Determination of hydrocarbons in marine organisms and sediments by thin layer chromatography. In: Marine Pollution Monitoring (Petroleum). Natl. Bur. Stand. Spec. Publ. 409, p. 213-6.
64. Hunter, L. (1975). Quantitation of environmental hydrocarbons by thin-layer chromatography. Gravimetry/densitometry comparison. Environ. Sci. Technol. 9:241-6.
65. Farrington, J.W. and G.C. Medeiros (1975). Evaluation of some methods of analysis for petroleum hydrocarbons in marine organisms. In: Proceedings of 1975 Conference on Prevention and Control of Oil Pollution, p. 115-21. American Petroleum Institute, Washington, D.C.
66. Blaylock, J.W., P.W. O'Keefe, J.N. Roehm, and R.E. Wildung (1973). Determination of *n*-alkane and methylnaphthalene compounds in shellfish. In: Proceedings of 1973 Joint Conference on Prevention and Control of Oil Spills, p. 173-7. American Petroleum Institute, Washington, D.C.
67. Warner, J.S. (1976). Determination of aliphatic and aromatic hydrocarbons in marine organisms. Anal. Chem. 48:578-83.
68. Anderson, J.W., R.C. Clark, Jr., and J.J. Stegeman (1974). Petroleum hydrocarbons. In: Marine Bioassays Workshop Proceedings (G.V. Cox, ed.), p. 36-75. Marine Technology Society, Washington, D.C.
69. McAuliffe, C.D. (1969). Solubility in water of normal C_9 and C_{10} alkane hydrocarbons. Science 163:478-9.
70. McAuliffe, C.D. (1971). GC determination of solutes by multiple phase equilibrium. Chem. Technol. 1:46-51.
71. McAuliffe, C.D. (1974). Determination of C_1-C_{10} hydrocarbons in water. In: Marine Pollution Monitoring (Petroleum). Natl. Bur. Stand. Spec. Publ. 409, p. 121-5.
72. Schink, D.R., N.L. Guinasso, Jr., S.S. Sigalov, and N.E. Cima (1971). Hydrocarbons under the sea - A new survey - Techniques. Paper OTC 1339 1. Offshore Technology Conference, Houston, Texas. Offshore Technology Conference, Dallas, Tex., p. 130-42.
73. Fort, E.R., B.O. Prescott, and A. Walters (1973). Mapping hydrocarbon seepages in water-covered regions. U.S. Patent 3,747,405.

74. U.S. Environmental Protection Agency (1974). Methods for Chemical Analysis of Water and Wastes: Oil and grease, total, recoverable. Methods Development and Quality Assurance Research Laboratory, Cincinnati, Ohio, p. 229-31.
75. Levy, E.M. (1971). The presence of petroleum residues off the east coast of Nova Scotia, in the Gulf of St. Lawrence, and the St. Lawrence River. Water Res. 5:723-33.
76. Thruston, A.D., Jr. and R.W. Knight (1971). Characterization of crude and residual-type oils by fluorescence spectroscopy. Environ. Sci. Technol. 5:64-9.
77. Zitko, V. and W.V. Carson (1970). The characterization of petroleum oils and their determination in the aquatic environment. Fish. Res. Board Can. Tech. Rep. 217, 29 p.
78. Hornig, A.W. (1974). Identification, estimation and monitoring of petroleum in marine waters by luminescence methods. In: Marine Pollution Monitoring (Petroleum). Natl. Bur. Stand. Spec. Publ. 409, p. 135-44.
79. Gordon, D.C., Jr. and P.D. Keizer (1974). Hydrocarbon concentrations in seawater along the Halifax-Bermuda section: Lessons learned regarding sampling and some results. In: Marine Pollution Monitoring (Petroleum). Natl. Bur. Stand. Spec. Publ. 409, p. 113-5.
80. Schwarz, F.P. and S.P. Wasik (1976). Fluorescence measurements of benzene, naphthalene, anthracene, pyrene, fluoranthene, and benzo[e]pyrene in water. Anal. Chem. 48:524-8.
81. Simard, R.G., I. Hasegawa, W. Bandaruk, and C.E. Headington (1951). Infrared spectrophotometric determination of oil and phenols in water. Anal. Chem. 23:1384-7.
82. Brown, R.A., J.J. Elliott, J.M. Kelliher, and T.D. Searl (1975). Sampling and analysis of nonvolatile hydrocarbons in ocean water. In: Analytical Methods in Oceanography (T.R.P. Gibbs, Jr., ed.). Am. Chem. Soc. Adv. Chem. Ser. 147, p. 172-87.
83. Bean, R.M. (1974). Suspensions of crude oils in sea water: Rapid methods of characterizing light hydrocarbon solutes. In: Marine Pollution Monitoring (Petroleum). Natl. Bur. Stand. Spec. Publ. 409, p. 127-30.
84. Kawahara, F.K. (1974). Recent developments in the identification of asphalts and other petroleum products. In: Marine Pollution Monitoring (Petroleum). Natl. Bur. Stand. Spec. Publ. 409, p. 145-8.
85. Gruenfeld, M. (1973). Extraction of dispersed oils from water for quantitative analysis by infrared spectrophotometry. Environ. Sci. Technol. 7:636-9.

86. Ehrhardt, M. and M. Blumer (1972). The source identification of marine hydrocarbons by gas chromatography. Environ. Pollut. 3:179-94.
87. Adlard, E.R., L.F. Creaser, and P.H.D. Matthews (1972). Identification of hydrocarbon pollutants on sea and beaches by gas chromatography. Anal. Chem. 44:64-73.
88. Kreider, R.E. (1971). Identification of oil leaks and spills. In: Proceedings of 1971 Joint Conference on Prevention and Control of Oil Spills, p. 119-24. American Petroleum Institute, Washington, D.C.
89. Zafiriou, O.C., M. Blumer, and J. Myers (1972). Correlation of oils and oil products by gas chromatography. Unpublished manuscript. Woods Hole Oceanogr. Inst. Tech. Rep. 72-55, 110 p.
90. Zafiriou, O.C., J. Myers, R. Bourbonniere, and F.J. Freestone (1973). Oil spill-source correlation by gas chromatography: An experimental evaluation of system performance. In: Proceedings of 1973 Joint Conference on Prevention and Control of Oil Spills, p. 153-9. American Petroleum Institute, Washington, D.C.
91. Zafiriou, O.C. (1973). Improved method for characterizing environmental hydrocarbons by gas chromatography. Anal. Chem. 45:952-6.
92. Grob, K. (1973). Organic substances in potable water and its precursor. Part I. Methods for their determination by gas-liquid chromatography. J. Chromatogr. 84:255-73.
93. Brown, R.A. (1965). Mass spectrometry of hydrocarbons. In: Hydrocarbon Analysis. ASTM Spec. Tech. Publ. 389, p. 68-102.
94. Bieri, R.H., A.L. Walker, B.W. Lewis, G. Losser, and R.J. Huggett (1974). Identification of hydrocarbons in an extract from estuarine water accommodated No. 2 fuel oil. In: Marine Pollution Monitoring (Petroleum). Natl. Bur. Stand. Spec. Publ. 409, p. 149-53.
95. Lao, R.C., R.S. Thomas, and J.L. Monkman (1975). Computerized gas chromatographic-mass spectrometric analysis of polycyclic aromatic hydrocarbons in environmental samples. J. Chromatogr. 112:681-700.
96. Farrington, J.W. and B.W. Tripp (1975). A comparison of analysis methods for hydrocarbons in surface sediments. In: Marine Chemistry in the Coastal Environment. Am. Chem. Soc. Symp. Ser. 18, p. 267-84.
97. Chesler, S.N., B.H. Gump, H.S. Hertz, W.E. May, S.M. Dyszel, and D.P. Enagonio (1976). Trace hydrocarbon analysis: The National Bureau of Standards Prince William Sound/Northeastern Gulf of Alaska baseline study. Natl. Bur. Stand. Tech. Note 889, 73 p.

98. Clark, R.C., Jr., J.S. Finley, B.G. Patten, D.F. Stefani, and E.E. DeNike (1973). Interagency investigations of a persistent oil spill on the Washington coast: Animal population studies, hydrocarbon uptake by marine organisms, and algal response following the grounding of the troopship *General M.C. Meigs*. In: Proceedings of 1973 Joint Conference on Prevention and Control of Oil Spills, p. 793-808. American Petroleum Institute, Washington, D.C.
99. Clark, R.C., Jr., J.S. Finley, B.G. Patten, and E.E. DeNike (1975). Long-term chemical and biological effects of a persistent oil spill following the grounding of the *General M.C. Meigs*. In: Proceedings of 1975 Conference on Prevention and Control of Oil Pollution, p. 479-87. American Petroleum Institute, Washington, D.C.
100. Clark, R.C., Jr. and M. Blumer (1967). Distribution of *n*-paraffins in marine organisms and sediment. Limnol. Oceanogr. 12:79-87.
101. Blumer, M., P.C. Blokker, E.B. Cowell, and D.F. Duckworth (1972). Petroleum. In: A Guide to Marine Pollution (E.D. Goldberg, ed.), p. 19-40. Gordon and Breach, New York.
102. Farrington, J.W. and J.G. Quinn (1973). Petroleum hydrocarbons in Narragansett Bay. I. Survey of hydrocarbons in sediments and clams (*Mercenaria mercenaria*). Estuarine Coastal Mar. Sci. 1:71-9.
103. Warner, J.S. (1975). Determination of sulfur-containing petroleum components in marine samples. In: Proceedings of 1975 Conference on Prevention and Control of Oil Pollution, p. 97-101. American Petroleum Institute, Washington, D.C.
104. Brown, R.A., J.J. Elliott, and T.S. Searl (1974). Measurement and characterization of nonvolatile hydrocarbons in ocean water. In: Marine Pollution Monitoring (Petroleum). Natl. Bur. Stand. Spec. Publ. 409, p. 131-3.
105. Blaylock, J.W., R.M. Bean, and R.E. Wildung (1974). Determination of extractable organic material and analysis of hydrocarbon types in lake and coastal sediments. In: Marine Pollution Monitoring (Petroleum). Natl. Bur. Stand. Spec. Publ. 409, p. 217-9.
106. Blumer, M. (1957). Removal of elemental sulfur from hydrocarbon fractions. Anal. Chem. 29:1039-41.
107. Wasik, S.P. and R.N. Boyd (1974). Determination of aromatic hydrocarbons in sea water using an electrolytic stripping cell. In: Marine Pollution Monitoring (Petroleum). Natl. Bur. Stand. Spec. Publ. 409, p. 117-8.
108. Hites, R.A. and K. Biemann (1972). Water pollution: Organic compounds in the Charles River, Boston. Science 178:158-60.

109. Iliffe, T.M. and J.A. Calder (1974). Dissolved hydrocarbons in the eastern Gulf of Mexico Loop Current and the Caribbean Sea. Deep-Sea Res. 21:481-8.

110. Clark, R.C., Jr., J.S. Finley, and G.G. Gibson (1974). Acute effects of outboard motor effluent on two marine shellfish. Environ. Sci. Technol. 8:1009-14.

111. Miles, D.H., M.J. Coign, and L.R. Brown (1975). The estimation of the amount of Empire Mix crude oil in mullet, shrimp, and oysters in liquid chromatography. In: Proceedings of 1975 Conference on Prevention and Control of Oil Pollution, p. 149-54. American Petroleum Institute, Washington, D.C.

112. Blumer, M., G. Sousa, and J. Sass (1970). Hydrocarbon pollution of edible shellfish by an oil spill. Mar. Biol. (Berl.) 5:195-202.

113. Zsolnay, A. (1974). Determination of aromatic and total hydrocarbon content in submicrogram and microgram quantities in aqueous systems by means of high performance liquid chromatography. In: Marine Pollution Monitoring (Petroleum). Natl. Bur. Stand. Spec. Publ. 409, p. 119-20.

114. Larson, R.A. and J.C. Weston (1976). Analysis of water extracts of crude petroleum by gel permeation chromatography. Bull. Environ. Contam. Toxicol. 16:44-52.

115. Schenck, P.A. and E. Eisma (1964). Quantitative determination of *n*-alkanes in crude oils and rock extracts by gas chromatography. In: Advances in Organic Geochemistry (U. Colombo and G.D. Hobson, eds.), p. 403-15. Macmillan Co., New York.

116. Schiessler, R.W. and D. Flitter (1952). Urea and thiourea adduction of C_5-C_{42}-hydrocarbons. J. Am. Chem. Soc. 74:1720-32.

117. Morris, L.J. (1966). Separations of lipids by silver ion chromatography. J. Lipid Res. 7:717-32.

118. Weiss, F.T. (1970). Determination of Organic Compounds: Methods and Procedures. Wiley-Interscience, New York, 475 p.

119. Leonard, R.O. (1974). A sampling procedure for lipids and other labile materials. Perkin-Elmer Instrument News 24(1):10-1.

120. Neff, J.M. and J.W. Anderson (1975). An ultraviolet spectrophotometric method for the determination of naphthalene and alkylnaphthalenes in the tissues of oil-contaminated marine animals. Bull. Environ. Contam. Toxicol. 14:122-8.

121. Zitko, V. (1975). Aromatic hydrocarbons in aquatic fauna. Bull. Environ. Contam. Toxicol. 14:621-31.

122. Keith, L.H. (1976). Identification and Analysis of Organic Pollutants in Water. Ann Arbor Science Publishers Inc., Ann Arbor, Michigan, 718 p.
123. John, P. and I. Soutar (1976). Identification of crude oils by synchronous excitation spectrofluorimetry. Anal. Chem. 48:520-4.
124. Wakeham, S.G. (1976). Synchronous fluorescence spectroscopy and its application to indigenous and petroleum-derived hydrocarbons in lacustrine sediments. Environ. Sci. Technol. In press.
125. National Bureau of Standards (1974). Marine Pollution Monitoring (Petroleum). Report of Symposium and Workshop, 13-17 May 1974. Natl. Bur. Stand. Spec. Publ. 409, p. 271-99.
126. Coakley, W.A. (1973). Comparative identification of oil spills by fluorescence spectroscopy fingerprinting. In: Proceedings of 1973 Joint Conference on Prevention and Control of Oil Spills, p. 215-22. American Petroleum Institute, Washington, D.C.
127. Hargrave, B.T. and G.A. Phillips (1974). Estimates of oil in aquatic sediments by fluorescence spectroscopy. Environ. Pollut. 8:193-215.
128. Brown, R.A. and H.L. Huffman, Jr. (1976). Hydrocarbons in open ocean waters. Science 191:847-9.
129. Kováts, E. and A.I.M. Keulemans (1964). The Kováts retention index system. Anal. Chem. 36:31A-41A.
130. Farrington, J.W., J.M. Teal, J.G. Quinn, T.L. Wade, and K.A. Burns (1973). Intercalibration of analyses of recently biosynthesized hydrocarbons and petroleum hydrocarbons in marine lipids. Bull. Environ. Contam. Toxicol. 10:129-36.
131. Quinn, J.G. and T.L. Wade (1974). Hydrocarbon analyses of IDOE intercalibration samples of cod liver oil and tuna meal. University of Rhode Island, Narragansett, Grad. Sch. Oceanogr., Mar. Memo Ser. 33, 8 p.
132. Burns, K.A and J.M. Teal (1971). Hydrocarbon incorporation into the salt marsh ecosystem from the West Falmouth oil spill. Unpublished manuscript. Woods Hole Oceanogr. Inst. Tech. Rep. 71-69, 14 p.
133. Morris, R.J. (1973). Uptake and discharge of petroleum hydrocarbons by barnacles. Mar. Pollut. Bull. 4:107-9.
134. Blumer, M., M. Ehrhardt, and J.H. Jones (1973). The environmental fate of stranded crude oil. Deep-Sea Res. 20:239-59.
135. Ehrhardt, M. (1972). Petroleum hydrocarbons in oysters from Galveston Bay. Environ. Pollut. 3:257-71.

136. Tissier, M. and J.L. Oudin (1973). Characteristics of naturally occurring and pollutant hydrocarbons in marine sediments. In: Proceedings of 1973 Joint Conference on Prevention and Control of Oil Spills, p. 205-14. American Petroleum Institute, Washington, D.C.
137. Youngblood, W.W. and M. Blumer (1975). Polycyclic aromatic hydrocarbons in the environment: Homologous series in soils and recent marine sediments. Geochim. Cosmochim. Acta 39:1303-14.
138. Blumer, M. and W.W. Youngblood (1975). Polycyclic aromatic hydrocarbons in soils and recent sediments. Science 188:53-5.
139. National Analytical Facility (1976). Unpublished data. Northwest and Alaska Fisheries Center, NMFS, NOAA, U.S. Dep. of Commerce, Seattle, Washington.
140. Blumer, M. (1969). Oil pollution of the ocean. In: Oil on the Sea (D.P. Hoult, ed.), p. 5-13. Plenum Press, New York.
141. Farrington, J.W., C.S. Giam, G.R. Harvey, P.L. Parker, and J.M. Teal (1972). Analytical techniques for selected organic compounds. In: Marine Pollution Monitoring: Strategies for a National Program (E.D. Goldberg, ed.), p. 152-76. Workshop, Santa Catalina, California Marine Laboratory, 25-28 October 1972. Sponsored by the National Oceanic and Atmospheric Administration, Rockville, Md.
142. Clark, R.C., Jr. (1966). Occurrence of normal paraffin hydrocarbons in nature. Unpublished manuscript. Woods Hole Oceanogr. Inst. Tech. Rep. 66-34, 56 p.
143. Baker, J.T. and V. Murphy (1976). Handbook of Marine Science. Compounds from Marine Organisms. Vol. I. CRC Press, Cleveland, Ohio, 226 p.
144. Goldacre, R.J. (1968). Effect of detergents and oils on the cell membrane. In: The Biological Effects of Oil Pollution on Littoral Communities (J.D. Carthy and D.R. Arthur, eds.), Suppl. to Field Studies, Vol. 2, p. 131-7. Obtainable from E.W. Classey, Ltd., Hampton, Middx., England.
145. Mair, B.J. (1964). Hydrocarbons isolated from petroleum. Oil Gas J. 62(3):130-4.
146. Mair, B.J., A. Ronen, E.J. Eisenbraun, and A.G. Horodysky (1966). Terpenoid precursors of hydrocarbons from the gasoline range of petroleum. Science 154:1339-41.
147. Smith, C.L. and W.G. MacIntyre (1971). Initial aging of fuel oil films of sea water. In: Proceedings of 1971 Joint Conference on Prevention and Control of Oil Spills, p. 457-61. American Petroleum Institute, Washington, D.C.

148. Meinschein, W.G. (1959). Origin of petroleum. Bull. Am. Assoc. Petrol. Geol. 43:925-43.
149. Youngblood, W.W., M. Blumer, R.L. Guillard, and F. Fiore (1971). Saturated and unsaturated hydrocarbons in marine benthic algae. Mar. Biol. (Berl.) 8:190-201.
150. Koons, C.B., G.W. Jamieson, and L.S. Ciereszko (1965). Normal alkane distribution in marine organisms; possible significance to petroleum origin. Bull. Am. Assoc. Petrol. Geol. 49:301-4.
151. Davis, J.B. (1968). Paraffinic hydrocarbons in the sulfate-reducing bacterium, *Desulfovibrio desulfuricaus.* Chem. Geol. 3:155-60.
152. Blumer, M. (1967). Hydrocarbons in digestive tract and liver of a basking shark. Science 156:390-1.
153. Avigan, J. and M. Blumer (1968). On the origin of pristane in marine organisms. J. Lipid Res. 9:350-2.
154. Whittle, K.J. and M. Blumer (1970). Interactions between organisms and dissolved organic substances in the sea. Chemical attraction of the starfish *Asterias vulgaris* to oysters. In: Organic Matter in Nature Waters (D.W. Hood, ed.), p. 495-507. Institute of Marine Science, University of Alaska, Fairbanks.
155. Hills, I.R. and E.V. Whitehead (1966). Triterpanes in optically active petroleum distillates. Nature 209:977-9.
156. Blumer, M., J.C. Robertson, J.E. Gordon, and J. Sass (1969). Phytol-derived, C_{19} di- and triolefinic hydrocarbons in marine zooplankton and fishes. Biochem. 8:4067-74.
157. Blumer, M., M.M. Mullin, and R.R.L. Guillard (1970). A polyunsaturated hydrocarbon (3,6,9,12,15,18-heneicosahexaene) in the marine food web. Mar. Biol. (Berl.) 6:226-35.
158. Youngblood, W.W. and M. Blumer (1973). Alkanes and alkenes in marine benthic algae. Mar. Biol. (Berl.) 21:163-72.
159. Lee, R.F. and A.R. Loeblich, III (1971). Distribution of 21:6 hydrocarbon and its relationship to 22:6 fatty acid in algae. Phytochemistry 10:593-602.
160. Lee, R.F., J.C. Nevenzel, G.A. Paffenhofer, A.A. Benson, S. Parron, and T.E. Kavanagh (1970). A unique hexaene hydrocarbon from a diatom (*Skeletonema costatum*). Biochim. Biophys. Acta 202:386-8.
161. Blumer, M. (1971). Scientific aspects of the oil spill problem. Environ. Aff. 1:54-73.
162. Mair, B.J. and J.M. Barnewall (1964). Composition of the mononuclear aromatic material in the light gas oil range, low refractive index portion, 230°-305°C. J. Chem. Eng. Data 9:282-92.

163. Mair, B.J. and J.L. Martinéz-Picó (1962). Composition of the trinuclear aromatic portion of the heavy gas oil and light lubricating distillate. Proc. Am. Petrol. Inst. 42(Sec. III), p. 173-85.
164. Pancirov, R.J. and R.A. Brown (1975). Analytical method for polynuclear aromatic hydrocarbons in crude oils, heating oils, and marine tissues. In: Proceedings of 1975 Conference on Prevention and Control of Oil Pollution, p. 103-13. American Petroleum Institute, Washington, D.C.
165. Gerarde, H.W. and D.G. Gerarde (1961). The ubiquitous hydrocarbons. Assoc. of Food and Drug Officials, U.S. Rep. 25/26, p. 1-47.
166. Borneff, J., F. Selenka, H. Kunte, and A. Maximos (1968). The synthesis of 3,4-benzpyrene and other polycyclic, aromatic hydrocarbons in plants. Arch. Hyg. Bakteriol. 152/3:279-82.
167. Borneff, J., F. Selenka, H. Kunte, and A. Maximos (1968). Experimental studies on the formation of polycyclic aromatic hydrocarbons in plants. Environ. Res. 2:22-9.
168. Gräf, W. and H. Diehl (1966). Über der naturbedingten Normalpegel kanzerogener polycyclischer Aromate und seine Ursache. Arch. Hyg. Bakteriol. 150:49-59.
169. Grimmer, G. and D. Düvel (1970). Investigations of biosynthetic formation of polycyclic hydrocarbons in higher plants. Z. Naturforsch. 25b:1171-5.
170. Lee, R.F. (1975). Fate of petroleum hydrocarbons in marine zooplankton. In: Proceedings of 1975 Conference on Prevention and Control of Oil Pollution, p. 549-53. American Petroleum Institute, Washington, D.C.
171. ZoBell, C.E. (1971). Sources and biodegradation of carcinogenic hydrocarbons. In: Proceedings of 1971 Joint Conference on Prevention and Control of Oil Spills, p. 441-51. American Petroleum Institute, Washington, D.C.
172. Andelman, J.B. and J.E. Snodgrass (1974). Incidence and significance of polynuclear aromatic hydrocarbons in the water environment. In: CRC Critical Reviews in Environmental Control (R.G. Bond and C.P. Straub, eds.), p. 69-83. CRC Press, Cleveland, Ohio.
173. Blumer, M. (1976). Polycyclic aromatic compounds in nature. Sci. Am. 234(3):34-45.
174. Martin, A. and M. Blumer (1975). Polycyclic aromatic hydrocarbons: Occurrence and analysis - a partial bibliography. Unpublished manuscript. Woods Hole Oceanogr. Inst. Tech. Rep. 75-22, 10 p.
175. Andelman, J.B. and M.J. Suess (1970). Polynuclear aromatic hydrocarbons in the water environment. Bull. W.H.O. 43:479-508.

176. Wilber, C.G. (1969). The Biological Aspects of Water Pollution. Charles C. Thomas, Springfield, Ill., 296 p.
177. Wynder, E.L. and D. Hoffmann (1968). Experimental tobacco carcinogenesis. Science 162:862-71.
178. Dastillung, M. and P. Albrecht (1976). Molecular test for oil pollution in surface sediments. Mar. Pollut. Bull. 7:13-15.
179. Teal, J.M. and J.W. Farrington (1976). A comparison of hydrocarbons in animals and their benthic habitats. Int. Counc. Explor. Sea, Rapports et Proces-Verbaux. In press.
180. Clark, R.C., Jr. and J.S. Finley (1974). Analytical techniques for isolating and quantifying petroleum paraffin hydrocarbons in marine organisms. In: Marine Pollution Monitoring (Petroleum). Natl. Bur. Stand. Spec. Publ. 409, p. 209-12.
181. Blumer, M. (1975). Organic compounds in nature: Limits to our knowledge. Angew. Chem. Int. Ed. Engl. 14:507-14.

Chapter 2

INPUTS, TRANSPORT MECHANISMS, AND OBSERVED CONCENTRATIONS OF PETROLEUM IN THE MARINE ENVIRONMENT

ROBERT C. CLARK, JR.
Environmental Conservation Division

WILLIAM D. MACLEOD, JR.
NOAA National Analytical Facility
Environmental Conservation Division

Northwest and Alaska Fisheries Center
National Marine Fisheries Service
National Oceanic and Atmospheric Administration
U.S. Department of Commerce
Seattle, Washington 98112

Hydrocarbons enter the environment by three general pathways: biosynthesis, geochemical processes, and anthropogenic activities. The following description of these pathways is drawn largely from the 1975 review by Farrington and Meyers [1]. Although an attempt was made to emphasize the arctic and subarctic environment, the majority of the data and information relates to the world oceans in the Northern Hemisphere, especially the North Atlantic Ocean and its margins where probably the greatest input of petroleum hydrocarbons has occurred.

POTENTIAL PETROLEUM HYDROCARBON INPUTS

Marine and land organisms biosynthesize hydrocarbons either *de novo* or from ingested precursor compounds. These hydrocarbons may be released during metabolism or upon death and decomposition of the organism. Estimates of the annual rate of biosynthesis of hydrocarbons by marine primary production are between 1 and 10 million metric tons [1].

The estimated annual rate of input of petroleum from submarine and coastal land seepages is between 0.1 and 10 million metric tons, with 0.6 million metric tons per year suggested as a reasonable estimate in a report by the National Academy of Sciences [2]. Additional, but probably relatively small, geochemical inputs occur from weathering of soils and sediments and from transport of some of the hydrocarbons in such materials to the marine environment. An unknown, but potentially significant, input of hydrocarbons may come from fallout of hydrocarbons from the atmosphere following injection from forest and range fires and incomplete combustion of fossil fuels. A fourth possible input is the production of geochemical hydrocarbons during diagenesis of organic matter [1], although this process may not be important until burial exceeds 1,000 m and the temperature exceeds 50°C.

Anthropogenic inputs discussed here include man's accidental and intentional discharges of fossil fuels that reach the marine environment regardless of the initial point of origin--water, land, or atmosphere. Oil pollution appears to be an inevitable consequence of the dependence of a growing population on an increasing oil-based technology. The widespread production, transportation, and use of petroleum leads to substantial losses which are widely, but not evenly, distributed over the earth. Oil transport is heavily concentrated in sea routes over continental shelves, in coastal areas, and in regions of upwelling water. Routes and estimates of worldwide movements of oil for 1974 are presented in Figure 1 [3].

The use of petroleum is generally concentrated in major industrial areas situated along sea coasts or beside rivers that empty into coastal waters. These coastal water areas have a great biological productivity [4]; thus, the potential for substantial damage to the marine ecosystem from oil pollution may well be greater than if petroleum production, transportation, and utilization were more evenly spread over the world [5].

In this chapter heavy reliance has been placed on two comprehensive publications concerning petroleum contamination of the marine environment: the report "Petroleum in the Marine Environment" [2] and the chapter "Petroleum," in the "Marine Bioassays Workshop Proceedings, 1974" [6]. Additional information was obtained from a report by McAuliffe [7] containing updated estimates of petroleum hydrocarbon inputs to the oceans, especially the North Atlantic.

Several major problems limit our ability to estimate the input and flux of petroleum hydrocarbons into the environment in general and into the oceans in particular. First, few reliable analytical data are available on the distribution of petroleum hydrocarbons in the atmosphere, oceans, earth, and biota [2]. Secondly, although some reliable data exist on sources of petroleum contamination in the areas of marine transportation, offshore petroleum production, and coastal oil refineries, the contamination from land runoff, natural seeps, and atmospheric fallout is quite difficult to quantify. Thirdly, considerable information exists on the contribution of hydrocarbons from large municipalities and industrial centers; however, the data obtained by the different analytical methods used are not always comparable. Furthermore, most of the data do not permit estimation of the contribution of hydrocarbons of biogenic origin [8]. Estimated amounts of petroleum hydrocarbons contributed by various sources are presented in Table 1.

LAND-BASED DISCHARGES

Refineries

Estimates of petroleum discharged into the marine environment from seaboard refineries and petrochemical plants have been made by multiplying the volume of the reported discharges by the petroleum content of these industrial discharges. These estimates range from 200,000 to 300,000 metric tons per year [12,13]. The rate is expected to diminish in the future,

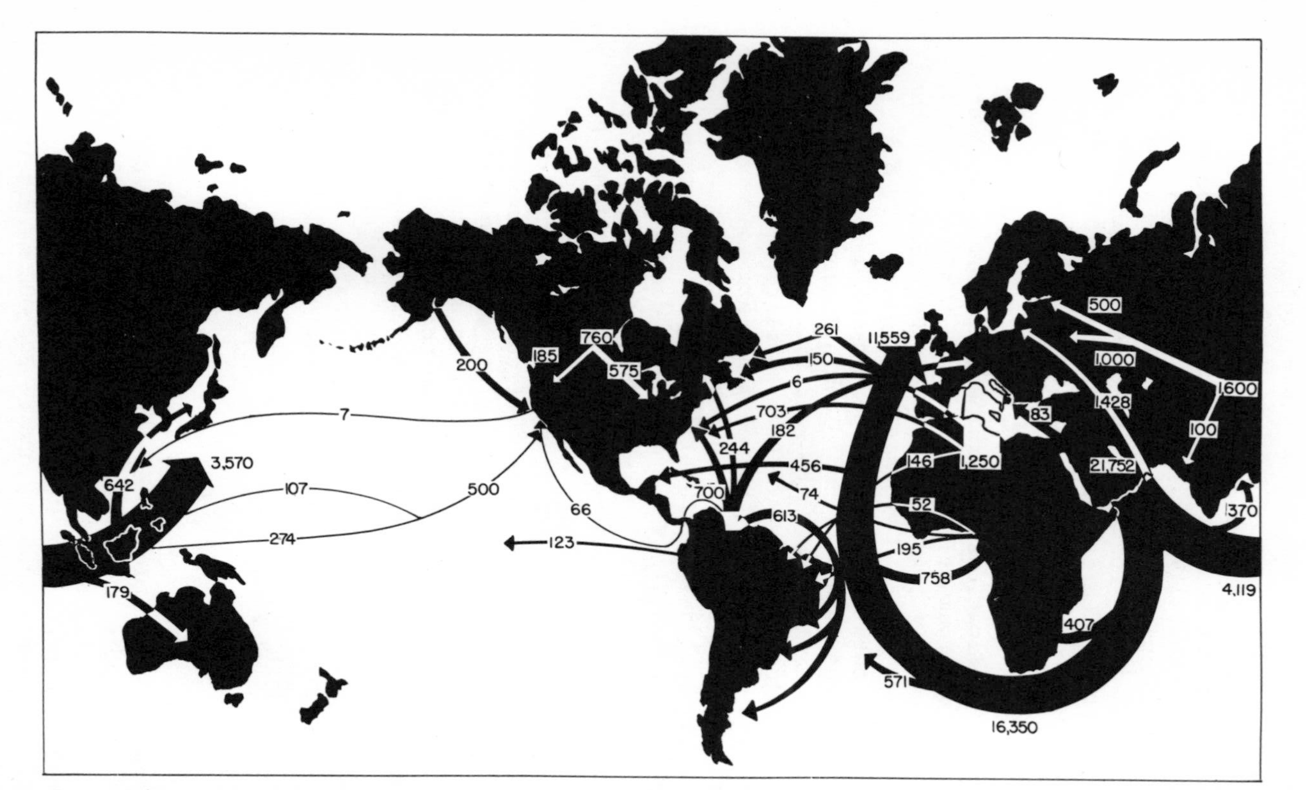

FIG. 1. Major petroleum transportation routes (1974). Values indicate thousands of barrels per day. (Adapted from International Petroleum Encyclopedia [3]. Used with permission of the Petroleum Publishing Company.)

TABLE 1

Sources of petroleum hydrocarbons found in the marine environment from worldwide inputs

Source	Estimated input of oil per year (metric tons)			
	1971[a]	1971[b]	1973[c]	1980[c]
Land-based discharges				
Refineries	2,000,000	300,000	200,000	20,000
Waste oils, runoff, sewage		500,000	2,500,000	2,400,000
Marine operations				
Tankers, using LOT[d]	3,000,000	100,000	310,000	200,000
Tankers, not using LOT	1,500,000	600,000	770,000	
Drydocking	—	—	250,000	300,000
Bilge discharges	500,000	50,000	500,000	
Marine accidents	—	200,000	300,000	150,000
Offshore production	—	150,000	80,000	200,000
Natural oil seeps	—		600,000	600,000
Atmospheric fallout	—	—	600,000	600,000
Total[e]	7,000,000	1,900,000	6,110,000	4,470,000

a Estimate by Blumer [9].

b Estimate by Wardley-Smith [10].

c Estimate by National Academy of Sciences [2].

d LOT: "Load-on-Top", see Nelson-Smith [11] for a description of this procedure for reducing tank washing discharges to the sea.

e For comparison, the total petroleum production and total tanker transportation in million metric tons were, respectively: 1,850 and 650 in 1967; 2,200 and 1,500 in 1970; and 4,000 and 2,000 estimated for 1980, according to Blumer et al. [5].

however, because of increasing awareness of the limitations of the environment to withstand exposure to petroleum materials. In the United States, for example, the petroleum industry is being encouraged to recycle refinery cooling waters and to lower the petroleum level in the discharged water from 20 ppm (parts per million, 10^{-6}) or more to less than 5 ppm.

Waste Oils, Runoff, and Sewage

About 2,000,000 metric tons of spent (used) lubricating oil is not accounted for each year in the United States alone. Assuming that (1) the production of spent oil is proportional to the population, (2) the ratio of United States to world consumption of petroleum is 3:1, and (3) the spent oil from coastal regions is discharged into the oceans, it is estimated that the world input of spent lubricating oil may be as high as 1,400,000 metric tons per year [2]. As an example of the problem of waste oils, the quantities of oil generated and the methods of waste oil disposal in the State of Washington during 1971 are given in Table 2. Approximately two-thirds of the waste oil from this coastal locality is either discharged directly into the environment or is unaccountable.

Significant amounts of petroleum hydrocarbons are deposited in urban areas from a variety of sources such as oil heating systems, fallout, and the operation of motor vehicles. Rainfall and runoff flush petroleum materials into storm drains and into coastal waters or rivers. It is estimated [2] that about 300,000 metric tons (range of 100,000 to 500,000) of petroleum hydrocarbons are deposited annually into the sea through urban storm drainage. The National Academy of Sciences [2] estimate of annual input of petroleum hydrocarbons into rivers was about 530,000 metric tons. Using this figure as a base, the total world input of petroleum hydrocarbons into the sea by river runoff is estimated at about 1,600,000 metric tons per year.

The world input of petroleum hydrocarbons from sewage discharges in coastal areas was estimated [2] at 600,000 metric tons annually. This estimate was based on the contribution of petroleum from a major coastal city of 8 g per capita per day from domestic sources and the same amount from industrial sources in a non-refinery region.

For example, Wakeham [15] demonstrated how the budget for aliphatic hydrocarbons could be established for a localized area, viz., Lake Washington. This urban lake is located between Seattle on the west and a major suburban population area on the east. There are no domestic or industrial sewage discharges into the lake. The total input of hydrocarbons was estimated at 57 tons per year of which 40% was contributed by rivers, 14% by direct fallout, and the remaining 46% by runoff (including the runoff from two floating bridges than span the

TABLE 2

The quantities of waste oil generated and the methods of waste oil disposal employed in Washington State during 1971.

	Amount of oil (tons)
Waste oils generated by type	
Automotive lubricating oils	35,500
Industrial oils	19,700
Tank cleanings	7,200
Total	62,400
Waste oils disposed of by method	
Returned to California refineries	2,800
Rerefined locally	8,600
Used as road oil	25,500
Dumped on ground surface	9,500
Disposed of at a sanitary landfill or dump	2,600
Reused as a lubricant or form oil	500
Used as fuel	9,200
Dumped into sewer or storm drain	100
Unaccountable	3,600
Total	62,400

From Washington State Department of Ecology [14].

lake). Disposition of aliphatic hydrocarbons was estimated at about 13% outflow through the ship canal to Puget Sound and about 87% deposited in the sediments of Lake Washington.

MARINE OPERATIONS

Operations associated with the marine transportation of petroleum contribute the second greatest amount of petroleum hydrocarbons to the marine environment. Inputs occur during the normal operations, dry-docking, and bunkering of all vessels. The total input of petroleum hydrocarbons from this general source was estimated at 1,830,000 metric tons based on 1973 data [2]; however, the amount is expected to decrease to about 500,000 metric tons by 1980.

Tankers

Of more than 50,000 merchant ships afloat in 1971, 6,000 tankers (total carrying capacity of 180 million metric deadweight tons) were used to transport the world's crude petroleum of 1,335 million metric tons. Two hundred and fifty million metric tons of refined petroleum products were moved across national borders, predominantly by tankers [2].

As a standard practice, the cargo tanks of tankers are usually washed out with seawater after delivery of the crude petroleum. These washings contain petroleum averaging about 0.35% (ranging from about 0.1% for refined oils to 1.5% for residual fuel oils) of the total capacity of the tanker. To avoid such waste, the industry has introduced a procedure called "Load-on-Top" (LOT). In the LOT procedure, the washings are retained on board the tankers as ballast and allowed to settle. The separated oil is then drawn off and included in the cargo of the next shipment; the seawater washings are then discharged at sea [11].

The estimated potential discharge of petroleum to the sea from tanker washings if the LOT procedure were not used would be 3.85 million metric tons per year. However, the procedure is currently being applied in about 80% of the world's tankers so that the estimated actual discharge of oil in tank washings is about 0.77 million metric tons per year [2].

The potential efficiency of the LOT oil recovery process is almost 99%; however, it will vary depending upon weather conditions and the experience, proficiency, and interest of the tanker crews. In practice, the overall efficiency is probably about 90%. The 10% loss would represent about 0.31 million metric tons of petroleum annually [2].

Additional petroleum may be discharged at sea during cleaning in preparation for inspection, maintenance, or refitting of vessels. The tankers are placed in dry-dock at intervals averaging 18 mo. For safety, all cargo tanks must be clean and free from gases and oil. Most of the time, the tankers are cleaned and the washings discharged at sea owing to lack of disposal facilities at the dock or shore. Thus, for a world tanker fleet having a deadweight capacity of 180 million metric tons, the amount of petroleum discharged to the sea from these periodic tank cleanings would amount to 250,000 tons annually [2].

Bilge Discharges and Bunkering

The rate of input of petroleum to the environment from bilge discharges and bunkering from all large vessels is difficult to estimate from the type of data available. Nevertheless, the National Academy of Sciences [2] estimated an average loss per vessel of about 10 metric tons per year for a total of 500,000 metric tons. Even though the volume of shipping will increase in the future, it is expected that petroleum losses from these sources will decrease because of the introduction of on-board separators and dockside waste receptor facilities.

Accidents

This category of petroleum input to the marine environment includes oil spills resulting from accidents involving vessels and accidents at port oil-transfer terminals and docks. Accidental losses from tankers and other vessels arise from collisions, groundings, explosions, fires, structural failure, and human error. A review of oil losses resulting from major vessel accidents [16] indicated that the amount of oil spilled at sea by accidental discharges from vessels varied considerably from year to year, depending upon the number and size of vessels involved. About 90% of the oil spilled was from tankers, half from vessels that had run aground. Eighty-five percent of the oil spills occurred within 90 km (50 nmi) of a port [16].

Worldwide accidental discharge of oil is estimated at 50,000 to 250,000 metric tons [2]. Keith and Porricelli [17] in 1973 suggested that the actual spillage is probably close to 250,000 metric tons per year; the National Academy of Sciences [2] estimated the annual spillage by all vessels to be 200,000 metric tons (range 120,000 to 250,000). A decline in accidental discharge of oil is expected by the late 1970's because of new international regulations for reducing operational discharges and because of modern technology in vessel construction and navigation.

Non-tanker vessels outnumber tankers by about nine times and are, on the average, much smaller. The bunker fuel oil carried in non-tanker vessels is limited to that required for operating the vessel. The estimated accidental loss of bunker fuel oil from non-tanker vessel accidents is about 100,000 metric tons per year [2]. In the United States, there is a direct relationship between the number of accidents and the number of ship movements in the major ports [18]. Although improvements in port navigation facilities and in vessel radar equipment are expected, accidents will continue because of human error. Thus, the amount of oil spilled accidentally from vessels is not expected to diminish in the future [2].

Milford Haven, the United Kingdom's major crude oil port, accommodates crude oil tankers up to 300,000 metric deadweight tons [3]. Brummage [19] indicated that spillage there during a nine year period was 0.00011% of all oil moved. This spillage is considered representative of a well-controlled oil terminal operation [2]. Applying this spill rate to world petroleum movements of 1,355 million metric tons gives an annual loss of 3,000 metric tons. An important consideration in oil terminal operation is the number of ship-and-shore transfers. The spill rate is likely to be greater in ports handling numerous small tankers carrying crude oil or refined products than in ports receiving a few very large crude oil tankers [2].

No change is expected in the amount of petroleum spilled at terminal facilities in the next several years. Although the quantity of petroleum handled at terminal ports will increase, some reduction in spillage can be expected as the result of increased surveillance and the establishment of a system of penalties. Furthermore, increasingly efficient technology and improved capability to cleanup small- and medium-sized petroleum spills will tend to reduce potentially adverse effects [2].

OFFSHORE PRODUCTION OPERATIONS

The worldwide input of petroleum into the marine environment from offshore drilling and production operations is estimated at 80,000 metric tons per year. Of this total, 20,000 metric tons are lost through minor spills (6.8 metric tons or less) and through discharges of oil field brines during normal drilling and production operations. The remainder is lost during major accidents that result from blowouts, ruptures of gathering lines and pipelines that carry the crude oil from production platforms to shore, and related unpredictable occurrences [2].

In 1965, Weeks [20] predicted that by 1978 total worldwide offshore petroleum production would increase to 1,200 million metric tons per year, or about two and one-half times that for 1971. If the rate of total hydrocarbon losses from offshore production remains the same, by the early 1980's the total input to the marine environment may reach 200,000 metric tons annually. In fact, there was no decrease in the rate of blowouts from offshore oil and gas wells during the period from 1964 to 1971. However, increasing concern for the environment, greater care in production operations, and improved equipment in all phases of offshore operations should reduce spillage below the projected level [2]. The projected reduction in oil spills in offshore production areas may be offset by the increasing potential for spills in newly-developing and proposed areas in arctic and subarctic waters where more severe weather and ice conditions can be expected. Development will involve deep water areas (e.g., down to 300 meters in the Santa Barbara Channel and the Gulf of Mexico) or shallow areas in arctic and subarctic waters (e.g., Gulf of Alaska, Bering Sea, Beaufort Sea, Barents Sea, Okhotsk Sea, and the Grand Banks) where ambient conditions make drilling and production extremely difficult. The risks of pollution incidents will undoubtably increase because of such environmental stresses.

OIL SEEPS

Wilson [21] found that the geological and geochemical factors affecting oil seepage allow the classification of land and near-shore seep areas into three groups: high seepage potential (35 metric tons/1,000 km^2 or about 100 bbl/1,000 mi^2); moderate seepage potential (1 metric ton/1,000 km^2 or about 3 bbl/1,000 mi^2); and low seepage potential (0.04 metric ton/1,000 km^2 or about 0.1 bbl/1,000 mi^2). Using this concept, the direct input of petroleum into the marine environment was estimated at 600,000 metric tons per year with a range of 0.2 to 1.0 million metric tons [2]. Blumer [22] suggested that this upper estimate was too high because such a seepage rate would have depleted all the petroleum reservoirs in offshore areas. Nevertheless, an estimated 50 to 100 times as much petroleum has been lost to the environment as now is known to exist underground. This estimate was based on observations of several major oil seep areas including the Athabasca Tar Belt of the western Canadian Basin [2]. Seeps tend to occur in tectonically-active areas containing source beds of petroleum. Since most petroleum is generated at depths below 1,000 m (ca. 3,000 ft), very little is lost through seepage until the geological areas containing the petroleum sources or reservoirs are fractured and the oil-bearing areas are uplifted or become exposed by erosion.

The contribution of petroleum to the marine environment from possible oil seeps beneath the ocean surface seaward of the continental shelf is open to question. The biological productivity of the oceanic areas is extremely low compared to shelf areas and the sedimentary deposits contain only small amounts of organic matter. It is unlikely, therefore, that such areas beyond the continental slope and shelf would contain large deposits of petroleum.

ATMOSPHERIC INPUT

Petroleum hydrocarbons are removed from the earth's surface and redistributed by atmospheric processes. Estimates of global removal by this mechanism are virtually impossible to quantitate because of the many types and sources of inputs. An estimate has been made by the National Academy of Sciences [2]: approximately two-thirds of the petroleum hydrocarbons entering the atmosphere emanate from transportation sources and the remainder arises from combustion of oil by stationary electrical power and heating facilities, industrial processes, and solvent and gasoline evaporation. The amount of petroleum hydrocarbons entering the oceans as direct aerial fallout (rainfall or dustfall) is estimated to be between 0.4 and 0.8 million metric tons per year. However, such estimates are subject to errors due to lack of data on: (1) reaction kinetics of hydrocarbons exposed to sunlight in the atmosphere; (2) resultant products; and (3) the petroleum hydrocarbon content of aerial particulate matter.

SPILL TRANSPORT AND RELATED MECHANISMS

Petroleum spilled at sea immediately undergoes weathering processes that alter its physical and chemical nature. These processes involve spreading of the petroleum on the sea surface to form slicks, evaporation of volatile components, dissolution of soluble components in the seawater beneath the slick, emulsification of fine particles of petroleum into the water column, adsorption of petroleum onto particles, compaction of the petroleum into tar balls, modification of petroleum mixtures by ingestion and excretion by bacteria and larger marine organisms, and photochemical modification. In arctic and subarctic regions the fate of petroleum spilled into the sea can be influenced by its interaction with ice and snow. The original physical characteristics and chemical composition of the petroleum affects its movement and destiny at sea.

A diagrammatic representation of the processes leading to the distribution and composition of petroleum at sea is given in Figure 2. Many of these weathering processes can occur simultaneously and are interrelated; therefore, their individual contributions to the weathering of an oil spill may be difficult to evaluate. The following discussion primarily addresses the fate of oil spilled into the marine environment; subsequent chapters deal with biological effects and ecological impacts (See Chapter 3 and Volume II).

SPREADING

Petroleum added to calm water tends to spread over the surface to form a thin continuous layer. This tendency to spread is the result of two physical forces acting together: the force of gravity which causes the oil to spread horizontally and the surface tension of pure water, which is usually greater than that of the oil film floating on water [23]. Theoretically, the oil could spread until it forms a monomolecular layer; however, spreading stops short of this point, most likely because of changes in the surface tension of the oil. Fay [23] stated that gravity and surface tension will promote the spreading of oil on a calm water surface while inertia and viscosity will retard spreading.

In the open ocean, spreading of oil is aided by wave motion, winds, and currents. These surface motions are often random in strength and direction so that the various elements

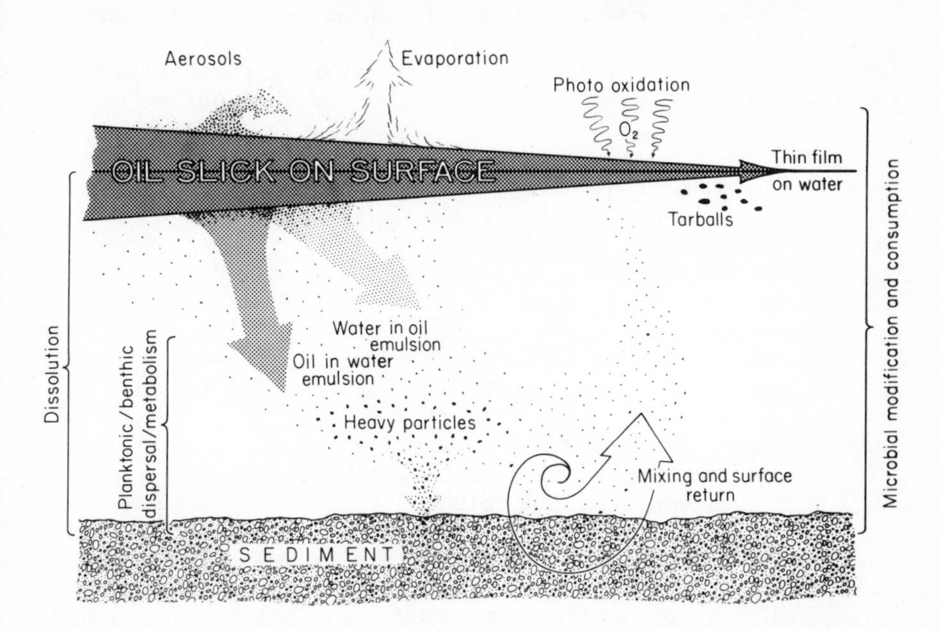

FIG. 2. Diagrammatic representation of the processes leading to the distribution and consumption of petroleum at sea.

of an oil slick are moved about indiscriminately, even when there is gradual drift of the entire slick in a specific direction. This two-dimensional drift of an oil slick, influenced by the random motions of waves, wind, and currents, may be analogous to the three-dimensional dispersion of a puff of smoke emitted into the atmosphere on a windy day [23].

The low molecular weight components of oil spilled at sea immediately begin to vaporize into the atmosphere or are leached into the seawater. Some oils eventually leave a fairly dense and viscous residue. Molecular diffusion of the volatile fractions within high specific gravity, viscous oils spilled on quiet water is slow; therefore, evaporation can also be expected to be slow [24]. As the slick spreads, the density, viscosity, and interfacial tension of the residue change [23]. As a result, the rate of spreading decreases with time; spreading, therefore, is a self-retarding process [11]. Eventually, wave action and other forces will tend to break up the slick into small patches.

Spreading of oil from a point source on a calm sea passes through three phases according to Fay [23]:

(1) The initial phase (up to an hour) in which only gravitational and inertial forces are important. On the basis of Fay's data, the diameter of a 10,000 ton spill of oil might increase by a factor of eight in the first hour.

(2) The intermediate phase (from one hour to a week) in which gravitational force and viscosity (flow-resistance) dominate. During this period the diameter of the theoretical spill would increase by five times that reached during the initial phase.

(3) The final phase (after one week if any surface oil remains) in which the surface tension is balanced by the viscosity. This condition usually occurs at a maximum critical oil thickness of 8 mm [11].

In the first phase, the spreading oil film drags with it a thin layer of water; when the thickness of this water layer is the same as that of the slick, viscosity of the oil becomes the dominant retarding force and the rate of spreading decreases markedly. Once the slick thins out to less than the critical thickness (8 mm), factors affecting spreading become independent of the volume of oil spilled.

The visual appearance of an oil slick on seawater at various oil thicknesses and volumes is given in Table 3.

The tendency of an oil to spread is also associated with the hydrostatic forces derived from differences in density of the oil and the water. A mathematical consideration of the relationship between hydrostatic and interfacial forces might lead to a hydrostatic spreading model for a given oil. Such a concept, however, would not be valid under open sea conditions because of the extremely complex composition of petroleum and the indeterminate effects of wind, waves, and currents. It has not been possible, therefore, to develop a predictive hydrostatic model that will define the spreading ability of an oil spill on a real ocean [26].

Only on a calm sea would an oil spill spread out in a circular patch. Most oil slicks become elongated, with wide and narrow areas that tend to move in many directions [26]. In the open ocean, wind coupling can provide sufficient energy to move a surface oil slick at a rate of about 3% of the wind velocity in the same direction as the wind [11,27].

Surface-active constituents of petroleum are important in the spreading and dispersion processes. Only aromatic and aliphatic hydrocarbons with a carbon number less than C_9 have positive spreading coefficients [2]. The spreading coefficient is defined as the surface tension of the water minus the sum of the surface tension of the oil and the interfacial tension between the oil and the water. For example, hexadecane (*n*-$C_{16}H_{34}$) on water at 20°C has a negative spreading coefficient (F_s) calculated as follows:

$$F_s = 72.8 - (27.5 + 53.8) = -8.5 \text{ dynes/cm.}$$

The negative value of F_s indicates that this compound will not be spread by surface forces on water. Most petroleum prod-

TABLE 3

Description of floating oil slicks

Appearance of oil slick	Approximate thickness		Quantity per unit area		
	(in)	(μm)	(gal/mi^2)	(l/km^2)	(ton/mi^2)
Barely visible under most favorable light conditions	1.5×10^{-6}	0.038	25	36	0.082
Visible as a silvery sheen on surface of water	3.0×10^{-6}	0.076	50	73	0.163
First trace of color may be observed	6.0×10^{-6}	0.152	100	146	0.326
Bright bands of color are visible	12.0×10^{-6}	0.305	200	292	0.652
Colors begin to turn dull	40.0×10^{-6}	1.016	666	973	2.17
Colors are much darker	80.0×10^{-6}	2,032	1,332	1,946	4.34

From American Petroleum Institute [25].

ucts, however, have a positive spreading coefficient because of the presence of surface-active NSO components (nitrogen-sulfur-oxygen-containing organic compounds). These surface-active constituents cause the otherwise non-spreading hydrocarbon oils to form expanded slicks surrounded by thinner films which approach monomolecular thickness at their outermost extremities [28].

According to Garrett [28], only small quantities of polar organic compounds are required to modify the behavior of oil on water. For example, 0.2, 0.5, and 1.0% of 1-dodecanol ($C_{12}H_{25}OH$) dissolved in nonspreading paraffin oil (F_s = -11.7 dynes/cm) gave positive spreading coefficients of 3.0, 11.0, and 17.5 dynes/cm, respectively. The orientation of surface-active molecules at the oil-water interface also lowers the interfacial tension and modifies the viscoelastic properties of this boundary, leading to enhanced formation of emulsions when the oil layer is agitated by wind and waves [28]. Spreading and dispersion speeds up the weathering process by increasing the surface area of the oil exposed to air, light, and seawater [2].

Fay [29] suggested that the cessation of spreading of an oil slick is largely due to the dissolving of surface-active NSO components. The NSO compounds are only slightly more soluble in seawater than hydrocarbons of comparable molecular weight and similar molecular configuration. Structural studies have indicated that NSO compounds are actually hydrocarbon-like; most petroleum NSO compounds contain only one (rarely two or more) heteroatoms [30] (See Chapter 1).

Crude petroleum and most types of refined products which spread across the surface of quiet or confined waters can be cleaned up, or at least dispersed. However, no satisfactory method has been found for dealing with the heavy fuel oils that tend to solidify when spilled onto cold seawater. Detergents do not have much effect on these heavy oils; they must be removed by physical means. Such a task, difficult enough in daylight and good weather, is virtually impossible in darkness, icy seas, or bad weather [31]. Self-mixed detergents are being developed to disperse heavy oils in cold water [24].

EVAPORATION

The first changes that take place in petroleum spilled on the sea are evaporation of the volatile components and alteration of the physical properties of the remaining slick. The rate and extent of these changes depend upon the chemical and physical nature of the components of the particular petroleum, the wave action, wind velocity and water temperature [32,33]. Simulated evaporative weathering tests under controlled conditions [34,35] indicated that virtually all hydrocarbons smaller than C_{15} (boiling point less than 250°C) will volatilize from the sea surface within 10 days; many lighter petroleum materials tend to disappear in hours. Hydrocarbons from C_{15} to C_{25} (boiling point range 250° to 400°C) will evaporate in more limited amounts and will; therefore, largely remain in the oil slick. Hydrocarbons above C_{25} will be retained.

Evaporation alone will remove about 30-50% of the hydrocarbons in a typical crude petroleum on the sea surface; a Bunker C fuel oil would probably lose only about 10% of its hydrocarbons. As evaporation of a crude petroleum proceeds, the hydrocarbon composition of the residue often approaches that of a Bunker C fuel oil; this is true for the majority of the world's crude oils. About 75% of the hydrocarbons of a No. 2 fuel oil and about 100% of the hydrocarbons of gasoline and kerosine will vaporize [2].

Field studies of a controlled oil spill in a ocean area (off Virginia) showed that the content of hydrocarbons less than C_{12} in a surface slick was reduced by more than 50% in 8 hr at 5°C water temperature and wind velocity of 17 to 22 km/hr (9 to 12 kn). The dissipation of this oil slick was presumed to be caused mostly by evaporation, although dispersion of the lower molecular weight aromatic compounds into seawater also occurred [36]. Analogous studies in warm water areas (off Santa Barbara [37] and off Grand Bahama Island [38]) showed similar loss of the more volatile hydrocarbons, although the rate of loss was higher, largely because of higher water temperatures.

If a petroleum spill occurs in the open sea, evaporation may well be complete before the slick reaches the shoreline. Rough seas would tend to increase evaporation rates because sea spray and bursting bubbles will eject petroleum components into the marine atmosphere. In the atmosphere these may be photochemically converted to less volatile oxygenated organic compounds, some of which may be reabsorbed into the sea [2]. Current assessment of the level of petroleum pollution of seawater is usually measured in terms of the total content of hydrocarbons. The contribution of NSO compounds to the total burden of oil pollutants in seawater may be significant [39]; however, their determination is not usually made; therefore their contribution is unfortunately ignored.

Petroleum hydrocarbons removed from the sea surface as aerosol particles are redistributed by atmospheric processes. Except in coastal areas with onshore winds, the removal of hydrocarbons from the ocean surface by spray and bursting bubbles is temporary. The airborne particles have relatively short residence times, ranging from seconds for particles larger than 0.1 mm up to a few days for micrometer-sized particles. Most of these particles will be redeposited in the ocean at distances ranging from a few meters to several hundred kilometers from injection [2].

Smith and MacIntyre [35] found a sharp increase in the evaporation of *n*-paraffins from a No. 2 fuel oil slick at sea after the wind velocity increased sufficiently to form whitecaps and breaking waves which increased the surface area of the film. They found that the major pathway of initial weathering of No. 2 fuel oil was evaporative loss of volatile hydrocarbons (boiling points less than 250°C). The initial losses appeared to be proportional to the amount of volatile components (alkyl-substituted benzenes and naphthalenes) present; the viscosity of the oil did not seem to be important [35]. Evaporation of the volatile hydrocarbons from an oil slick was probably controlled primarily by the amount of mixing by wave action rather than by molecular diffusion of the volatile components through the oil slick.

Simulated weathering of oil at elevated temperatures (e.g., distillation of the oil) may not produce the same results as environmental weathering because the vapor pressures of the component petroleum hydrocarbons are temperature dependent [40]. Reports in the literature [41,42] show that low-boiling (below 200°C) aromatic hydrocarbons have a lower vapor pressure at ambient temperatures than *n*-paraffins of the same boiling points. For hydrocarbons boiling above 200°C, the *n*-paraffins have a greater vapor pressure than aromatics of the same boiling points. Hence, the low-boiling aromatics should be depleted more slowly by evaporation under environmental conditions than the *n*-paraffins of the same boiling point [40].

The air bubbles caused by breaking wave action mentioned earlier inject considerable amounts of dissolved hydrocarbons [43] and surface hydrocarbon-rich films [34] into the marine atmosphere. Baier [44] suggested that the removal of "thin" oil films or "sheens" can be largely attributed to the bursting of these bubbles and the formation of spray. Solar radiation can cause the formation of polar, surface-active molecules that promote further spreading and weathering of the oil. Furthermore, the hydrocarbons may be solubilized by naturally-occurring surfactants such as humic substances, fatty acids, and fatty alcohols that are concentrated in the sea surface microlayer.

Hydrocarbons are present in the surface microlayer of the sea in non-spill areas. Duce and coworkers [45] reported slight hydrocarbon enrichment of the surface microlayer (top 150 to 200 μm) in non-spill areas in Narragansett Bay, Rhode Island, compared with water collected 20 cm below the surface. Wade and Quinn [46] found hydrocarbon enrichment up to 26-fold in 12 of 17 samples of surface microlayer from the open Atlantic Ocean in non-spill regions. However, they concluded that hydrocarbon enrichment in the surface microlayer of non-spill areas is apparently not a universal feature of open ocean water.

DISSOLUTION

In the context of this review, dissolution is arbitarily defined as the disintegration and dispersal of the bulk oil into water. Dissolution can include such phenomena as the formation of macroparticles (droplet dispersions), the formation of microparticles (colloidal dispersions and oil-in-water emulsions), and the formation of a single phase, homogeneous mixture (dissolved fractions) of hydrocarbons in water. There are no definitive demarkations between these states of dissolution. Frequently, authors in the literature reviewed in this chapter did not distinguish which of the above states was being considered in their investigations of the weathering processes of spilled petroleum.

Besides evaporation, the low molecular weight aliphatic and aromatic hydrocarbons and non-hydrocarbon organic compounds are removed from an oil slick by dissolution into the underlying seawater. Dissolution depends on ambient weather conditions (e.g., air and water temperatures, wind, sea state, currents, and waves) and on the physical characteristics and chemical composition of the petroleum. Dissolution starts immediately upon contact of the oil with seawater. The process may be self-sustaining; biological and photochemical oxidation of the components of the oil produce additional surface-active, polar compounds which are more soluble in seawater. For example, naphthalene, with a solubility of about 31 ppm, can be oxidized to α-naphthol, with a solubility of 740 ppm [2].

The solubility of the low molecular weight *n*-paraffins in fresh water is inversely proportional to their molecular weight. Solubilities of various paraffins, cycloparaffins, and aromatics in fresh and salt water are given in Table 4. McAuliffe [47] found that for various hydrocarbons from C_1 to C_9 dissolved in fresh water:

(1) The logarithms of the solubilities of the members of each homologous series is a linear function of the molar volumes of the hydrocarbons.

(2) Branched-chain paraffins, olefins, and acetylenes are more soluble than their straight-chain counterparts.

(3) Branched-chain substituted cyclic hydrocarbons (cycloparaffins, cycloolefins, and aromatic compounds) show about the same solubility as their straight-chain counterparts.

(4) For a given carbon-number, ring hydrocarbons show greater solubility than non-ring hydrocarbons.

TABLE 4

Solubility of petroleum hydrocarbons in water

Compound	Carbon number	Solubility[a] (ppm)	Reference
PARAFFINS			
Methane	1	24	47
Ethane	2	60	47
Propane	3	62	47
n-Butane	4	61	47
n-Pentane	5	39	47
n-Hexane	6	9.5	47
2-Methylpentane	6	13.8	47
3-Methylpentane	6	12.8	47
2,2-Dimethylbutane	6	18.4	47
n-Heptane	7	2.9	47
n-Octane	8	0.66	47
n-Nonane	9	0.220	48
n-Decane	10	0.052	48
n-Undecane	11	0.0041	48
n-Dodecane	12	0.0037 0.0029 (SW)	49
n-Tetradecane	14	0.0022 0.0017 (SW)	49
n-Hexadecane	16	0.0009 0.0004 (SW)	49
n-Octadecane	18	0.0021 0.0008 (SW)	49
n-Eicosane	20	0.0019 0.0008 (SW)	49
n-Hexacosane	26	0.0017 0.0001 (SW)	49
n-Triacontane	30	0.002	2
n-Heptacontane	37	10^{-8}[b]	50

Compound	Carbon number	Solubility[a] (ppm)	Reference
CYCLOPARAFFINS			
Cyclopentane	5	156	47
Cyclohexane	6	55	47
Cycloheptane	7	30	47
Cyclooctane	8	7.9	47
AROMATICS			
Benzene	6	1,780	47
Toluene	7	515	47
o-Xylene	8	175	47
Ethylbenzene	8	152	47
1,2,4-Trimethylbenzene	9	57	47
iso-Propylbenzene	9	50	47
Naphthalene	10	31.3 22.0 (SW)	51
1-Methylnaphthalene	11	25.8	51
2-Methylnaphthalene	11	24.6	51
2-Ethylnaphthalene	12	8.00	51
1,5-Dimethylnaphthalene	12	2.74	51
2,3-Dimethylnaphthalene	12	1.99	51
2,6-Dimethylnaphthalene	12	1.30	51
Biphenyl	12	7.45 4.76 (SW)	51
Acenaphthene	13	3.47	51
Phenanthrene	14	1.07 0.71 (SW)	51
Anthracene	14	0.075	2
Chrysene	18	0.002	2

a In distilled water, except where noted by (SW), indicating filtered seawater, usually corrected to a salinity of 35°/oo (parts per thousand); ppm = parts per million - micrograms per gram.

b Extrapolated.

(5) For a given carbon number, unsaturated ring or open chain hydrocarbons are more soluble than saturated hydrocarbons; the solubility increases with an increase in the number of double bonds in the molecule.

(6) For a given carbon number, hydrocarbons containing a terminal triple bond (α-acetylenes) are more soluble than hydrocarbons containing two double bonds.

Generalizations regarding the solubility of various petroleum hydrocarbons in natural seawater are difficult to make because the published data on solubility are not always comparable. This is true because such data were obtained in laboratory tests involving different experimental parameters, different petroleum substances, and different seawater compositions. Variations in experimental parameters include: type, degree, and duration of agitation of the petroleum and seawater mixture, the ratio of petroleum to seawater, the filtration system used on the mixture (type of filter, pore diameter, and filtering rate or pressure), and the methods of chemical analyses (See Chapter 1). The petroleum substances varied as to type (crude petroleum or a refined product), fraction (whole oil, weathered residue, or distillation fraction), and source. The composition of seawater varied in terms of salinity, pH, source (estuarine, coastal, open-ocean, artificial), and content of naturally-occurring organic matter (dissolved and particules) [52-54].

McAuliffe [55] determined the hydrocarbon content of subsurface brines in contact with reservoir crude petroleum and found that the major hydrocarbon components consisted of low molecular weight (C_1 to C_7) alkanes, cycloalkanes, and aromatic compounds (e.g., benzene and toluene). This compositional pattern of hydrocarbons in the brines reflected their general solubility values as determined in laboratory studies [47]. Later research on seawater extracts of crude petroleum [56,57] and refined products [58] confirmed results of the earlier studies [47,48] and showed, in addition, that the seawater extracts also contained high molecular weight compounds such as indanes, naphthalenes, acenaphthenes, fluorenes, phenanthrenes and traces of higher molecular weight aromatic compounds [59]. Nevertheless, high molecular weight aromatic compounds (C_{14} plus) are not much more soluble in seawater than paraffins of comparable molecular weight, so that preferential leaching of these aromatic hydrocarbons into seawater from an oil spill is probably insignificant [2].

In laboratory studies, Boylan and Tripp [56] accommodated suspended and dissolved petroleum and refined products in filtered seawater using several different modes of mixing: (1) slow stirring for 12 hr to produce a vortex of less than 25% of the depth of the solution; (2) moderate, but not turbulent, mixing to the point where the magnetic stirring bar and oil come into contact; and (3) vigorous shaking. Reproducible extracts were obtained by the slow stirring method; the extracts contained mostly aromatic compounds (e.g., benzenes and naphthalenes) in the boiling point range of 80° to 260°C. The seawater extracts from the mixtures that were moderately stirred and vigorously shaken contained relatively high levels of NSO compounds, probably due to the incorporation of suspended oil droplets. Seawater extracts of Kuwait crude oil contained saturated and aromatic hydrocarbons as well as polar cyclic and high-molecular weight aromatic compounds. Boylan and Tripp also found that 1 liter of seawater at 23°C could accommodate 1.453 μg of aromatic organic compounds consisting of 657 μg of benzenes and naphthalenes and 796 μg of polar aromatics. At the same temperature, 1 liter of seawater extract of kerosine contained 796 μg of polar aromatic organic compounds which included 135 μg of alkylated benzenes and 362 μg of naphthalenes, principally naphthalene [56].

In field studies, Harrison and coworkers [38] determined the disappearance of certain aliphatic and aromatic compounds from a slick of South Louisiana crude oil in a tropical sea. The study was carried out off Grand Bahama Island where the seawater temperature was 23.6°C (75°F) and the wind velocity was 0 to 29 km/hr (0 to 16 kn). To serve as an internal standard, 4.2% by weight of cumene (isopropyl benzene) was added to the crude oil. The purpose of the test was to develop a guide for estimating the relative rates of evaporation and dissolution of the various hydrocarbon classes in petroleum following an oil spill. The concentrations of cumene and certain alkanes in the slick were measured up to 5 hr after the experimental spill. Cumene disappeared from the slick faster than *n*-nonane (the two compounds have similar vapor pressures but different solubilities in seawater at ambient temperature). An interpretative model developed by the authors [38], indicated that the dissolution rate for cumene could be approximately two orders of magnitude lower than the evaporation rate. Other measurements on the composition of the oil slick indicated that the dissolution rate of the crude oil slick was less than half the evaporation rate. The analytical techniques used by Harrison and coworkers were not sufficiently sensitive to allow a closer comparison between the field studies and their model.

Smith and MacIntyre [35] also found that the total loss of components from three refined oil products, No. 2, No. 4 and No. 6 (Bunker C) fuel oils, in artificial seawater was negligible compared to evaporation. Dissolution was 0.2 to 1.1% of evaporation. They used a laboratory system for their tests in which 2 l/min of air was bubbled through a mixture containing 10 liters of artificial seawater and 30 ml of the test oil.

In laboratory bioassay experiments, Anderson and coworkers [60] prepared what was claimed to be "truly soluble fractions of test oils" by placing one part of oil over nine parts of artificial seawater (20 $^o/oo$ salinity at 20°C) in a 19-liter glass bottle and stirring the mixture with a magnetic stirring bar for 20 hr. The speed of stirring was adjusted until the vortex extended 25% of the distance to the bottom of the container; the bottle was capped to minimize evaporation. The mixture was allowed to stand 1 to 6 hr; then the seawater phase was siphoned off and used immediately. Such mixtures yielded 19.8 ppm total hydrocarbons (by infrared analysis) from a South Louisiana crude oil, 10.4 ppm from a Kuwait crude oil, 8.7 ppm from a No. 2 fuel oil, and 6.3 ppm from a Bunker C fuel oil. These oils were American Petroleum Institute reference oils--See Tables 4 and 6 in Chapter 1.

Under the test conditions, Bunker C fuel oil tended to form persistent suspensions of oil droplets in the artificial seawater; it appeared likely that a significant portion of the petroleum remained as very fine droplets suspended in all the mixtures.

Roubal and coworkers [61] designed a sophisticated flow-through system for exposing aquatic organisms to seawater extracts of petroleum. Petroleum was metered into filtered seawater where it was churned into a froth by a mechanical stirrer in a specially-designed glass chamber. The non-homogenous mixture of oil and water then flowed into a second compartment where large droplets rose to the surface and were discarded. A third chamber allowed the further removal of surface oil before the mixture flowed into a baffle box containing 16 glass baffle plates. A test of the system gave an uninterrupted flow of seawater extracts of Prudhoe Bay crude oil for 5 weeks. The total hydrocarbon content of the extracts (mostly benzenes and naphthalenes) varied from 5 ppm initially to about 6 ppm at the end of the experiment. The diluted seawater extracts exhibited no Tyndall Effects and thus appeared to be "water clear." The composition of the water-soluble fraction (WSF) was essentially free of alkanes, which implied the virtual absence of suspended droplets.

It is difficult to compare WSF of petroleum reported on bioassay and exposure studies because different techniques were used in their preparation. Moreover, insufficient information was given on their chemical composition and physical characteristics. Some researchers attempting to define the WSF of petroleum have reported chemical analyses of the major hydrocarbon compounds but they provided only limited data on oil-particle sizes or made only visual examinations of the clarity of the fractions. Such information is inadequate because (1) oil particles of 1 μm diameter or less are not readily discernable to the human eye [11] and (2) oil droplets smaller than 1-2 μm in diameter remain suspended in seawater for hours or days [50], which is much longer than the settling periods used routinely in preparing water-soluble fractions. Because it is difficult to establish experimentally that an oil-in-seawater mixture is a single-phase solution, perhaps one should speak of "accommodated" oil fractions rather than "water-soluble" fractions.

Other methods have been proposed for preparing seawater extracts of petroleum [49,52,56,62] involving various combinations of mixing, shaking, filtering, and settling. In all cases, fine suspensions of oil droplets formed in the seawater which could not be removed except by filtration [52,54,62]. Filtering pressure appears important; when gentle suction was employed, apparent solubilities were higher presumably because of rupturing of oil droplets trapped on the filter [54].

Parker, Freegarde, and Hatchard [50] suggested that suspended oil particles in seawater beneath an oil slick have greater impact on the marine environment than dissolved hydrocarbons. A volume of ocean water one kilometer square and one meter deep containing 1 ppm of evenly suspended oil droplets would contain one ton of petroleum [50]. In laboratory studies, these investigators found that 1 ppm of suspended oil in seawater persisted for 5 weeks. Therefore, in rough seas significant amounts of oil may remain suspended long enough to become dispersed far beyond the original contamination. Such a transport mechanism may considerably increase the exposure of pelagic organisms to petroleum. In field studies, the level of dissolved paraffin hydrocarbons in seawater from a low viscosity South Louisiana offshore crude oil dropped rapidly from 60 to 70 ppm to 2 ppm in a few hours within 300 to 1,500 m of the source [63].

The natural ocean environment has a distribution of energy (e.g., winds, waves, tides, heat, light) virtually impossible to duplicate in the laboratory; hence a full understanding of the degree of partitioning of petroleum components into the various weathering and transport compartments will be difficult to achieve.

EMULSIFICATION

An emulsion is a dispersion of one liquid (in the form of fine droplets or suspension) in a second immiscible liquid (continuous phase). Petroleum and seawater emulsions can be of two types: oil droplets in seawater (oil-in-water) and seawater droplets in oil (water-in-oil). The tendency of oil and seawater to form emulsions is important in the transport of certain types of spilled petroleum in the ocean.

Oil-in-seawater emulsions are not stable; however, continuous agitation, such as occurs in most arctic and subarctic exposed surface waters will tend to maintain emulsions and facilitate dispersion [50]. Milne [64] reported that oil droplets in an oil-in-water emulsion have an average diameter of 0.5 μm, a volume of 6 x 10^{-17} l, and a surface area of 8 x 10^{-13} m^2 (at 20°C). Thus, one milliliter of oil can form 16 x 10^{12} droplets with a total surface area of about 13 m^2. The oil droplets tend to coalesce and return to the sea surface to form a slick. The emulsion can be stabilized, however, by natural or added emulsifiers (detergents or dispersants) [1]. Where seawater is the continuous phase of emulsified oil-in-seawater, there is no limit to the dispersion of oil droplets.

Following the *Arrow* incident in Chedabucto Bay, Nova Scotia, Canada, the size and distribution of Bunker C fuel oil particles in the water column were assessed [62]. The effective length (cube root of the volume) of the droplets to a depth of 80 m generally ranged from 5 μm to 1 mm; occasionally a droplet was found as large as 2 mm. The droplets were formed primarily by surf and waves. However, some droplets may have formed in the vicinity of surface sea ice here the ice formed a barrier to floating oil and when the water currents under the oil slick and the ice barrier exceeded a certain velocity (usually 2 to 3 m/s or 4 to 6 kn), droplets of oil on the underside of the windward edge of the slick were probably torn free from the bulk oil and carried under the slick and the ice barrier [66].

Suspended oil droplets about 1 mm in diameter, separated by only a few centimeters, were found in the shallow water near oil-contaminated beaches after the *Arrow* incident. The concentration of the larger oil droplets decreased with increasing depth [65]. A large production of small oil particles (<1 mm diameter) during the first fifteen days after the spill were attributed to: (1) stormy weather conditions, (2) presence of detergents used during the first few days in an unsuccessful attempt to emulsify the oil, and (3) unweathered condition of the oil. Forrester [65] stated that complete erosion of a Bunker C fuel oil particle, having 1 mm^3 volume, would require 30 days at 0° to 2°C water temperatures off the Nova Scotia coast in late winter, making no allowance for erosion of oil from the particle by microbial degradation or by solution.

Suspended oil droplets apparently can serve as an excellent tracer for near-surface currents and drift [65]. The oil particles from the *Arrow* travelled about 250 km southwestward from Chedabucto Bay in a band extending up to 25 km off shore. Two weeks after the wreck, particles of oil were found 70 km to the east of Nova Scotia in a 10 km wide tongue, probably carried there by wind-driven currents.

Gordon, Keizer, and Prouse [62] determined the dissolution of three different oils (Venezuelan crude oil, No. 2 fuel oil, and Bunker C fuel oil) in natural seawater. The oils were dissolved individually in *n*-hexane, the mixture added to seawater in separatory funnels, and shaken in a series of tests in which the amount of oil, the duration of shaking, temperature, and degree of exposure to air varied. The size distribution of the dispersed droplets was determined with a Coulter counter. In addition, estimates were made of total oil and subparticulate oil by fluorescence spectrometry; particulate oil concentrations were determined as total carbon [67]. Other determinations included paraffin hydrocarbons by gas-liquid chromatography, pH, dissolved oxygen, and bacterial counts. Particulate and subparticulate oils were defined as those oil droplets in the extracts that were, respectively, retained or passed through a silver membrane filter of 0.8 μm porosity using a pressure of 1 kg/cm^2 (about 15 psi or 1 atmosphere).

Total oil content (dissolved oil plus suspended oil particles) in the seawater at 19-21°C for each of the three oils (5 mg/l, added) was about the same (mean: 0.3 mg/l) but the ratio of particulate oil content to subparticulate oil content varied considerably. The ratio of particulate to subparticulate fractions was directly related to the viscosity of each oil. The Bunker C fuel oil showed a particulate to subparticulate ratio of 40.7; the No. 2 fuel oil, a ratio of 9.8; and the Venezuelan crude oil, a ratio of 7.5 [62].

Gordon and coworkers [62] also observed that about two-thirds of the oil retained on the filters consisted of particles greater than 2 μm in diameter. Only 2 to 13% of the oil in the seawater passed through the filter; thus, the filtrate contained oil in solution plus suspended oil particles less than 0.8 μm in diameter. Nevertheless, the values for total dissolved and subparticulate oil did not reach theoretical levels possible for saturated solutions of individual hydrocarbons (acetylenes, diolefins, and aromatics) based on published solubility data and laboratory and field studies [47, 49,51,55].

Ordinarily, the residue of a crude oil after loss of volatile components has a specific gravity greater than the original material. Kuwait crude oil, for example, has a specific gravity of about 0.869; its residue (after removal of fractions boiling below 520°C) has a specific gravity of 1.023. Iranian heavy crude oil has a specific gravity of 0.869; whereas its residue has a specific gravity of 1.027 [32]. Because seawater has a specific gravity near 1.025, residues of these two crude oils could tend to disperse readily, either by emulsification or by sinking.

Chemical detergents tend to increase the capacity of the seawater to accommodate petroleum by facilitating the dispersion of the oil and by stabilizing the emulsions produced. In 1970, a 34° API gravity crude petroleum blew out of a production platform in the Gulf of Mexico over a 3-week period. An estimated 25 to 30% of the oil was lost by evaporation of low molecular weight hydrocarbons, 10 to 20% of the oil was removed from the sea surface by skimming, less than 1% dissolved in the seawater, and less than 1% was deposited in the sediment within an 8 km radius of the blowout. Detergents were used in the water sprayed on the burning platform. The remaining oil presumably became emulsified and dispersed to undetectable levels (less than 1 ppm), or became biodegraded or photooxidized [63]. Observations revealed that the oil developed three major modes of appearance once clear of the platform. The most noticeable was a bright reddish-brown band of thick oil roughly 3 m in width extending for many kilometers on the sea surface; the band took on various ropey shapes. It was suggested that this was "mousse" or seawater-in-oil emulsion, induced by either the detergents or the high pressure water streams, or both. The second mode appeared to vary from a thin dull gray through an irridescent to a silvery sheen; it became a widening surface plume or slick downwind from the platform. The third mode was a creamy yellow subsurface plume emanating from the platform; it was thought to be an oil-in-seawater emulsion with the oil dispersed into very fine droplets by the seawater-detergent mix [68].

The formation of seawater-in-oil emulsions usually involves the more viscous oils, such as high asphalt crude petroleums and residual oils. Seawater-in-oil emulsions formed rapidly during the cleanup operations following the *Torrey Canyon* spill of Kuwait crude oil. Such emulsions tend to form semi-solid, gel-like masses often referred to as "chocolate mousse" or "mousse" in reports following the *Torrey Canyon* incident. These emulsions are fairly stable and may contain up to 80% water. Those containing 30-50% water are fluid and look like crude oil. Mousses containing 50-80% water have a grease-like consistency and are more pale than those containing less water. Crude petroleums containing high levels of asphaltenes, such as those from Venezuela and Kuwait, can produce a mousse that is stable for several months. Crude petroleums containing low levels of asphaltenes (e.g., from Lybia and Nigeria) produce a fairly unstable mousse [11].

Naturally-occurring surfactants in the oil or the addition of detergents will help stabilize the seawater-in-oil emulsions. Dodd [69] showed that the rate of formation of emulsions varied dramatically with the nature of the oil. Under comparable laboratory conditions a Kuwait crude oil took up to 40% seawater in 2 to 4 hr; a Venezuelan Tia Juana crude oil took up 3% seawater within 2 days and 57% within 8 days. Seawater-in-oil emulsions eventually break up in seawater or combine with particulate matter, sand, or other detritus; subsequently they are biodegraded or incorporated into sediment [2].

During large spills, thick layers of oil persist for long periods and large aggregates of mousse can be produced. The mousse is fairly stable because of the reduced surface exposed to the elements; however, anaerobic bacteria may degrade some of the oil (See Chapter 3). The mousse washed ashore will incorporate sand and debris and eventually form compact tarry lumps after the water evaporates. Further degradation of the tarry lumps seems to occur slowly [50].

Davis and Gibbs [70] exposed a high-boiling (>350°C) fraction of Kuwait crude oil to the environment for two years in enclosures floating on the sea surface (Portsmouth, U.K.). One enclosure was open to the seawater below the surface oil slick and the other was closed off from the surrounding seawater by a bottom panel; both were exposed to the atmosphere. The initially fluid oil fraction (0.7 cm thick layer) rapidly formed a viscous seawater-in-oil emulsion (50% water content which remained constant after the first 2 months) which completely covered the surface of the enclosures for much of the 2-yr period. The thick (1.4 cm) chocolate mousse displayed a similar increase in asphaltene content in both enclosures, but the increase in viscosity and specific gravity was greater in the enclosure open to seawater beneath the oil slick. The *n*-alkane content of the emulsion from the "open" enclosure decreased significantly while the alkane content of the "closed" enclosure showed no change. Davis and Gibbs discounted (1) direct solution of the oil components into the water, (2) build-up of organic products which might inhibit biodegradation, or (3) lack of nutrients as the causes for the differences between the closed and the open enclosures. The low oxygen tension (20% of saturation) because of the thick mousse layer on the closed enclosure may be the most probable explanation for retention of the alkanes.

Although Kuwait crude oil residues can be substantially oxidized by microorganisms with adequate oxygen and nutrients, the field experiments of Davis and Gibbs [70] indicated that large globules of seawater-in-oil emulsions decompose only slowly. The metabolic activity of any aerobic microorganisms within the mousse could be limited by the amount of oxygen and nutrients in the aqueous phase. The aqueous phase within the mousse is not continuous with the bulk seawater so a rapid exchange of oxygen and nutrients between the continuous and discontinuous aqueous phases could not occur. Breaking waves would be expected to enhance weathering through mixing of emulsions into the water column. Davis and Gibbs suggested that effective degradation of mousse would occur only after its physical dispersal into small oil particles [70].

Zajic and Supplisson [71] added Bunker C fuel oil to a buffered mineral salt medium containing unidentified microorganisms isolated from sewage and exposed the system to mild continuous stirring. An oil-in-water emulsion formed during the first 24 hr; after 48 hr the mixture was converted to a water-in-oil emulsion. Apparently, compounds in Bunker C fuel oil and those formed during microbial action contributed to this shift in the type of emulsion. The functional groups involved in this shift were attributed to: -COOH (carboxyl), -OH (alcohol), -CHO (aldehyde), $-OSO_3$ (sulfate), and $-SO_3H$ (sulfonate) [71].

Microorganisms contribute to the emulsification of oil particles by forming biochemical intermediates which act as surface-active agents and by metabolizing certain constituents of the Bunker C fuel oil. For example, the paraffinic and low-molecular weight aromatic hydrocarbons were preferentially removed from the emulsion [71]. The oil "pellets" in the emulsion observed by Zajic and Supplisson became progressively smaller and sank to the bottom of the flasks. The sinking of oil-bearing material was attributed to the bio-utilization of the low molecular weight hydrocarbons in the Bunker C fuel oil emulsion, to increasing entrapment of water, and to the development of a microorganism-rich, oil-and-water mass denser than the test medium. The deposited pellets were sticky and soft and could easily entrap particulate matter and sediment. It was suggested that microorganisms may not be an absolute requirement for forming mousse, but that microorganisms were definitely involved [71]. Some of the pellets resurfaced; Zajic and Supplisson speculated this resulted from leaching of the denser microorganisms from the emulsion or because of biosynthesis and entrapment of gases (e.g., methane, hydrogen, or hydrogen sulfide) formed under partial or strict anaerobic conditions. Zajic and Supplisson [71] suggested that the biochemical mechanisms involved need verification. A critical review of microbial interactions with petroleum is given in Chapter 3.

During the research cruise of the RSS *Discovery* in the eastern Mediterranean in 1970, Morris and Culkin [72] found large-scale distributions of oil-in-water emulsions off the coast of Libya and Egypt. These emulsions probably developed from the cargo tank washings of crude oil tankers; the tank washings are discharged at sea before the vessels enter port. Using a 54-mesh (about 0.25 mm) net, the authors found 0.1 to 0.5 mg of oil per square meter of sea surface near the coast. The droplets within the emulsions contained both natural organic components (about 15%) and a complex mixture of pollutant hydrocarbons (greater than 75% of the total extract). The two sources were assigned on the basis of chromatographic analyses for fatty acids (free and esterified), polycyclic aromatic compounds, and saturated alkanes (straight-chain and branched).

AGGLOMERATION AND SINKING

The processes, as suggested in previous discussions, that tend to increase the specific gravity and cause petroleum to sink are (1) evaporation and dissolution of the low molecular weight hydrocarbons, (2) degradation and oxidation of the components of the oil, (3) formation of dispersed particles and subsequent agglomeration of some of these particles, (4) absorption and adsorption of the oil by particulate matter [2], and (5) uptake of seawater during emulsification [11].

Formation and recombining of liquid petroleum particles occurred following the *Arrow* spill of Bunker C fuel oil in Chedabucto Bay [64]. Evaporation and dissolution, combined with other processes including oxidation, lead to formation of tar balls. Morris and Butler [73] suggested that degradation of the surface of such tar balls leads to the formation of many small, dense particles which can then sink. There appears to be a downward flux of such particles in seawater [2].

Indirect evidence for sinking as a transport mechanism was provided by Stehr [74] who noted that 20,000 tons of crude petroleum lost from the tanker *Anne Mildred Brøvig* in 1966 never appeared on nearby beaches of northern Germany, presumably because the oil had sunk in the open sea. However, some 6,400 tons of a heavy fuel oil sank in cold water off the eastern United States and reappeared at the surface later under warmer conditions [75].

Conomos [76] reported more direct evidence for the sinking of oil globules to near-bottom waters where they moved horizonally with the bottom currents. In 1971, two tankers collided at the center of the Golden Gate, the entrance to San Francisco Bay, California, causing a major spill of Bunker C fuel oil. Some of the oil globules were neutrally-buoyant in the near-bottom layer of water in the bay. This layer moved independently of the surface floating oil slicks, depositing oil ashore on different beaches. The trajectory of this near-bottom oil closely followed the movements of experimental seabed drifter devices used in current flow studies of the local estuarine circulation system. Oil residues appeared along the beaches of eastern San Pablo Bay in the Northern Bay region; they had been transported landward from the Golden Gate spill site by the near-bottom currents. These oiled beaches were 16 km east (upriver direction) of the eastward limit of beaches stained by floating oil. Oil also drifted under the wharves of San Francisco and along the bottom in Bolinas Bay up the ocean coast. Shrimp trawl nets dislodged oil apparently suspended at the pycnocline (density gradient) in Bolinas Bay [76].

This type of perennial, river-induced, estuarine circulation is typical of estuaries in arctic and subarctic regions where the land runoff and precipitation exceed the evaporation. Runoff from rivers, mixed with underlying ocean water, forms a low-salinity, low-density upper water layer which generally moves seaward. Saline oceanic water at depth moves into the estuary toward the river mouths to replace the ocean water which was entrained or mixed upward into the brackish surface outflow. In San Francisco Bay, this upwelling or entrainment of landward-flowing deep seawater takes place in San Pablo and Suisun Bays near the beaches where the submerged oil was deposited. The average surface seaward flow exceeds 5 km/day and the near-bottom landward drift from the Golden Gate to San Pablo Bay averages 4 km/day [76]; hence, both surface and subsurface movement of spilled oil in estuaries can be rapid.

Sinking of petroleum components to the ocean floor depends, at least in part, on the availability of other particulate material which can physically or chemically sorb the petroleum. Such materials include organics, clays, calcite, aragonite, glacial flour, and siliceous sand grains. Rough seas mix this particulate matter with petroleum, especially in shallow waters; the particles with sorbed petroleum eventually settle to the bottom during calm seas. Silt particles have a density of about 2.65, so that only small amounts need to be incorporated with the petroleum to produce an oil-silt combination denser than seawater [2].

Large quantities of suspended silt and other particles can occur in coastal waters, particularly in tidal estuaries. When an oil slick is agitated by wind and wave action in such an area, the particulate matter becomes incorporated into a petroleum-water emulsion or is coated by dispersed oil droplets. The resulting emulsion and droplets may then sink [2]. Powdered chalk was used successfully to sink Kuwait crude petroleum spilled by the *Torrey Canyon* [27].

Theoretically, oil droplets should settle more rapidly in freshwater than in seawater because of the lower specific gravity of the freshwater. Salt water, however, alters the ionic field surrounding the oil droplets increasing their tendency to flocculate and settle. An example of this phenomenon is the presence of soft, flocculant deposits of mud near the mouth of fresh water rivers entering oceans [11].

Laboratory studies [77] showed that fine-grained (less than 44 μm in diameter) clays absorbed test hydrocarbons or

No. 2 fuel oil components in the following decreasing order of efficiency: bentonite, kaolinite, illite, and montmorillonite. Natural sediments sorbed about one-third less No. 2 fuel oil components than the same sediments that were pretreated with hydrogen peroxide to remove organic coatings (presumed to be humic substances). Meyers and Quinn [77] stated that the organic matter probably masked the sorption sites in the sediment or bound sedimentary particles together, reducing the available surface area for sorption of the oil components. However, Suess [78] suggested that 3 to 5% organic material on clay will enhance the sorption process. Obviously, the physical and chemical processes involved in the sorption of petroleum by particulate matter are not yet fully understood.

Sorption apparently increases with salinity of the seawater and decreases with temperature. Fuel oil sorbed on clay is not readily released by repeated washings with clean seawater; Meyers and Quinn [77] found that only 15% of the fuel oil sorbed onto clay was removed after three rinses.

Although the rate of sedimentation of petroleum cannot be calculated in the environment, a considerable amount of sinking can take place in just a few weeks [79,80]. Significant lateral spreading of sedimented petroleum can occur for at least several months after spills [79,80,81]. Bottom deposits of oil, bound by fine sediments, might be expected to occur in sluggish estuaries and shallow coastal waters at the approaches to major ports and industrial regions. In most large estuaries, especially in arctic and subarctic regions, the net water circulation (beyond the tidal oscillations) consists of a seaward movement of low-salinity water at the surface which is replaced by the inflow of more saline oceanic water along the bottom. Sedimented petroleum is thus likely to be concentrated in estuaries rather than dispersed. When petroleum residues are spread out over a large surface area in contact with marine muds they are readily attacked by microorganisms. Masses of petroleum can sink in shallow offshore areas and be rolled along the bottom by waves and currents, accumulating larger particles of sand, shell fragments, and small pebbles; if these are moved shoreward they may form hard, tarry masses described as "coquina oil" [11,75].

The horizontal transport of sunken oil particles can have a serious impact on certain intertidal and benthic organisms. For instance, oil incorporated into the top centimeter of a fine sediment may have an adverse effect on small deposit feeders such as the clam, *Macoma balthica*, which siphons the slurry at the sediment-seawater interface [82].

The persistence of oil in intertidal areas depends upon the wave and surf forces acting on the area and on the type of sedimentary substrate present. On high energy shores (exposed to a high degree of wave and surf action), petroleum deposited during one tidal cycle can be removed on the next cycle [83]. Seasonal erosion and accretion of beaches can alternately expose and cover sedimented oil [84]; oxidation and dissolution of oil occurs during these cycles depending partly on the flow of seawater through the beach and partly on the exposure lengths and conditions [85]. Tarry globules on beaches tend to accumulate sediment and become rounded like pebbles [86]. The globules also tend to concentrate on beach areas above the high tidal level where degradation of the large masses is very slow [84,87].

Low energy shores (protected areas) tend to contain fine-grained sediments and show little or no seasonal cycles of erosion and accretion. The rate of sedimentation of geological particulate matter (e.g., sand, silt, clay) can be quite slow; several months may be required to cover an oil layer on such shores. Microbial degradation of oil buried in a low-energy beach involves primarily anaerobic microorganisms because of lack of oxygen replenishment at sediment depths greater than a few centimeters [2,88]. Three and one-half years after the Bunker C fuel oil spill in Chedabucto Bay, Rashid [89] found oil residues in nearly all the initially contaminated beaches located in protected areas of low- and moderate-energy intertidal environments. Some of the residues were still quite fluid. In the high-energy beaches contaminated by the original spill, the oil residues were highly viscous and adhered to the sand and pebble substrate. These residues were formed through loss of *n*-alkanes and subsequent concentration of the asphaltenes, resins, and non-hydrocarbon organic compounds from the original oil. The chemical compositions of the original Bunker C fuel oil, a sample of the same oil stored for several years, and the residues from low-, moderate-, and high-energy coastlines are presented in Table 5.

On rocky shores, deposition of slicks and finely dispersed petroleum is often confined to the algal felt, or canopy, in the lower intertidal zone and to pores in the rock surface. Tidal flushing removes considerable petroleum over periods as long as several months. Additional residues are removed by erosion of coated dead algae, grazing of intertidal animals [90], or biological incorporation into plant and animal cellular material [84,91]. Permanent deposits of particulate petroleum are usually limited to shore areas of relatively low slope where the oil becomes stranded and bonds to the substrate by partial melting [2].

Oil deposited by wind and wave action on shore above the normal tidal excursions may persist a long time [87,92,93]. Immediately after the grounding of the *General M.C. Meigs*, one rock in the high splash zone was coated with liquid oil. Three years later small deposits of a fluid residue covered by

TABLE 5

Chemical and physical characteristics of Bunker C oil in sediments at Chedabucto Bay

Characteristics	Bunker C oil		Bunker C oil in sediments		
	Fresh sample	Stored sample	Low energy coast	Moderate energy coast	High energy coast
HYDROCARBONS (%)					
Saturated	—[a]	26	25	23	18
Aromatic	—	25	24	24	16
Total hydrocarbons	73.1	51	49	47	34
Ratio: $\frac{\text{saturated}}{\text{aromatic}}$	—	1.04	1.04	0.96	1.12
NONHYDROCARBONS (%)					
Asphaltenes	16.3	20	22	23	22
Resins and NSO	10.6	29	29	30	44
Total nonhydrocarbons	26.9	49	51	53	66
Ratio: $\frac{\text{hydrocarbons}}{\text{nonhydrocarbons}}$	2.72	1.04	0.96	0.88	0.52
PHYSICAL PROPERTIES					
Specific gravity	0.950	0.963	0.9953	0.9765	0.9823
Viscosity (cP)	—	19,584	28,600	1,210,000	3,640,000

Adapted from Rashid [89].

a — indicates no data.

a tough weathered skin were still found in crevices of the rock; the fluid petroleum portion had retained the paraffin hydrocarbon distribution pattern of the original Navy Special Fuel Oil (similar to a Bunker C fuel oil) [84].

Petroleum deposited in the sediment of the intertidal zone can be redistributed into the environment depending upon the nature of the petroleum and the shore area. Light oilings will be carried by successive tides to the strandline at the high-water mark. Heavy, viscous oils and weathered petroleum may mix with sand, gravel, and shell fragments to form hard, tarry, oil cakes or flat patties on the beach. Liquid petroleum does not readily penetrate wet or fine-grained sands, but wave and surf action may throw fresh sand over the oil deposits. In this way, a badly polluted beach may appear to be clean shortly after a spill, but the subsurface oil layers can become exposed during storms or the shifting of sand to again reveal evidence of oil pollution [94]. Oil is readily sorbed by the byssus-threads of mussels, the horny outer layer of shells, and by certain upper-shore seaweeds and plants [84].

Rocky tide pools within an intertidal zone are a refuge for plants and animals unable to withstand the rigors of the exposed open ocean. Because oxygen is as soluble in oil as it is in seawater, the change in the rate of oxygen replenishment for a tide pool covered with a thin layer of oil may be slight [11]. However, if the oil layer is thick enough to substantially decrease the amount of light reaching the tide pool organisms, photosynthetic production of oxygen by the algae will diminish or cease. If sufficient oxygen cannot be provided by diffusion through the thick oil layer, the pool may not support the contained plant and animal marine life. A dark, thick layer of oil will absorb heat from the sun. Some of this heat will be transferred to the water in the pool, lowering the dissolved oxygen capacity of the seawater and increasing the respiratory rate of the animals. The indirect effect (alteration of the oxygen balance) of an oil slick on the marine organisms in an intertidal pool, however, is probably insignificant compared to the direct effect of toxic petroleum components [24].

In temperate climates, oil remaining on a beach after one year retains about 10% of the original n-C_{17}-C_{18} hydrocarbons, about 50% of the n-C_{19}-C_{20} hydrocarbons, and practically all of the hydrocarbons above n-C_{23}-C_{24} [87]. In comparison, oil incorporated into nearshore or marsh bottom sediments retains most of the original hydrocarbons above n-C_{12} [80,81,87].

Blumer, Ehrhardt, and Jones [87] studied the weathering of two light paraffinic crude oils which stranded on a coastal subarctic beach (Martha's Vineyard, Massachusetts) and on a tropical island beach (Bermuda). Climatic differences between the two locations had only minor effects on the degradation rates during the first year. Physical disintegration of the weathered crude oil at Bermuda provided an enlarged surface area and may have facilitated evaporation during the thirteenth month. At Martha's Vineyard, this stage of weathering was not reached in 16 mo after stranding, based on gas chromatographic analyses. In both areas, the original paraffins above n-C_{22} (boiling points of 320°C and above) were retained in the residues, implying the retention of high boiling aromatic hydrocarbons such as phenanthrene, anthracene, and their substituted homologs, and the higher molecular weight aromatics [87].

After 13-1/2 mo exposure at Bermuda, the residue showed little evidence of purely chemical alteration. The initial ratio of saturated to aromatic hydrocarbons was preserved, the hydroxyl content was not increased, and only a modest rise in ester or acid content was found. After 13-1/2 mo at Bermuda and 16 mo at Martha's Vineyard, the spill residues were still far from being inert asphalts [87].

Vandermeulen and Gordon [95] noted the long-term re-entry of stranded oil into interstitial water environments of oiled beaches following the *Arrow* incident. A stranded Bunker C fuel oil was markedly altered, with almost total loss of the *n*-alkane portion, according to Vandermuelen, Keizer, and Ahern [96]. They suggested that natural beach sediments appeared to function as a filter for removing alkanes, perhaps through microbial activity. The loss of *n*-alkanes was nearly complete in the oil contained within the top few centimeters of the beach, but a sizable fraction of aromatics and cycloalkanes remained which could eventually re-enter the water column [95].

MICROBIAL MODIFICATION

A detailed discussion of the chemical, physical, and biological parameters that affect the microbial modification of petroleum is given in Chapter 3. Two references broadly covering this problem are the Symposium proceedings "The Microbial Degradation of Oil Pollutants" edited by Ahearn and Meyers [97] and the section on microbial degradation in the National Academy of Sciences report "Petroleum in the Marine Environment" [2].

In the marine environment, hydrocarbon-utilizing bacteria [98-100], yeast, and fungi [67,99] occur in greater numbers in those areas exposed to recurring petroleum contamination. Petroleum modification by these microorganisms is selective and relatively slow; different microorganisms appear to prefer different chemical components of petroleum [99-101]. Apart from temperature the simultaneous availability of water, petroleum, oxygen, and nutrients appears to control the rate of degradation [32].

The rates of microbial degradation of fossil hydrocarbons and their derivatives vary with the chemical complexity of the petroleum, the number and types of microbial populations, and environmental conditions [102]. With this multivariable system, it is impossible to predict the rate of microbial degradation with ease or accuracy [2]. Few experiments have been carried out in the marine environment. Floodgate [102] demonstrated that, even under laboratory conditions, utilization of petroleum hydrocarbons by marine bacteria is measured on the scale of weeks. Microbial modification of petroleum in a natural marine system may be considerably different from that in the laboratory because of variations in environmental factors (e.g., temperature, solar radiation, salinity of seawater, nutrients) which could effect the action of the microorganisms and alter the physical and chemical nature of the petroleum substrate [2].

The turbulence is ordinarily sufficient to supply oxygen for microorganisms living in surface layers and facilitate emulsification and dispersion of petroleum [2]. ZoBell [103] calculated that the theoretical amount of oxygen needed by microorganisms to oxidize 1 liter (0.26 gal) of crude oil to carbon dioxide and water would require all the dissolved oxygen in 3.2×10^5 l (84,000 gal) of average air-saturated seawater (20°C). Before oxygen depletion becomes a serious problem, the rate of microbial consumption must greatly exceed the rate of oxygen renewal in the seawater; however, such a condition is not likely to occur in the open ocean [24].

Anaerobic degradation of petroleum by marine microorganisms seems insignificant [104]. Dean [32] suggested that petroleum would probably remain indefinitely on the sea bottom where the oxygen content is low. In support of this view, Blumer and Sass [80,81] demonstrated the persistence of a fuel oil incorporated into the sediments of an active estuarine environment.

Temperature may influence the rate of microbial utilization of petroleum in the marine environment in several ways. Within certain limits, depending upon the types of bacteria present, raising culture temperatures (e.g., by 5° to 10°C) increases the degradation rates and activities of microorganisms. Also, higher temperatures reduce the viscosity of the petroleum and increase the tendency of oil and water to form emulsions, thus enhancing the availability of the petroleum to microbial action [101]. The temperature effect on the activity of microorganisms involves different ranges depending upon the microorganisms. For example, the psychrophilic (low temperature) microorganisms commonly found in the sea, are more active at 5° to 10°C and may not survive at 20°C, while the mesophilic (medium temperature) microorganisms are more active around 20°C. Since the temperature of ocean water below the upper layers rarely rises above 5°C, most of the microorganisms associated with the marine system are psychrophiles. However, many of the laboratory enrichment studies carried out so far involve incubation temperatures above 15°C, approaching the temperatures for mesophilic microorganisms [105]. Laboratory experiments by ZoBell [101] indicate that certain psychrophilic microorganisms can degrade saturated hydrocarbons at 0° to 2°C. Cundell and Traxler [98] found a diverse microbial population capable of degrading low-molecular weight saturated, cyclic, and aromatic hydrocarbons in Chedabucto Bay (Nova Scotia) water and sediments and well as in soil adjacent to a natural oil seep at Cape Simpson (Alaska). The environmental temperatures at these two sites were 16° and 8°C, respectively. The enrichment flasks were incubated at 0°, 8°, 16°, and 24°C; the majority of the microorganisms were isolated from the samples incubated at 16° and 24°C. Growth occurred between 8° and 24°C within 14 days of incubation, suggesting that the microorganisms were tolerant to a range of temperatures. Certain petroleum components, such as certain long-chain alkanes, polynuclear aromatic hydrocarbons, and asphaltenes, seem to be resistant to rapid microbial degradation [98]. Many polynuclear aromatic hydrocarbons may be both biosynthesized and biodegraded [106]. Nevertheless, Cundell and Traxler [98] suggested that microorganisms could play a significant role in the degradation of petroleum in the marine environment even at temperatures as low as 0°C.

Although psychrophilic bacteria can grow at low temperatures, their proliferation in arctic environments will be curtailed at the freezing point. The combination of high pressure and low temperature may mean that petroleum hydrocarbons reaching deep-sea sediments may not be degraded rapidly. This possibility was dramatically exemplified by the recovery of almost perfectly preserved food in a waterfilled lunchbox from the salvaged research submersible *Alvin* of the Woods Hole Oceanographic Institution after one year below 1,500 m of seawater [107,108].

PHOTOCHEMICAL MODIFICATION

Oxidation of crude oils can be initiated thermally or photochemically [50]. Light of wavelengths shorter than 400 nm (ultraviolet light and part of the visible spectrum) tend to accelerate the oxidation process. Certain metallo-organic compounds, often found in petroleum, promote oxidation of hydrocarbons. Sulfur-containing organic compounds, normally present in petroleum, tend to inhibit the rate of oxidation;

however, the sulfur compounds may be oxidized and converted to water-soluble components [32,59].

Paraffins, unconjugated olefins, and many aromatic compounds can undergo a variety of chemical reactions resulting in the formation of ketones, aldehydes, carboxylic acids, esters, and polymeric gums and tars. Oxidation of petroleum hydrocarbons generates reactive free radical intermediates; the reactions are initiated by free radicals produced by processes such as the absorption of light [50]. Further reactions can be self-sustaining or even accelerating because the reaction products (hydroperoxides) can decompose to form more free radicals [50]. In a particular hydrocarbon system the reaction rate can be limited by inhibitors (e.g., sulfur compounds [50], phenolic compounds and organic amines [109]).

The compounds formed by hydroperoxide decomposition may further react (e.g., dehydrogenation, oxidation) to yield aldehydes, ketones, or carboxylic acids having a lower molecular weight [50]. Higher molecular weight products may also form by oxygen addition to conjugated dienes or by the decomposition of polymeric alkoxy radicals [109]. For example, the gum formed in degraded gasoline largely consists of peroxides with higher molecular weights than those of the precursor olefins [109]; degradation products of lubricating oils contain alcohols, esters [50], aldehydes, polyermic material, and acids with lower molecular weights than those of the precursor peroxides [109].

The oxidation rate depends on the chemical nature of the hydrocarbons and non-hydrocarbons (NSO compounds) in petroleum [110]. For example, alkyl-substituted cycloalkanes tend to oxidize faster than the normal alkanes [2]. Parker et al. [50] found that photooxidation of petroleum in fresh water proceeded more rapidly when the mixture was exposed to light of wavelength shorter than 300 nm (i.e., shorter than the ultraviolet limit of sunlight) than when exposed to direct sunlight. Furthermore, the optical density of the petroleum seemed to be an important variable in the photooxidation processes, particularly when light in the ultraviolet region was used [2].

Freegarde and Hatchett [111] estimated that an oil slick 2.5 μm thick (2 metric tons per km^2) could be degraded by photooxidation in 100 hr of continuous exposure to sunlight. Thus, assuming 8 hr of sunlight per day, an average oil slick in the open ocean could decompose in a few days [2]. Under ordinary environmental conditions, destruction of the oil slick would be more rapid because other factors, such as dissolution of photolysis products (e.g., carboxylic acids or phenolic acids), biological modification, evaporation, and emulsification, would enter into the process. Parker, Freegarde, and Hatchard [50] tentatively concluded that photochemical degradation of oil in a thin film or dispersed droplets near the sea surface would be slow, but probably appreciable on long exposures. Photooxidation of chocolate mousse and the resulting tarry lumps must be extremely slow because of their low surface area to volume ratio.

In laboratory simulation tests, Burwood and Speers [59] exposed surface slicks of crude petroleum to artificial light to determine the dissolution of petroleum hydrocarbons in seawater. They postulated that indigenous autooxidizable hydrocarbons react photolytically with indigenous thiacyclanes to form a complex water-soluble mixture of thiacyclane oxides (sulfoxides). Such compounds were first detected in seawater extracts following prolonged equilibrium with a medium-sulfur content Middle East crude oil. These compounds showed up in gas chromatograms of high-boiling, water-soluble components as an unresolved envelope (C_{15} to C_{23} range). The area of this unresolved envelope increased conspicuously as the photooxidation progressed. Burwood and Speers [59] suggested that such a process might explain the loss of sulfur materials during weathering of Kuwait crude oil in the *Torrey Canyon* spill [112].

Laboratory studies on the spreading properties of several crude oil slicks (0.15 ml of oil on a circular chamber 5 cm in diameter containing seawater) irradiated with ultraviolet light revealed that viscous oils (greater than 40,000 centipoise at 25°C) behave differently than non-viscous oils (less than 500 cP) [113]. As photooxidation proceeds, a viscous oil slick may actually contract owing to polymerization of the petroleum components and to the resulting increased viscosity which restricts diffusion of oxidation products to the oil-water interface. Photooxidation could thereby help generate intractable tarry residues and perhaps water-in-oil emulsions.

The impact of photolysis products on the marine environment may extend beyond the apparent dissipation of the oil slick. Lacaze and Villedon de Naïde [114] found that the deleterious effect of Kuwait crude oil on the primary production of a microalga in water-oil mixtures was much greater in the light than in the dark.

BIOLOGICAL INGESTION AND EXCRETION

Bioassimilation of petroleum from some marine environments is aided by the dispersion of the oil into fine droplets by wave action and subsequent dispersal of these droplets by currents [59]. Some of the ingested material is not incorporated into tissues but is eliminated in the feces. Following the *Arrow* incident in Chedabucto Bay, Nova Scotia, Conover [115] noted that zooplankton ingested significant amounts of Bunker C fuel oil droplets. Most of the ingested oil was eliminated in the feces which contained up to 7% by weight of oil. Copepods voided ingested oil within 24 hr and generally showed no adverse effects, as determined by microscopic examinations. Conover estimated that under the conditions following the *Arrow* spill, zooplankton could remove as much as 20% of the oil droplets (less than 0.1 mm diameter) from the water column. The density of copepod fecal matter is greater than that of seawater, so encapsulated oil will sink.

In 1970, Parker [116] found oil droplets in the gut of copepods and barnacle larvae and in their fecal pellets. The oil apparently passed through the organisms unchanged chemically. The plankton were not adversely affected by the oil droplets although long-term impact studies had not been conducted to verify this conclusion. Parker, Freegarde, and Hatchard [50] estimated that one copepod (*Calanus finmarchicus*) could encapsulate up to 1.5×10^{-4} g of oil per day. At this rate, a population of 2,000 individuals per cubic meter of seawater, covering an area of 1 km^2 to a depth of 10 m, could remove as much as 3 tons of oil daily if the oil concentration were 1.5 μg/l or greater. Although the copepods did not seem directly harmed by the oil, oil-containing plankton and fecal matter could be ingested by other members of the food web [2].

In laboratory studies, Alyakrinskaya [117] found that the mussel, *Mytilus galloprovincialis*, from the Black Sea could tolerate mixtures containing up to 20 ml of oil per liter of seawater. The mussels formed mucous-encased pseudofeces from the oil droplets. The encapsulation of oil by these mussels is similar in principal to the response observed in plankton [115,116].

Many investigators [e.g., 60,84,91,118-122] have determined the hydrocarbon content of filter-feeding organisms exposed to oil contamination. The hydrocarbon patterns were similar to those in petroleum. Furthermore, the hydrocarbons in the organisms occurred at levels well below the odor and taste thresholds of humans [9,120]. When placed in unpolluted water or when the pollutant source is removed, many organisms tended to eliminate most the absorbed petroleum within several weeks [2]. See Chapter 3 of Volume II for a detailed discussion of physiological effects resulting from exposing marine organisms to petroleum.

TAR BALL FORMATION

The occurrence of petroleum residues on beaches and in the open ocean is well-documented [2,72,73,83]. The residues vary considerably in composition, but most of them contain paraffinic hydrocarbons up to C_{40}, typical of weathered whole crude petroleum or waxy crude petroleum sludge from the cargo tank washings of tankers [83]. Few residues containing hydrocarbons of less than 15 carbon atoms have been found. About 20 to 40% of a typical residue (tar ball) from the North Atlantic Ocean is not volatilized in 10 min at 200°C in the injection port of a gas chromatograph [123], indicating the presence of high molecular weight components.

Tar balls collected along the Texas Gulf coast east of Galveston Bay had a high NSO compound content (up to 70%, with an average of 45%) [124]. The level of NSO compounds in tar balls is frequently higher than would be expected from simple evaporation and dissolution of weathered crude oil. This enrichment apparently arises from the formation of NSO compounds from petroleum hydrocarbons by photooxidation; aromatic compounds can be photochemically, or catalytically, oxidized in the presence of trace metals (See the preceding section on photochemical weathering). Many of the polynuclear aromatic hydrocarbons and trace metals [125] present in spilled petroleum would remain after the low-molecular weight hydrocarbons are lost. Some of the NSO compounds formed during the weathering process would be retained in the residue and would be lost extremely slowly from the semi-solid matrix of the resulting tar balls [2].

Chloroform-insoluble material from tarry residues collected off the Florida coast [124] consisted of pulpy organic matter and dense mineral-like particles. Some of these particles were insoluble in acid and seemed to be quartz. The ferric oxide content of some residues was as high as 18% on a dry weight basis, substantially greater than the usual iron content of crude petroleum (100 ppm, total iron). The increased levels of iron probably originated from rust particles picked up by the oil from steel tanks or other steel apparatus (e.g., offshore oil production, wastes from engines either on land or in ships, or rinsings from cargo tanks of crude oil tankers) [124].

Tar balls found on California beaches near seep areas were porous and were probably produced by the agglomeration of smaller lumps in the surf zone [86]. In contrast, most tar lumps collected at sea appeared to have been formed as a single dense piece, probably from a single oil sludge (tank washings) or slick residue [2]. Since natural oil seeps contribute crude petroleum to the environment at a slow steady rate, the resulting oil slick would be expected to be quite thin [126]. Tar lumps from seeps, when formed, would be expected to be different to tarry residues formed from tanker washings [2].

Oil slicks and tar balls on the ocean surface are transient phenomenon. Slicks may last only a few weeks, whereas tar balls may last for a year or more [127]. McAuliffe [7] suggested that the half-life of oceanic tar may be as short as one to four months.

INTERACTION OF ICE WITH PETROLEUM

Oil spilled in arctic regions would be subject to the previously described processes of spreading, evaporation, dissolution, emulsification, sinking, microbial modification, photochemical modification, and biological degradation. However, considerable differences can be expected between the rate of these processes in arctic regions and those in temperate regions about which we have more data and understanding. The colder climate of the arctic and subarctic increases the longevity of spilled oil by decelerating natural degradative processes. Moreover, climatic conditions impose severe constraints on men and machinery [128]. Depending on the sea and ice conditions, spilled petroleum may spread over the ice, under the ice, on the water, or in several modes at once [129, 130].

Campbell and Martin [131,132] have speculated about serious complications which might occur in the Arctic from a major oil spill. They suggested that three processes are involved in the dispersal of oil in the Arctic: (1) lead-matrix pumping where the oil could be dispersed by the continuous opening and closing of leads in the open ice pack (the ice pack can move at rates as high as 10 km/day); (2) oiled hummock melting where oil particles incorporated into ice hummocks formed during closing of oil-filled leads would have a different absorption of solar radiation; and (3) under-ice transport in the form of oil-in-water and water-in-oil emulsions. A hypothetical oil spill of 2 million barrels (about 270,000 metric tons), as an extreme case, might cover an open water area of 16,000 km^2 possibly affecting as much as 8% of the Arctic Ocean. The wide dispersal of oil combined with its slow rate of degradation under arctic conditions and the dynamics of the Arctic Ocean could increase the areal distribution of oil on ice surfaces. Campbell and Martin [131] speculated extensive coatings of oil could lower the natural albedo and significantly change the heat balance of the Arctic Ocean, thereby causing massive ice melting. Extensive dispersion of oil and the inability to clean up a major spill in the highly mobile and deformable ice pack during the nine-month arctic winter increases the potential for long-term environmental damage.

In response to Campbell and Martin's paper, Ayers, Jahns, and Glaeser [133] stated that a significant alteration of the Arctic heat balance from a major spill would probably not occur. They suggested that the processes described by Campbell and Martin involved in the dispersion of oil in the ice areas of the Arctic would compete with other processes (e.g., "herding" by wind and ice, and the closing of oil-covered leads) that tend to concentrate the oil. Ayers and coworkers predicted that the hypothetical spill would cover only 100 km^2 or 0.001% of the pack ice resulting in no significant change in the heat balance of the entire Arctic Ocean. They also stated that heating effects would occur only from oil on the ice surface and only when the oil was not covered with snow. Results from the NORCOR report [134] indicated, however, that solar radiation penetrates snow and that underlying oil will absorb heat through a snow covering. Obviously, more research on the interaction of oil with sea ice is needed to make accurate predictions.

Although ice packs can confine spilled oil, open leads in the pack ice are often interconnected and serve as convenient escape paths from wind-driven floating oil [133,135,136]. Opening and closing of the leads (matrix-lead pumping) may force some of the oil to the surface of the ice pack or into more distant leads. The diffusion or spreading of large quantities of oil by the ice pack may contaminate large areas over a period of several years. Any oil in open leads could be bound in crushed ice formed during the continuous opening and closing of the ice pack. Because of repeated collisions of ice masses, large accumulations of ice, called pressure ridges, are often formed. Oil-contaminated crushed ice could form oily pressure ridges (hummocks) above the water and large masses of contaminated ice could make up the pressure ridge keel. This contained oil would be practically inaccessible for cleanup. The melting of the oily hummock could decrease the albedo of the surface ice; the abrasion of the contaminated ridge keel on the ocean bottom could slowly release oil into the seawater. Since keel depths range from 5 to 40 m, some of the oil could be released into the currents below the ice-seawater boundary allowing the oil to slowly disperse for years over large ocean areas [132].

Field tests [137] indicated that Prudhoe Bay crude oil has a viscosity low enough to spread easily over seawater in the arctic summer but at winter temperatures the oil will solidify. The rough upper surface of the ice often acts to contain the spilled oil. The oil is trapped in depressions or is absorbed into the coarse upper ice layers; about 10 to 24% by volume of the oil can be absorbed [137]. The transition point where an oil slick on the sea surface would stop spreading by surface forces was determined [23]. Glaeser [129] found that a Prudhoe Bay crude oil spilled on the Arctic Ocean (0°C water temperature) reached a stable spreading state at about 5 mm thickness; further spreading was effected only by the wind. During the winter, McMinn and Golden [33] spilled the same oil over arctic ice and found that the crude oil never reached a point where surface tension became an important factor in spreading. They suggested that the rough, uneven surface of the ice and the subsequent pocketing of the oil limited the spreading so that the transition point thickness (calculated to be about 3 mm) was never reached.

Using laboratory experiments, Martin, Kauffman, and Welander [138] investigated the formation and growth of sea ice in the presence of waves and spilled oil. Two kinds of sea ice grew in the wave field--grease ice which appeared as a thick soupy, flexible layer on the sea surface and pancake ice which consisted of solid, circular pieces of ice with raised rims. The rims were formed from the ice cakes striking each other in the wave field. These two kinds of ice occur in the transition zone where new ice forms between the open waters of the arctic and subarctic seas and the more rigid pack ice. Cold, high velocity winds generate waves in this transition zone.

Under laboratory conditions (-20°C) in the presence of waves (0.6 to 1.3 m wavelengths), the first ice to form was the grease ice which had a high porosity. Heat transfer occurred by convection so that the surface temperature of the grease ice (6 cm thick) was only a few tenths of a degree lower than temperatures of the water below. Diesel oil released 20 cm under the grease ice rose rapidly and spread out on the sea surface; no signs of oil absorption below the grease ice surface were evident [138]. The grease ice grew until it reached a thickness of about 10 cm, at which time pancake ice began to form. These cakes grew from the grease ice and initially floated on the much thicker grease ice layer. Grease ice was pumped onto the ice surface of the pancakes by the convergent-divergent motion of the wave field where the grease ice froze forming the raised rims on the edges.

Cold Prudhoe Bay crude oil (between -2° and +0.5°C) released under the pancake ice (dispensed as viscous globules at 30 cm) moved through the grease ice and appeared on the surface between the rims of the pancakes. Wave oscillations pumped the oil laterally and onto the depressed pancake ice surfaces where it tended to remain because of the raised rims. Most of the crude oil in this experiment appeared to penetrate the pancake ice surface (58% by volume released) or remained in the grease ice cracks between the cakes; however, Martin and coworkers observed a very light film of oil on parts of the undersides of the pancakes [138].

Petroleum discharged below the ice cover (such as from an oil well blowout) rises to the underside of the ice in a cone-shaped plume and becomes trapped in the uppermost irregular pockets at the ice-seawater interface. The effect of a gas-rich oil well blowout might be the formation of a strong central buoyant jet of bubbles which could entrain a large quantity of bottom water. Water transported to the surface could then form a radially-spreading surface current under the ice. In general, the entrained deep water is denser than the surface seawater; the denser upwelled water would sink at some distance from the vertical plume. Some melting of the ice could occur either from the heat transfer from the hot crude oil or from entrainment of warmer bottom water (0.5°C warmer than the surface water) [139].

A rising oil plume tends to be unstable, breaking up into small spherical particles, 1 cm in diameter or less. On striking the underside of the ice the oil radiates outward, progressively filling the depressions in the underside of the ice sheet. Most crude oils form sessile drops at an ice-water interface; the minimum thickness of a continuous oil film is about 0.8 cm for Norman Wells (Northwest Territories, Canada) crude oil. Smaller spherical drops can exist, but usually they occur only near the periphery of a plume. The maximum thickness of the entrapped oil is determined by the depth of depressions, or variations in ice thickness; typical depressions are about ±20% of the average ice thickness [134]. Studies also suggest that oil trapped under the ice is virtually indistinguishable from fresh crude oil with regard to viscosity, density, gas chromatographic hydrocarbon profiles, and other parameters [134].

If turbulence is low, the petroleum spilled under ice can remain unchanged for a long time. Sea ice will, in general, have a specific gravity of 0.85 compared to 0.91 for salt-free ice. Sea ice has a lower density because of its porous structure caused by the migration of brine during the freezing process. Most Arctic crude oils are denser than sea ice and will tend to flow under the ice [129].

The liquids (crude oil and production waters) from an oil well blowout in quiet ice-free waters will have a radial surface flow pattern with a ring of waves concentric with the plume center. This stationary ring of waves marks a division in the directions of the surface radial currents--outward flowing currents inside this ring and inward beyond the ring. Such a flow system might provide a limited amount of natural containment for oil in quiescent periods [139].

Regardless of the season, the temperature of the ice at the ice-seawater interface will be at the freezing point. At this temperature, a typical North Slope (Alaska) crude oil will have a viscosity low enough to flow easily. Multi-year ice in general has a very rugged underside, with pressure ridges extending vertically downward as much as 50 m. These features will tend to severely restrict the horizonal spreading of a large volume of oil under the ice; the oil will be trapped or caused to flow around the obstructions [129].

Oil enclosed in brine channels and trapped under areas of first-year ice will remain essentially in its original state through the winter. During the spring and summer, after the first-year surface ice melts and the brine channels open up, the oil may migrate to the surface of the ice. As the dark oil absorbs heat from the sun, the rate of migration increases and the oil forms a melt pool on top of the ice. Multi-year ice contains no brine channels so the oil may remain trapped for up to 3-4 yr [132,140].

Following the grounding of the tanker *Arrow* in Chedabucto Bay, Nova Scotia, oil droplets were found in the bay ice near the spill area. Apparently, the oil particles were contained in the sea surface layers of the bay water and had become entrapped in the bay ice during the freezing process. Later, as the ice melted, subsurface areas (often 15 m in diameter) of the ice appeared to be "dirty" as a result of incorporated oil particles; near-surface area accumulations of fluid oil were from 5 to 15 cm in diameter. Barber [137] suggested that this pattern resulted from a redistribution of particulate oil caused by the intermingling of oil with melt water; the ice cover was 10 cm thick and had been 20 cm thick previously.

A Norman Wells crude oil released under sea ice early in the ice growth season penetrated between 5 and 10 cm into the loose skeletal layer on the underside of the sea ice. The oil remained in the ice until the ice sheet warmed in the spring. Then, melting of the brine channels intensified and the oil began to migrate upward. Initially the movement was slow, typically in the range of 15 to 20 cm during February and March. The rate of migration increased with more sunshine and warmer air temperatures. Oil released in late April under 150 cm of ice was detected on the surface within one hour [134].

The pour point temperature may also influence the behavior of an oil in seawater [141]. Divers observed that the Bunker C fuel oil leaking from small openings in the wrecked tanker *Arrow* generally surfaced in a "discrete piece like a rope 1 to 2 feet long" [142]. This behavior was undoubtedly related to the fact that the pour point (-1°C) of the escaping oil was close to the temperature of the surrounding water (0° to 2°C).

In those arctic areas where ice continually moves and breaks up annually, alteration of the density, viscosity, and surface tension of the oil may influence the methods required to contain a major oil spill. The density of undeformed sea ice depends on its salinity and porosity and averages about 0.91 g/cm^3; the density of the underlying seawater is approximately 1.030 g/cm^3. If the oil is allowed to weather, it will be less buoyant and may migrate under the ice or into the seawater [33]. Oil denser than ice but less dense than seawater will remain on the water surface until forced under the ice by waves or currents.

In laboratory experiments, Keevil and Ramseier [141] observed what took place when warm petroleum (20°C) was injected under ice (-15°C) in a fresh water (0°C) basin without horizonal currents. Initially, the petroleum (Norman Wells crude) separated into hundreds of small globules or particles, 0.1 to 2.0 mm in diameter, which immediately rose to the underside of the floating ice sheet. Some particles coalesced to form lenses which did not adhere to the ice-water interface under quiet water conditions. The ice continued to grow, first down around the sides of the petroleum lens and then finally underneath, sandwiching the oil lens between ice layers. The oil lens may have insulated the water beneath from the cold air above but did not significantly alter the rate of ice growth.

In a laboratory test, Wolfe and Hoult [143,144] injected a 1 to 2.6 cm thick layer of North Slope (Alaska) crude petroleum under a 12 to 16 cm thick sea ice layer in a brine solution. Little mixing of the oil and ice in the brine occurred although the oil rose in small pockets under the ice surface. The ice did not grow through the oil, although a second layer of ice formed beneath. Wolfe and Hoult found further that the oil had an insulating effect which slowed the formation of the lower ice layer. After a single experimental discharge of oil beneath a sea-ice sheet during the depth of winter, a similar lip of ice formed around the oil lens within a matter of hours, preventing horizonal movement of the spilled oil. A new layer of ice formed beneath the oil within a few days. Once entrapped in this fashion, the oil became stabilized until spring [134]. If an oil spill were to occur in winter, most of this entrained oil would be entrapped by the growing ice mass making cleanup virtually impossible.

Martin and Campbell [132] speculated that multi-year ice containing oil might move thousands of kilometers in the one to four years that it takes for the oil in the ice to reach the surface of pack ice. Because the ice is also continuously converging and diverging, the width of the band of oil-contaminated ice will progressively increase [134]. Hence, it is difficult to predict where ice-entrapped oil might surface. If the oil was released after one year, oil from a blowout in Mackenzie Bay in the Canadian Arctic, for example, might be transported as far west as Barrow, Alaska; if released after two years, the oily ice could move as far as Vamkaren in eastern Siberia.

Prudhoe Bay crude oil exposed in the arctic marine environment during the summer lost volatile components (fraction boiling below 100°C) in two to five days with a resulting increase in viscosity and specific gravity. No marine microorganisms were found that were able to degrade the crude oil at normal summertime arctic temperatures under the ice. However, two fungi and one bacterium isolated from the test area were able to grow at elevated temperatures in laboratory enrichment cultures. The supply of oxygen for microbial degradation under the ice may have been limited by either the surface covering of ice or the layer of low salinity seawater just beneath which restricted oxygen replenishment. The oil had also a relatively small surface area due to pocketing at the ice-seawater interface, further reducing the potential for microbial activity [129].

In situ microbial degradation of Prudhoe Bay crude oil under arctic conditions was studied using three types of miniature, contained oil slicks: (1) oil plus a poison to kill indigenous bacteria, (2) oil alone, and (3) oil plus a fertilizer. There were large increases in the number of bacteria underlying the unpoisoned slicks with the greatest numbers associated with the fertilized slick. After 5 weeks, non-biological loss of oil (poisoned slick) was 31% of the added oil by weight, loss of untreated (no fertilizer) oil was 60%, and loss of fertilizer-enriched oil was 80% [145]. While it is difficult to relate the results of this type of controlled field experiment to large-scale natural spill conditions, the following tentative conclusions might be drawn: (1) microbial degradation of oil is possible under summer arctic conditions near Prudhoe Bay, and presumably in other arctic localities; (2) microbial degradation in the contained system was faster than non-biological degradation processes; (3) inorganic nutrients may limit the overall rate of degradation; and (4) even under conditions favorable for effective microbial degradation of spilled petroleum, periods of time ranging from weeks to months may be needed.

The dissolution of petroleum into seawater and the absorption of petroleum into ice may substantially affect the fate of spilled oil; however, little is known about the rates of these processes in the arctic. Photochemical oxidation is not expected to be a major factor because of low ambient light intensities in the winter, insulation of oil by the ice cover, and low oil surface areas of exposure [129]. When Prudhoe Bay crude oil was exposed to an arctic winter environment, the more volatile components evaporated. Although the rate of evaporation was much less in the winter [35] than in the summer [129], the oil underwent weathering in the winter months when exposed to the atmosphere and approached the density of sea ice and possibly even that of the seawater.

The evaporation of oil poured onto snow in the arctic was not affected by wind because the oil was protected by the oil-absorbing snow cover. However, the evaporation rate on snow-free ice increased directly with ambient wind velocity. Snow-protected oil evaporated much slower despite a 10° to 25°C increase in temperature [33]. When crude petroleum is spilled onto snow at the same temperature, the oil will penetrate the snow and completely fill the voids in the snow structure until there is a decrease in the hydrostatic pressure of the oil. When the hydrostatic pressure of the oil drops, gravitational flow assisted by capillary forces will occur, partially draining the oil from the snow. Ultimately, the amount of oil remaining on the drained snow crystals will depend on the physical structure of the snow; however, Mackay and coworkers [146] believed that the oil coating may amount to about 20% of the void volume.

When hot oil is spilled onto snow, the snow melts and an oil-water mixture is formed which penetrates the snow in advance of the hot oil. The whole process is very complex, involving fluid flow, heat transfer, phase changes, and liquid-liquid separations [35,146]. Warm Prudhoe Bay crude oil (15°C) poured on snow (-15° to -25°C) melted the surface of the snow layer as expected, and then the oil-water mixture flowed down 2 mm into the snow column due to gravity and capillary action. At this point the water under the oil layer refroze which effectively blocked the pore channels and prevented further downward migration of the petroleum. The layer of petroleum on the snow was quickly covered with wind-driven snow. This fresh snow partially diffused into the petroleum layer and formed a petroleum-crystalline mulch, containing as much as 80% water by volume in the form of snow. When the temperature increased above the pour point of the petroleum, the mixture became more fluid and the petroleum began to separate from the dry-appearing mulch [35].

An ice surface covered with black crude oil absorbs more solar radiation than a clean, white ice surface. Under experimental summer field conditions, ice covered with Prudhoe Bay crude oil absorbed approximately 30% more radiation than clear ice. As a result, oil-covered ice melted approximately 2 cm per day faster than clear ice [129]. Oil released under ice migrated to the surface through melting brine channels. On reaching the surface, the oil saturated the snow cover and substantially reduced the albedo; the albedo of oiled snow is about 4 times that of an oil film on water (Table 6). Increased absorption of sunlight accelerated the melting of the snow and oiled melt pools quickly developed. Oil was splashed into the surrounding snow by wind and wave action and the pools gradually enlarged until interconnected. New oil was continually released until the melt reached the initial level of the oil lens which had originally formed at the bottom of the ice sheet. Once melt holes develop and surface drainage patterns are established, the ice sheet rapidly deteriorates. Studies in the Canadian Arctic indicated that, depending on

TABLE 6

Measured albedos - Canadian Arctic oil tests

Surface type	Albedo
Fresh snow	0.85
Wet snow	0.65
Oil on snow	0.30 - 0.70
Thin oil film on water	0.10
Thick oil film on water	0.06 - 0.10
Water	0.08 - 0.15
Global earth average[a]	0.34

From NORCOR Engineering Research Ltd. [134] .

a Von Arx [147].

the physical structure and geographic location of the oiled ice, the contaminated areas were likely to be free of ice one to two weeks earlier than areas of clean sea ice [134].

Most of the research to date on the behavior of spilled oil in and under sea ice has involved experiments in laboratory cold-rooms or in stationary first-year sea ice. The results of such research can further our understanding of the nearshore, fast ice zone which undergoes annual melting. However, oil drilling activities have been proposed for offshore seasonal and polar pack ice zones in the Alaskan and Canadian Arctic where ice conditions are very dynamic and where a large proportion of the ice is multi-year ice. According to the NORCOR report [134], a great deal of research is necessary simply to delineate composition, physical characteristics, and movements of ice in these offshore regions. For instance, without data on the roughness of the underside of the pack ice, it is impossible to assess the probable depth and size of pools of spilled oil and to estimate the area of the possible contamination. Data are required on the frequency, orientation, size, and dynamics of leads to evaluate properly the significance of "lead matrix pumping" and the spread of oil between flows. Studies on the entrainment and migration of oil in multi-year ice are necessary. The brine channels, by which oil reaches the surface of first-year ice, do not exist in multi-year ice; therefore the oil could remain in the ice until the surface melt actually reaches the level of the entrapped oil, except where cracks, "worm" holes, and other flaws permit the oil to escape. Surface oil contamination and solar irradiaton of oil within the ice sheet may well shorten the natural ice pack depletion rate, commonly estimated to be between four and seven years [134].

PETROLEUM HYDROCARBON LEVELS IN THE MARINE ENVIRONMENT

Data on the hydrocarbon content of the several compartments of the marine environment were collected from the literature and are summarized in the following seven tables. Information is presented on sample description, sampling location, method of analysis, hydrocarbon fraction, hydrocarbon content, and reference source. The descriptive terms for the various hydrocarbons appearing in the tables are those used in the referenced reports; however, other hydrocarbons may have been present in the samples but were not determined or reported. Original data on hydrocarbon content of biological samples presented on a dry-weight basis were converted to wet-weight basis by dividing the values by 10, unless some other conversion factor was specified by the author(s).

ORGANISMS

Unexposed Baseline Locations

Marine organisms can acquire hydrocarbons from their surroundings [84,91,118,120,122]; Mackie, Whittle and Hardy [148] speculated that the levels in such organisms could be increasing. Analyses for petroleum-type hydrocarbons in marine organisms have been initiated in several regions (the coastal margins of North America, around the British Isles, along the German coast, and in the northern Mediterranean) to establish existing (baseline) levels and to determine the relative amounts derived from petroleum and from biological sources. The baseline levels of hydrocarbons in the tissues of marine organisms from areas apparently not exposed to petroleum contamination are presented in Table 7.

A striking feature of the data is that most of the organisms from all the coastal regions show consistently low levels of hydrocarbons. The levels usually range from 0.1 to 10 ppm; these are about the minimum levels routinely detectable by current laboratory methods and, for most of the samples, may represent hydrocarbons of biogenic origin. Generally, hydrocarbons produced by organisms encompass only a few compounds of a hydrocarbon type or sub-class [8] (See Chapter 1).

TABLE 7

Baseline levels of hydrocarbons in unexposed marine organisms

Marine organism[a]	Location	Hydrocarbon type, sub-class, or class	Method of analysis[b]	Hydrocarbon content, wet weight (µg/g)[c]	Reference
Bacteria					
Vibrio marinus	Pure culture	Alkanes & alkenes	GC	10*	149
Unicellular Algae					
Algal mats	Gulf of Mexico	Alkanes & alkenes	GC	1-10*	149
Anacystis nidulans	Blue-green	Alkanes & alkenes	GC	6*	149
Syracosphaera carterae	Pure culture	*n*-Paraffins	GC	3.3*	150
Skeletonema costatum	Pure culture	*n*-Paraffins	GC	12*	150
Undetermined cryptomonad	Pure culture	*n*-Paraffins	GC	3.4*	150
Plants					
Phyllospadix scouleri	Washington coast	*n*-Paraffins	GC	110*	84

a All analyses were made on the whole organism unless otherwise noted.

b GC: gas chromatography
TLC: thin-layer chromatrography
UV: ultraviolet spectrometry
FS: fluorescence spectrometry
MS: mass spectrometry
GR: gravimetric

c Values marked with an asterisk (*) were converted to wet weight by dividing the published dry weight value by 10, unless another conversion factor was given by the author(s).

Marine organism[a]	Location	Hydrocarbon type, subclass, or class	Method of analysis[b]	Hydrocarbon content, wet weight (μg/g)[c]	Reference
Spartina alteriflora					
Dormant growth period	NW Florida	Aliphatics	GC	23*	151
Active growth period	NW Florida	Aliphatics	GC	49*	151
Juncus romerianus					
Dormant growth period	NW Florida	Aliphatics	GC	2.2*	151
Active growth period	NW Florida	Aliphatics	GC	22.2*	151
Macroalgae					
Nereocystis sp.	Puget Sound, Wash.	*n*-Paraffins	GC	0.74	152
Ulva sp.	Puget Sound, Wash.	*n*-Paraffins	GC	20.3–23.0	152
Ulva lactura	Cape Cod, Mass.	Saturates	GC	2.7*	153
		Unsaturates	GC	543*	153
Fucus distichus	Cape Cod, Mass.	Saturates	GC	8.9*	153
		Unsaturates	GC	0.009*	153
Fucus spiralis	Cape Cod, Mass.	Saturates	GC	12.5*	153
		Unsaturates	GC	0.45*	153
Fucus gardneri	Washington coast	*n*-Paraffins	GC	2.0*–3.7*	84
Fucus sp.	New Hampshire	*n*-Paraffins	GC	3.2*	150
	Falmouth, Mass.	*n*-Paraffins	GC	13.8*	150
	Woods Hole, Mass.	*n*-Paraffins	GC	12.4*	150
	Puget Sound, Wash.	*n*-Paraffins	GC	3.0–56	152
Bossiella sp.	Puget Sound, Wash.	*n*-Paraffins	GC	1.8*	84
Polysiphonia sp.	New Hampshire	*n*-Paraffins	GC	6.2*	150
P. urceolata	Cape Cod, Mass.	Saturates	GC	30*	153
		Unsaturates	GC	1.0*	153

Marine organism[a]	Location	Hydrocarbon type, subclass, or class	Method of analysis[b]	Hydrocarbon content, wet weight (μg/g)[c]	Reference
Rhodymenia palmata	New Hampshire	*n*-Paraffins	GC	11*	150
	Cape Cod, Mass.	Saturates	GC	34*	153
		Unsaturates	GC	0.24*	153
Chondrus crispus	New Hampshire	*n*-Paraffins	GC	3.2*	150
	Falmouth, Mass.	*n*-Paraffins	GC	6.2*	150
	Cape Cod, Mass.	Saturates	GC	82*	153
		Unsaturates	GC	0.83*	153
Chaetomorpha linum	New Hampshire	*n*-Paraffins	GC	1.4*	150
Ascophyllum nodosum	New Hampshire	*n*-Paraffins	GC	6.5*	150
	Cape Cod, Mass.	Saturates	GC	16*	153
		Unsaturates	GC	12*	153
Agarum cribrosum	New Hampshire	*n*-Paraffins	GC	1.1*	150
Laminaria digitata	New Hampshire	*n*-Paraffins	GC	1.5*	150
	Cape Cod, Mass.	Saturates	GC	1.3*	153
		Unsaturates	GC	0.63*	153
L. agardhii	Cape Cod, Mass.	Saturates	GC	26*	153
		Unsaturates	GC	1.8*	153
Enteromorpha compressa	Cape Cod, Mass.	Saturates	GC	1.1*	153
		Unsaturates	GC	30*	153
Spongomorpha arcta	Cape Cod, Mass.	Saturates	GC	0.30*	153
		Unsaturates	GC	27*	153
Codium fragile	Cape Cod, Mass.	Saturates	GC	51*	153
		Unsaturates	GC	4.0*	153
Ectocarpus fasciculatus	Cape Cod, Mass.	Saturates	GC	2.2*	153
		Unsaturates	GC	23*	153

Marine organism[a]	Location	Hydrocarbon type, subclass, or class	Method of analysis[b]	Hydrocarbon content, wet weight (μg/g)[c]	Reference
Pilayella littoralis	Cape Cod, Mass.	Saturates	GC	0.0004*	153
		Unsaturates	GC	0.054*	153
Chordaria flagelliformis	Cape Cod, Mass.	Saturates	GC	5.4*	153
		Unsaturates	GC	0.20*	153
Leathesia difformis	Cape Cod, Mass.	Saturates	GC	11*	153
		Unsaturates	GC	0.01*	153
Punctaria latifolia	Cape Cod, Mass.	Saturates	GC	77*	153
		Unsaturates	GC	11*	153
Scytosiphon lomentaria	Cape Cod, Mass.	Saturates	GC	1.9*	153
		Unsaturates	GC	1.6*	153
Chorda filum	Cape Cod, Mass.	Saturates	GC	0.88*	153
		Unsaturates	GC	1.3*	153
C. tomentosa	Cape Cod, Mass.	Saturates	GC	1.2*	153
		Unsaturates	GC	2.6*	153
Porphyra leucosticta	Cape Cod, Mass.	Saturates	GC	8.2*	153
		Unsaturates	GC	42*	153
Dumontia incrassata	Cape Cod, Mass.	Saturates	GC	33*	153
		Unsaturates	GC	0.36*	153
Ceramium rubrum	Cape Cod, Mass.	Saturates	GC	56*	153
		Unsaturates	GC	1.3*	153
Sargassum sp.	Sargasso Sea	*n*-Paraffins	GC	0.95*	150
S. natans (narrow blade)	Open Atlantic	*n*-Paraffins	GC	0.5-4.3	154
(wide blade)		*n*-Paraffins	GC	11.8	154

Marine organism[a]	Location	Hydrocarbon type, subclass, or class	Method of analysis[b]	Hydrocarbon content wet weight (μg/g)[c]	Reference
Sponge					
Tethya sp.	Central California	Alkanes	TLC	0.5	155
Plankton (mixed tows)	Cape Cod, Mass.	*n*-Paraffins	GC	10*	150
	Firth of Clyde, Scotland	*n*-Paraffins	GC	32.4	148
	British Isles (inshore)	*n*-Paraffins	GC	0.57–159*	156
	British Isles (offshore)	*n*-Paraffins	GC	0.32–26.3*	156
	Southern Beaufort Sea	Non-polar HC	GC	1.5–203	157
		Paraffins	GC	0.1–198	157
	South Atlantic	Saturates	GR	2.0–31.8	158
Barnacles					
Mitella polymerus	Washington coast	Saturates	GC	1.4*	91
	Puget Sound, Wash.	Saturates	GC	1.2–4.5	152
Balanus cariosus	Puget Sound, Wash.	Saturates	GC	0.66	152
Lepas sp.	Open Atlantic	Saturates	GC	6	159
Shrimp					
Pandalis borealis	North Atlantic	Saturates	GR	43.6	158
Palaemonetes pugio	Galveston Island, Tex.	Saturates	GC	24.8	160
	Marsh Island, Tex.	Saturates	GC	10.9	160
Penaeus setiferus	Dow mariculture, Tex.	Saturates	GC	15.0	161
		Aromatics	GC	8.0	161

Marine organism[a]	Location	Hydrocarbon type, subclass, or class	Method of analysis[b]	Hydrocarbon content wet weight (μg/g)[c]	Reference
P. schmitti	Venezuela	Saturates	GC	2.8-5.7	162, 163
		Aromatics	GC	0.1-0.3	162, 163
Leander tenuiformis	Open Atlantic	*n*-Paraffins	GC	0.16	154
Sargassum shrimp	Open Atlantic	*n*-Paraffins	GC	3	2
Unidentified species	Arctic Ocean	*n*-Paraffins	GC	0.37-21.6	152
(muscle)	Texas coast	*n*-Paraffins	GC	0.04	164
Lobsters					
Homarus americanus					
(intestine)	Pictou, Nova Scotia	Aromatics: Bunker C	FS	57	165
(stomach)	Pictou, Nova Scotia	Aromatics: Bunker C	FS	19	165
(claw muscle)	Pictou, Nova Scotia	Aromatics: Bunker C	FS	4	165
(abdominal muscle)	Pictou, Nova Scotia	Aromatics: Bunker C	FS	5	165
Nephrops norvegicus					
(head)	Scotland	*n*-Paraffins	GC	2.62	148
(tail)	Scotland	*n*-Paraffins	GC	5.98	148
Crabs					
Pagurus (hermit)	Eastern Canada	Aromatics: Bunker C	FS	2	165
Cancer antennarius	San Francisco, Calif.	Alkanes	TLC	1*	155
C. irroratus	Eastern Canada	Aromatics	FS	7	166
Hemigrapsus nudus	Washington coast	*n*-Paraffins	GC	0.12-0.24*	84
	Puget Sound, Wash.	*n*-Paraffins	GC	0.082-3.7	152
Planes minuta	Open Atlantic	Paraffins	GC	<.01	154
Portunus sayi	Open Atlantic	Paraffins	GC	<.01	154

Marine organism[a]	Location	Hydrocarbon type, subclass, or class	Method of analysis[b]	Hydrocarbon content, wet weight (μg/g)[c]	Reference
Chiton					
Mopalia sp.	Puget Sound, Wash.	*n*-Paraffins	GC	0.50	152
Limpets					
Notoacmea scutum	Puget Sound, Wash.	*n*-Paraffins	GC	2.5	152
Snails					
Buccinum undatum	Scotland	*n*-Paraffins	GC	0.650	148
Thais lamellosa	Puget Sound, Wash.	*n*-Paraffins	GC	0.06-1.5	152
	Dungeness, Wash.	Saturates	GR	0.15-4.6*	167
		Unsaturates	GR	23-69*	167
Littorina littorea	Eastern Canada	Aromatics: Bunker C	FS	11	165
Littorina sp.	Valdez, Alaska	*n*-Paraffins	GC	16.1	152
Mussels					
Mytilus edulis	Valdez, Alaska	*n*-Paraffins	GC	0.40-0.95	152
	Puget Sound, Wash.	*n*-Paraffins	GC	0.37-21.6	152
	Puget Sound, Wash.	*n*-Paraffins	GC	0.21*-0.26*	120
	Dungeness, Wash.	Saturates	GR	2.1-34*	167
		Unsaturates	GR	34-100*	167
	Newport, Ore.	*n*-Paraffins	GC	0.34*	168
	Eastern Canada	Aromatics: Bunker C	FS	3	166
	Valdez, Alaska	n-C_{16-28}	GC	1.9	169
	Northern California	Alkanes	TLC	0.9*	155
M. galloprovincialis	Venice, Italy	*n*-Paraffins	GC	8-10	170
m. californianus	Washington coast	*n*-Paraffins	GC	0.45*	120
	Puget Sound, Wash.	*n*-Paraffins	GC	0.088-0.58	152
	Vancouver Is.,Canada	Benzo[a]pyrene	FS	$\leq 0.2 \times 10^{-3}$	171

Marine organism[a]	Location	Hydrocarbon type, subclass, or class	Method of analysis[b]	Hydrocarbon content, wet weight (μg/g)[c]	Reference
M. californianus	Cape Mendocino, Calif.	Alkanes	TLC	0.60*	172
M. edulis	Drake's Estero, Calif.	Alkanes	TLC	0.75*	172
	Bolinas Lagoon, Calif.	Alkanes	TLC	1.35*	172
Modiolus modiolus	Eastern Canada	Aromatics	FS	3	165
Scallops					
Aequipecten irrandians	Waquoit Bay, Mass.	Total hydrocarbons	GR	2.3-5.5	118
	Central California	Alkanes	TLC	2*	155
Oysters					
Ostrea lurida	Newport, Ore.	*n*-Paraffins	GC	0.35*	168
Crassostrea virginica	Redfish Reef, Tex.	Saturates	GC/MS	1.5	173
	Aransas Bay, Tex.	Saturates	GC/MS	1	173
	Quissett, Mass.	Total hydrocarbons	GC	1	174
	Galveston Is., Tex.	Total hydrocarbons	GC	2.0	175
	Eight Mile Road Reef, Tex.	Total hydrocarbons	GC	2.0	175
		Saturates	GC	0.1	175
		Aromatics	GC	0.1	175
Unidentified species	Vancouver Island, Canada	Polycyclic arom.	GC	0.123-.154	157
Clams					
Mercenaria mercenaria	Narragansett, R.I.	Saturates	GC	2.9-10.1	119
	S. Mass. coast	Paraffins & olefins	GC/GR	0.46-1.8	176
		Aromatics	GC	1.9-6.8	176
Mya arenaria	Maine	Total hydrocarbons	GC	2-12	93
	Eastern Canada	Aromatics: Bunker C	FS	8	166

Marine organism[a]	Location	Hydrocarbon type, subclass, or class	Method of analysis[b]	Hydrocarbon content, wet weight (μg/g)[c]	Reference
Mya sp.	Valdez, Alaska	C_{16-28}	GC	1.1	169
	Sippewissett, Mass.	Total hydrocarbons	GR	1.7	177
Starfish					
Asterias vulgaris	Nova Scotia, Canada	Aromatics: Bunker C	FS	11	165
A. rubens	Scotland	*n*-Paraffins	GC	0.403	148
Pisaster ochraeceus					
(digestive)	Central California	Alkanes	TLC	1*	155
(gonads)	Central California	Alkanes	TLC	0.7*	155
Unidentified	Valdez, Alaska	C_{16-28}	GC	0.80-12.4	169
	Alaska	*n*-Paraffins	GC	4.0	178
Sand dollar	Alaska	*n*-Paraffins	GC	2.6	178
Urchin					
Strongylocentrotus droebachiesis	Nova Scotia, Canada	Aromatics	FS	22	165
S. purpuratus	Washington coast	*n*-Paraffins	GC	0.038*	84
Atlantic Salmon					
Salmo salar	Eastern Canada	Aromatics: Bunker C	FS	10	166
Pacific Salmon	Ocean Station PAPA	Polycyclic arom.	GC	0.020-0.082	157
	North Pacific	Non-polar HC	GC	0.90	157
		Paraffins	GC	0.8	157
Barracuda					
Sphyraena barracuda	Texas	*n*-Paraffins	GC	22.6	178

Marine organism[a]	Location	Hydrocarbon type, subclass, or class	Method of analysis[b]	Hydrocarbon content, wet weight (μg/g)[c]	Reference
Bocachico					
Prochilodus r. reticulatus	Lake Maracaibo, Venezuela	Saturates	GC	16.2	162
		Aromatics	GC	22.2	162
	Freshwater pond, Venezuela	Saturates	GC	6.2-9.1	162
		Aromatics	GC	3.4-4.9	162
Cod					
Gadus callarias (livers)	North Atlantic	Saturates	GC	128-345	158
G. morhua (liver)	North Atlantic	Saturates	GC	332	158
(liver)	Scotland	*n*-Paraffins	GC	4.67	148
(liver)	British Isles(inshore)	*n*-Paraffins	GC	0.9-3.1	156
(liver)	British Isles(offshore)	*n*-Paraffins	GC	1.4-4.6	156
(flesh)	Scotland	*n*-Paraffins	GC	0.24	148
(muscle)	British Isles(inshore)	*n*-Paraffins	GC	.005-.03	156
(muscle)	British Isles(offshore)	*n*-Paraffins	GC	0.1-0.3	156
(codling) (liver)	Scotland	n-C_{15-33}	GC	3.4-4.9	179
(codling) (muscle)	Scotland	n-C_{15-33}	GC	0.1-0.2	179
Boreogadus esmarki	Arctic Ocean	*n*-Paraffins	GC	1.26	152
	Iceland	Saturates	GR	116.6	158
Curvina					
Cynoscion maracaiboensis	Lake Maracaibo, Venezuela	Saturates	GC	0.0-1.8	162
		Aromatics	GC	0.0-1.4	162
Dogfish					
Squalus (flesh)	Scotland	*n*-Paraffins	GC	2.77	148
acanthias (liver)	Scotland	*n*-Paraffins	GC	13.7	148

Marine organism[a]	Location	Hydrocarbon type, subclass, or class	Method of analysis[b]	Hydrocarbon content wet weight (μg/g)[c]	Reference
Flounder					
Pseudopleuronectes					
americanus (gut)	Eastern Canada	Aromatics:Bunker C	FS	21	166
(skin and flesh)	Eastern Canada	Aromatics:Bunker C	FS	0	166
Syncium gunteri	Gulf of Mexico	*n*-Paraffins	GC	8.7	178
Unidentified species	Alaska	*n*-Paraffins	GC	8.0	178
Flying fish	Open Atlantic	C_{14-20}	—	0.3	159
Greenland halibut					
Reinhandtius					
hippolossoides (liver)	North Atlantic	Saturates	GR	230	158
Hake					
Merluccius merluccius	Scotland	*n*-Paraffins	GC	6.00	148
Herring					
Clupea harengus (flesh)	Scotland	*n*-Paraffins	GC	11.8	148
C. pallasii (eggs)	Puget Sound, Wash.	*n*-Paraffins	GC	3.1	152
Haddock					
Gadus aeglefinus (liver)	North Atlantic	Saturates	GR	120	158
(liver)	Georges Bank	Saturates	GR	252	158
Lisa					
Mugil curema	Lake Maracaibo, Venezuela	Saturates	GC	2.0-2.1	162
		Aromatics	GC	0.8-3.2	162

Marine organism[a]	Location	Hydrocarbon type, subclass, or class	Method of analysis[b]	Hydrocarbon content, wet weight (μg/g)[c]	Reference
Mackerel					
Scomber scombrus (muscle)	British Isles(inshore)	*n*-Paraffins	GC	0.4-2.8	156
	British Isles(offshore)	*n*-Paraffins	GC	0.6-1.4	156
(liver)	British Isles(inshore)	*n*-Paraffins	GC	0.4-17.3	156
	British Isles(offshore)	*n*-Paraffins	GC	1.3-47.6	156
Scomberomorus cavalla	Gulf of Mexico	*n*-Paraffins	GC	11.3	178
Manamama					
Anodus laticeps	Lake Maracaibo, Venezuela	Saturates	GC	2.8-9.2	162
		Aromatics	GC	2.1-3.3	162
Perch					
Sebastes marinus (liver)	North Atlantic	Saturates	GR	110	158
(liver)	Georges Bank	Saturates	GR	30.6	158
Unidentified species	Grays Harbor, Wash.	Saturates	GC	0.4	162
		Aromatics	GC	0.4-0.7	162
Pipefish					
Syngnathus pelagicus	Open Atlantic	n-C_{14-20}	GC	0.27	154
Plaice					
Pleuronectes platessa					
(flesh)	Scotland	*n*-Paraffins	GC	1.12	148
(muscle)	British Isles(inshore)	*n*-Paraffins	GC	.005-1.8	156
	British Isles(offshore)	*n*-Paraffins	GC	1.5	156
(liver)	Scotland	*n*-Paraffins	GC	1.99	148
	British Isles(inshore)	*n*-Paraffins	GC	1.1-118	156
	British Isles(offshore)	*n*-Paraffins	GC	8.9	156

Marine organism[a]	Location	Hydrocarbon class, sub-class, or class	Method of analysis[b]	Hydrocarbon content, wet weight (µg/g)[c]	Reference
Pollock					
Pollachius verins (liver)	Georges Banks	Saturates	GR	262	158
Robalo					
Centropomus ensiferus	Lake Maracaibo, Venezuela	Saturates	GC	0.3-2.4	162
		Aromatics	GC	0.6-1.2	162
Sargassum fish					
Histrio histrio	Open Atlantic	n-C_{14-20}	GC	0.05	154
Salt marsh minnow					
Fundulus heteroclitus	Cape Cod, Mass.	Paraffins	GC	7	180
Triggerfish					
Canthidermis sp.	Open Atlantic	n-C_{14-20}	GC	0.08	154
Trout					
Spotted sea trout	Galveston Beach, Tex.	Saturates	GC	0.7	162
		Aromatics	GC	1.0-1.7	162
Unidentified species	Merida Hatchery, Venezuela	Saturates	GC	0.7-2.5	162
		Aromatics	GC	0.4-0.9	162
Wenchmen	Texas coast	n-Paraffins	GC	1.71	164
Whitefish (flesh)	Alberta, Canada	Steam distillables	GC	4-14	181
Yellow Sole					
Lamanda sp.	Valdez, Alaska	C_{16-28}	GC	0.51-.97	169

Marine organism[a]	Location	Hydrocarbon type, subclass, or class	Method of analysis[b]	Hydrocarbon content, wet weight (μg/g)[c]	Reference
Arctic cisco	Southern Beaufort Sea	Paraffins	GC	1.3-2.6	157
		Non-polar HC	GC	2.0-4.0	157
		Polycyclic arom.	GC/MS	0.014-.031	157
Least cisco	Southern Beaufort Sea	Paraffins	GC	1.4-6.0	157
		Non-polar HC	GC	2.0-7.8	157
		Polycyclic arom.	GC/MS	0.019-.026	157
Pomfret	Southern Beaufort Sea	Polycyclic arom.	GC/MS	0.009	157
Tuna	Ocean Station PAPA North Pacific	Paraffins	GC	49	157
		Non-polar HC	GC	55	157
		Polycyclic arom.	GC/MS	0.184	157
Silky shark *Carcharinus calciformis* (liver)	Gulf Stream	Saturates	GR	1,128-1,440	158
Herring Gull *Larus argentatus* (brain)	Cape Cod, Mass.	Total hydrocarbons	GC	15.3	182
(fatty tissue)	Cape Cod, Mass.	Total hydrocarbons	GC	10.0	182

Exposed Locations

Contamination of organisms (Table 8) by petroleum in the environment was attributed to (a) spill incidents, (b) the pattern of hydrocarbons present in the organisms, or (c) the conclusions by the referenced author(s) [8]. Marine organisms of various trophic levels may be contaminated with petroleum hydrocarbons by a variety of means. Based on laboratory experiments and field observations detailed information on the uptake, metabolism, and depuration of petroleum hydrocarbons by marine organisms is presented in Chapter 3 of this volume and in Chapter 3 of Volume II.

Tissues of organisms exposed to petroleum contained from 10 to 100 times more hydrocarbons than the tissues of marine animals not exposed to contamination (See comparative data in Tables 7 and 8) [8]. ZoBell [106] indicated that certain carcinogenic aromatic hydrocarbons occur in some marine organisms; it is not possible at this time to estimate the extent to which these compounds are petroleum derived or biosynthesized. In studies on marine sediments, Youngblood and Blumer [190] were able to discriminate between polynuclear aromatic hydrocarbons originating from crude petroleum and those from other pollutants. Polynuclear aromatic hydrocarbons from petroleum have long-chain alkyl substituents; polycyclic aromatic hydrocarbons produced at medium temperatures (e.g., forest fires) and at high temperatures (e.g., fossil fuel combustion products) have shorter and fewer alkyl side chains. Data on the hydrocarbon levels in plankton may be subject to question because there are several inherent problems associated with the use of collecting nets and contamination of samples may have occurred [191].

Laboratory Exposed Organisms

Table 9 summarizes the results of studies on the maximum accumulation of petroleum hydrocarbons by marine organisms under controlled laboratory conditions. In most of the tests, the organisms were exposed to relatively high levels of contaminants for relatively short periods under static test conditions. Detailed discussions of uptake and depuration of petroleum hydrocarbons by marine organisms is presented in Chapter 3 of Volume II.

SEDIMENT

Total hydrocarbon concentrations in a variety of surface sediments are presented in Table 10. These data are drawn in large part from the NAS report "Petroleum in the Marine Environment" [2]. In polluted coastal areas, the hydrocarbon content of sediments ranges from 100 to 12,000 ppm, with most levels less than 1,000 ppm. The hydrocarbon content of sedi-

TABLE 8

Hydrocarbon levels in tissues of marine organisms exposed to petroleum materials in the environment.

Marine organism[a]	Probable contaminant or contamination[b]	Hydrocarbon type, subclass, or class[c]	Method of analysis[d]	Hydrocarbon content, wet weight (μg/g)[e]	Reference
Plants					
Salicornia sp.	No. 2 F.O., spill	Total hydrocarbons	GC	13.2	182
Spartina alterniflora	No. 2 F.O., spill	Total hydrocarbons	GC	15.2	182
Phyllospadix scouleri	N.S.F.O., spill	*n*-Paraffins	GC	390*	84
Zostera marina	Bunker C, spill	Aromatics: Bunker C	FS	17	95

a All analyses were made on the whole organism unless otherwise noted.

b Oil types: F.O. = fuel oil; N.S.F.O. = Navy Special Fuel Oil.

c Only *n*-paraffins were indicated in some cases; the probable presence of other petroleum-type hydrocarbons, such as aromatic hydrocarbons, is not to be excluded.

d GC: gas chromatography
TLC: thin-layer chromatography
UV: ultraviolet spectrometry
FS: fluorescence spectrometry
MS: mass spectrometry
GR: gravimetric

e Values marked with an asterisk (*) were converted to wet weight by dividing the published dry weight value by 10, unless another conversion factor was given by the author(s).

Marine organism[a]	Probable contaminant or contamination[b]	Hydrocarbon type, sub-class, or class[c]	Method of analysis[d]	Hydrocarbon content, wet weight (μg/g)[e]	Reference
Macroalgae					
Enteromorpha clathrata	No. 2 F.O., spill	Total hydrocarbons	GC	429	182
Fucus gardneri	N.S.F.O., spill	*n*-Paraffins	GC	20.3*	84
Polysiphonia fibrillosa	No. 2 F.O., spill	Total hydrocarbons	GC	6.28	182
Sargassum natans	Chronic, N. Atl.	*n*-Paraffins	GC	0.05-.44	154
		Unresol. envelope	GC	0.99-4.72	154
Sponge					
Ophlatispongia sp.	Chronic, harbor	Alkanes	TLC	17*	155
Barnacles					
Mitella polymerus	N.S.F.O., spill	*n*-Paraffins	GC	16.1*	91
Plankton					
(mixed tows)	Chronic, coastal	Benzo[a]pyrene		0.4	183
Shrimp					
Leander tenuiformis	Open Atlantic	*n*-Paraffins	GC	0.51	154
		Unresol. envelope	GC	2.35	154
Lobsters					
Homarus americanus					
(intestine)	Bunker C F.O., spill	Aromatics: Bunker C	GS	103-130	165
(stomach)	Bunker C F.O., spill	Aromatics: Bunker C	FS	15-230	165
(claw muscle)	Bunker C F.O., spill	Aromatics: Bunker C	FS	2-3	165
(abdominal muscle)	Bunker C F.O., spill	Aromatics: Bunker C	FS	1-4	165

Marine organism[a]	Probable contaminant or contamination[b]	Hydrocarbon type, subclass, or class[c]	Method of analysis[d]	Hydrocarbon content, wet weight (μg/g)[e]	Reference
Crabs					
Cancer antennarius					
(internal)	Chronic, harbor	Alkanes	TLC	23*	155
(muscle)	Chronic, harbor	Alkanes	TLC	3*	155
(internal)	Chronic, harbor	Aromatics	TLC	8*	155
(muscle)	Chronic, harbor	Aromatics	TLC	1*	155
C. irroratus	Bunker C F.O., spill	Aromatics	FS	7-118	165
Hemigrapsus nudus	N.S.F.O., spill	*n*-Paraffins	GC	0.39*	84
Lady crab	Chronic, harbor	C_{14-20}	-	4	159
Planes minuta	Open Atlantic	*n*-Paraffins	GC	0.28	154
		Unresol. envelope	GC	10.60	154
Portunus sayi	Open Atlantic	*n*-Paraffins	GC	8.22	154
		Unresol. envelope	GC	25.48	154
Snails					
Buccinum undatum	Creosote, harbor	(As pyrene)	FS	3.0-5.3	184
Littorina littorea	Bunker C F.O., spill	Aromatics	FS	27-604	165
	Creosote, harbor	(As pyrene)	FS	6.9-49	184
Thais lamellosa	No. 2 F.O., spill	*n*-Paraffins	GC	5.4	152
	Chronic, harbor	Saturates	GR	2.2-48*	167
		Unsaturates	GR	18-27*	167
Mussels					
Modiolus demissus	No. 2 F.O., spill	Total hydrocarbons	GC	218	182
Modiolus modiolus	Bunker C F.O., spill	Aromatics	FS	21-372	165
Mytilus edulis	Bunker C F.O., spill	Aromatics	FS	77-103	166
	No. 2 F.O., spill	*n*-Paraffins	GC	1.4*	119

Marine organism[a]	Probable contaminant or contamination[b]	Hydrocarbon type, subclass, or class[c]	Method of analysis[d]	Hydrocarbon content, wet weight (μg/g)[e]	Reference
Mytilus edulis	Chronic, harbor	*n*-Paraffins	GS	0.97*	120
	Chronic, harbor	Alkanes	TLC	7.0-33*	172
		Aromatics	TLC	3.5-38*	172
	Chronic, estuarine	Alkanes	TLC	3.4-7.5*	172
		Aromatics	TLC	2.9-3.2*	172
	Chronic, coastal	Saturates	GC	13.8	185
		Aromatics	GC	10.8	185
	Chronic, harbor	Saturates	GR	22-46*	167
		Unsaturates & arom.	GR	36-120*	167
	Chronic, harbor (flushed)	Benzo[a]pyrene	FS	$0.0\text{-}8.3 \times 10^{-3}$	171
	Chronic, harbor (poorly flushed)	Benzo[a]pyrene	FS	$27\text{-}63 \times 10^{-3}$	171
	Creosote, harbor	(As pyrene)	FS	16	184
M. galloprovincialis	Chronic, harbor	*n*-Paraffins	GC	90-220	170
M. californianus	N.S.F.O., spill	*n*-Paraffins	GC	0.87*	120
Scallops					
Aequipecten irradians (muscle)	Chronic, harbor	Total hydrocarbons	TLC	5.5*	155
	No. 2 F.O., spill	Total hydrocarbons	GR	7.4-14	117
Oysters					
Crassostrea virginica	Chronic, estuarine	Total hydrocarbons	GC/MS/TLC	236	122
	Harbor	Total hydrocarbons	GC	10	186
	Chronic, harbor	Aromatics	UV	1	187
	Chronic, harbor (California)	Saturates	GC/MS	13-39	173

Marine organism[a]	Probable contaminant or contamination[b]	Hydrocarbon type, subclass, or class[c]	Method of analysis[d]	Hydrocarbon content, wet weight (μg/g)[e]	Reference
Crassostrea virginica	Chronic, harbor (Tex.)	Saturates	GC/MS	15	173
	Chronic, harbor	Total hydrocarbons	GC	160	175
	Chronic, harbor	Saturates, C_{12-24}	GC	11.2	175
	Chronic, harbor	Dimethyl-naphthalenes	GC	0.6	175
		Trimethyl-naphthalenes	GC	0.6	175
	No. 2 F.O., spill	Total Hydrocarbons	GR	69	118
	No. 2 F.O., spill	Total hydrocarbons	GR	38-126	177
Clams					
Mercenaria mercenaria	Sewage effluent	C_{16-32}	GC	14.3-16.0	119
	Chronic, harbor	Total hydrocarbons	TLC	16*	155
	Estuarine	Paraffins	GR/GC	69	176
		Olefins & aromatics	GR/GC	21	176
	Coastal	Paraffins	GR/GC	6.6	176
		Olefins & aromatics	GR/GC	3.0	176
	No. 2 F.O., spill	Total hydrocarbons	GR	8.3-12	177
Mya arenaria	Crude oil, spill	Total hydrocarbons	GC	64-88	93
	No. 2 F.O., spill	Total hydrocarbons	GR	21-36	177
	Creosote, harbor	(As pyrene)	FS	6.9	184
	Bunker C F.O., spill	Aromatics: Bunker C	FS	12-14	95
Starfish					
Asterias vulgaris	Bunker C F.O., spill	Aromatics	FS	14-400	165
Luidia sp.	Chronic, coastal	C_{14-20}	–	3.5	159

Marine organism[a]	Probable contaminant or contamination[b]	Hydrocarbon type, subclass, or class[c]	Method of analysis[d]	Hydrocarbon content, wet weight (μg/g)[e]	Reference
Pisaster ochraceus					
(digestive)	Chronic, harbor	Alkanes	TLC	8*	155
		Aromatics	TLC	7*	155
(gonad)	Chronic, harbor	Alkanes	TLC	3*	155
		Aromatics	TLC	3*	155
Urchins					
Strongylocentrotus droebachiensis	Bunker C F.O., spill	Aromatics	FS	17-94	165
S. purpuratus	N.S.F.O., spill	*n*-Paraffins	GC	0.75*	84
Eel					
Anguilla rostrata					
(liver)	No. 2 F.O., spill	Paraffins	GC	50.9-84.9	182
(muscle)	No. 2 F.O., spill	Paraffins	GC	23.5-89.4	182
Minnow					
Fundulus heteroclitus	No. 2 F.O., spill (11 months)	Paraffins	GC	75	182
	No. 2 F.O., spill (5 years)	Paraffins	GC	8	180
Mullet					
Mugil cephalus (flesh)	Chronic, harbor	Kerosene taint	GC/MS	100	188
Pipefish					
Syngnathus pelagicus	Open Atlantic	*n*-Paraffins	GC	1.41	154
		Unresol. envelope	GC	7.39	154

Marine organism[a]	Probable contaminant or contamination[b]	Hydrocarbon type, subclass, or class[c]	Method of analysis[d]	Hydrocarbon content, wet weight (μg/g)[e]	Reference
Sargassum fish					
Histrio histrio	Open Atlantic	*n*-Paraffins	GC	0.13	154
		Unresol. envelope	GC	1.45	154
Smelt	Chronic, harbor	Benzo[a]pyrene		0-0.5	189
Triggerfish					
Canthidermis sp.	Open Atlantic	*n*-Paraffins	GC	0.26	155
		Unresol. envelope	GC	1.41	155
Whitefish (flesh)	Diesel oil, spill	Steam-distillables	GC	29-88	181
Flatfish	Chronic, coastal	C_{14-20}	-	4	159
Herring gull					
Larus argentatus					
(brain)	No. 2 F.O., spill	Total hydrocarbons	GC	584	182
(fatty muscle)	No. 2 F.O., spill	Total hydrocarbons	GC	535	182

TABLE 9

Hydrocarbon levels in tissue of marine organisms exposed to petroleum materials under laboratory conditions

Marine organism[a]	Exposure conditions			Hydrocarbon type, class, subclass, or measurement	Method of analysis[c]	Hydrocarbon content, wet weight (μg/g)[d]	Reference
	Test material[b]	Level	Duration				
Copepods							
Calanus helgolandicus	^{3}H-benzo[a]pyrene	1.2 ppb	48 hr	Total radioactivity	^{3}H	0.19	192
	WSF No. 2 F.O.	50 ppb					
Calanus hyperboreus	^{3}H-benzo[a]pyrene	1.2 ppb	96 hr	Total radioactivity	^{3}H	0.03	192
	WSF No. 2 F.O.	50 ppb					
Calanus plumchrus	^{3}H-benzo[a]pyrene	1.2 ppb	72 hr	Total radioactivity	^{3}H	0.22	192
	WSF No. 2 F.O.	50 ppb					
	^{3}H-methylcholanthene	0.2 ppb	24 hr	Total radioactivity	^{3}H	0.052	192
	^{14}C-1-octadecane	6.0 ppb	96 hr	Total radioactivity	^{14}C	0.022	192

a All analyses were made on whole organisms unless otherwise noted and represent hydrocarbons incorporated above biogenic or baseline levels.

b WSF: water-soluble fraction; OWD: oil-water dispersion; F.O.: fuel oil.

c GC: gas chromatography
UV: ultraviolet spectrometry
FS: fluorescence spectrometry
^{3}H: radioactive tritium-labelled compound
^{14}C: radioactive carbon-14 labelled compound

d Values marked with an asterisk (*) were converted to wet weight by dividing the published dry weight value by 10 unless another conversion factor was given by the author(s).

Marine organism[a]	Exposure conditions			Hydrocarbon type, class, subclass, or measurement	Method of analysis[c]	Hydrocarbon content, wet weight (μg/g)[d]	Reference
	Test material[b]	Level	Duration				
Polychaetes							
Neanthes arenacoedentata	WSF No. 2 F.O.	100%	4 hr	Total paraffins	GC	5	193
				C_4 benzenes	GC	1.9	193
				Naphthalene	GC	0.9	193
				1-Methylnaphthalene	GC	3.0	193
				2-Methylnaphthalene	GC	3.7	193
				Dimethylnaphthalenes	GC	5.8	193
				Trimethylnaphthalenes	GC	1.5	193
				Other aromatics	GC	3.2	193
				Olefins	GC	3	194
	WSF South Louisiana crude oil	100%	4 hr	Total paraffins	GC	6	193
				C_3 benzenes	GC	2.5	193
				C_4 benzenes	GC	1.0	193
				Naphthalene	GC	1.1	193
				1-Methylnaphthalene	GC	1.4	193
				2-Methylnaphthalene	GC	0.7	193
				Dimethylnaphthalenes	GC	1.5	193
				Other aromatics	GC	1.8	193
Shrimp							
Palaemonetes pugio	OWD No. 2 F.O.	0.9 ppm	4 hr	Total naphthalenes	UV	6.2	195
			6 hr	Total naphthalenes	UV	9.2	195
			10 hr	Total naphthalenes	UV	10.2	195
Panaeus aztecus	WSF No. 2 F.O.	20%	24 hr	Saturates, each peak	UV	<0.1	161,196
				Naphthalene	UV	<0.1	161,196
				1-Methylnaphthalene	UV	<0.1	196
				2-Methylnaphthalene	UV	0.1	196
				Dimethylnaphthalenes	UV	2.7	196
				Trimethylnaphthalenes	UV	0.6	196
	No. 2 F.O. in open pond	≈40 ppm	24 hr	Total saturates	GC	6.2	161
				C_4 benzenes	GC	1.2	161,196
				Naphthalene	UV	3.3	161,196
				1-Methylnaphthalene	UV	8.0	161,196
				2-Methylnaphthalene	UV	8.9	161,196
				Dimethylnaphthalenes	UV	19.2	161,196
				Trimethylnaphthalenes	UV	4.2	161,196
				Phenanthrenes	UV	12.7	161,196

Marine organism[a]	Exposure conditions: Test material[b]	Level	Duration	Hydrocarbon type, class, subclass, or measurement	Method of analysis[c]	Hydrocarbon content, wet weight (μg/g)[d]	Reference
Penaeus setiferus	Total naphthalenes	0.7 ppm	24 hr	Total naphthalenes	UV	3	197
	No. 2 F.O. in open pond	40 ppm	24 hr	Naphthalenes	UV	50	196
			72 hr	Naphthalene	UV	0.45	196
				Methylnaphthalenes	UV	3.61	196
				Dimethylnaphthalenes	UV	14.7	196
Lobster							
Homarus americanus							
intestine	Bunker C F.O.	10,000 ppm	6½ da	Aromatics	FS	1,810	165
stomach				Aromatics	FS	2,840	165
abdominal muscle				Aromatics	FS	137	165
claw muscle				Aromatics	FS	33	165
Crab							
Uca minax (fiddler)	No. 2 F.O. in open pond	40 ppm	72 hr	Naphthalene	UV	0.75	196
				Methylnaphthalenes	UV	4.52	196
				Dimethylnaphthalenes	UV	16.8	196
Sesarma cinereum (wharf)	No. 2 F.O. in open pond	40 ppm	72 hr	Naphthalene	UV	0.93	196
				Methylnaphthalenes	UV	6.05	196
				Dimethylnaphthalenes	UV	24.7	196
Mussels							
Mytilus edulis	^{14}C-heptadecane	6.2 ppm	24 hr	Total radioactivity	^{14}C	400*	198
	^{14}C-naphthalene	32 ppb	4 hr	Total radioactivity	^{14}C	0.47*	198
	No. 2 F.O.	slick	48 hr	*n*-Paraffins	GC	7.7*	121
	No. 2 F.O.	dissolved	24 hr	*n*-Paraffins	GC	2.5*	121
	No. 5 F.O.	slick	32 hr	*n*-Paraffins	GC	7.4*	121
	Outboard motor effluent	10%	24 hr	*n*-Paraffins	GC	0.63*	168
Oysters							
Crassostrea gigas	No. 2 F.O.	50 ppm	11 da	n-C_{14-20}	GC	1.3	199
				Alkylnaphthalenes	GC	26	199
Crassostrea virginica	No. 2 F.O. - OWD	400 ppm	8 hr	*n*-Paraffins	GC	235	197
				Naphthalene	GC	14.7	197
				1-Methylnaphthalene	GC	8.7	197
				2-Methylnaphthalene	GC	15.0	197
				Dimethylnaphthalenes	GC	21.8	197
				Trimethylnaphthalenes	GC	9.1	197

Marine organism[a]	Exposure conditions			Hydrocarbon type, class, subclass, or measurement	Method of analysis[c]	Hydrocarbon content, wet weight (μg/g)[d]	Reference
	Test material[b]	Level	Duration				
Crassostrea virginica				Biphenyls	GC	0.8	197
				Fluorenes	GC	2.2	197
				Phenanthrenes	GC	4.1	197
	No. 2 F.O.	1,000 ppm	24 hr	Saturates	GC	4	200
				Mono & diaromatics	GC	121	200
				Triaromatics	GC	5	200
	No. 2 F.O.	1,000 ppm	96 hr	*n*-Paraffins	GC	3.1	200
				Naphthalenes	GC	84.1	200
				Higher aromatics	GC	9.5	200
	No. 2 F.O.	1,000 ppm	96 hr	Saturates	GC	0.1-3.1	175
				Mono & diaromatics	GC	9.6-84.1	175
				Triaromatics	GC	9.5	201
	No. 2 F.O.	106 ppb	7 wk	Non-polar HC	GC	334	174
	No. 2 F.O. in open pond	40 ppm	48 hr	Naphthalenes	GC	27	196
		0 ppm	0 hr	Naphthalene	GC	0.2	197
				1-Methylnaphthalene	GC	0.1	197
				2-Methylnaphthalene	GC	0.3	197
				Dimethylnaphthalenes	GC	1.0	197
				Trimethylnaphthalenes	GC	0.8	197
	Bunker C F.O.	1,000 ppm	96 hr	*n*-Paraffins	GC	1.1	200
				Naphthalenes	GC	41.3	200
				Higher aromatics	GC	5.0	200
	Bunker C F.O.	1,000 ppm	96 hr	Saturates	GC	0.1-1.1	175
				Mono & diaromatics	GC	40.4-41.3	175
				Triaromatics	GC	5.0	201
	Kuwait crude oil	1,000 ppm	96 hr	*n*-Paraffins	GC	46.0	200
				Naphthalenes	GC	55.1	200
				Higher aromatics	GC	6.0	200
	Kuwait crude oil	1,000 ppm	96 hr	Saturates	GC	0.1-46.0	175
				Mono & diaromatics	GC	55.1-83.1	175
				Triaromatics	GC	6.0	201
	South Louisiana crude oil	1,000 ppm	96 hr	*n*-Paraffins	GC	13.8	200
				Naphthalenes	GC	44.0	200
				Higher aromatics	GC	8.0	200
	South Louisiana crude oil	1,000 ppm	96 hr	Saturates	GC	0.1-13.8	175
				Mono & diaromatics	GC	44.0-55.5	175

Marine organism[a]	Exposure conditions			Hydrocarbon type, class, subclass, or measurement	Method of analysis[c]	Hydrocarbon content, wet weight (μg/g)[d]	Reference
	Test material[b]	Level	Duration				
Crassostrea virginica	South Louisiana crude oil	1,000 ppm	96 hr	Triaromatics	GC	8.0	201
Ostrea lurida	Outboard motor effluent	10%	5 da	*n*-Paraffins	GC	0.59*	168
		10%	10 da	*n*-Paraffins	GC	0.98*	168
Clams							
Rangia cuneata	No. 2 F.O.	1,000 ppm	48 hr	Total saturates	GC	26	200
				Mono & diaromatics	GC	481	200
				Poly-aromatics	GC	34	200
	Naphthalene	9 ppm	6 hr	Naphthalene	UV	64	202
					GC	50	202
	No. 2 F.O. - WSF	-	24 hr	Naphthalenes	GC	34	197
	2-Methylnaphthalene	11 ppm	24 hr	2-Methylnaphthalene	UV	70	202
					GC	83	202
	No. 2 F.O.	40 ppm	96 hr	Naphthalenes	GC	43	196
	^{14}C-benzo[a]pyrene	30.5 ppb	24 hr	Total radioactivity	^{14}C	5.7-7.2	203
Mya arenaria	Bunker C F.O.	-	-	Aromatics	FS	87	166
Atlantic salmon							
Salmo salar	Bunker C F.O.	-	-	Aromatics	FS	113	166
Flounder							
Pseudopleuronectes americanus (skin)	Bunker C F.O.	-	-	Aromatics	FS	182	166
(gut)	Bunker C F.O.	-	-	Aromatics	FS	622	166
(flesh)	Bunker C F.O.	-	-	Aromatics	FS	7	166
Perch							
Cymatogaster aggregata	No. 2 F.O.	50 ppm	96 hr	*n*-C_{12-20}	GC	2.3	199
				Alkylnaphthalenes	GC	19	199
Salt marsh minnow							
Fundulus similus	No. 2 F.O. - WSF	-	2 hr	Naphthalenes	GC	43.2	197
Sheepshead minnow							
Cyprionodon variegatus	1-Methylnaphthalene	1 ppm	4 hr	1-Methylnaphthalene	UV	205	204

Marine organism[a]	Exposure conditions			Hydrocarbon type, class, subclass, or measurement	Method of analysis[c]	Hydrocarbon content, wet weight (μg/g)[d]	Reference
	Test material[b]	Level	Duration				
Mudsuckers							
Gillichthys mirabilis							
(flesh)	^{3}H-benzo[a]pyrene	1 ppb	96 hr	Total radioactivity	^{3}H	0.02*	189
(liver)						0.10*	189
(flesh)	^{14}C-naphthalene	32 ppb	2 hr	Total radioactivity	^{14}C	0.2*	189
(liver)						4.1*	189
Tide pool sculpins							
Oligocottus maculosus							
(flesh)	^{3}H-benzo[a]pyrene	1 ppb	1 hr	Total radioactivity	^{3}H	0.13*	189
(liver)						0.12*	189
(flesh)	^{14}C-naphthalene	32 ppb	2 hr	Total radioactivity	^{14}C	0.5*	189
(liver)						2.6*	189
Sand dabs							
Citharichthys stigmaeus							
(flesh)	^{3}H-benzo[a]pyrene	1 ppb	1 hr	Total radioactivity	^{3}H	0.03*	189
(liver)						0.13*	189
(flesh)	^{14}C-naphthalene	32 ppb	2 hr	Total radioactivity	^{14}C	0.8*	189
(liver)						1.4*	189

TABLE 10

Hydrocarbon levels in marine sediments

Description of sediment samples					Hydrocarbon type, sub-class, or class	Hydrocarbon content, dry weight (μg/g)	Reference
Location and type (Sampling period)	Water depth (m)	Sample depth (cm)	Condition of area[a]	Number of samples			
Chedabucto Bay, Nova Scotia	Upper beach	Surface	P	≧4	Aromatics: Bunker C	$(6.7-72.1)\times10^6$	95
	Middle beach	Surface	P	≧6	Aromatics: Bunker C	$(0.03-1.5)\times10^6$	95
(Over 2 yr. period)	3	5	N-P	7	Aromatics: Bunker C	34-778	165
	12	5	N-P	7	Aromatics: Bunker C	11-1,240	165
(After 6 yr. period)	Upper beach	5	P	1	Saturates	67	96
					Aromatics: Bunker C	77	96
		10	P	1	Saturates	620	96
					Aromatics: Bunker C	2,400	96
		15	P	1	Saturates	1,200	96
					Aromatics: Bunker C	5,500	96
	Middle beach	5	P	1	Saturates	110	96
					Aromatics: Bunker C	380	96
		10	P	1	Saturates	1,300	96
					Aromatics: Bunker C	8	96
		15	P	1	Saturates	27	96
					Aromatics: Bunker C	65	96
	Lower beach	5	P	1	Saturates	160	96
					Aromatics: Bunker C	660	96
		10	P	1	Saturates	10	96
					Aromatics: Bunker C	36	96
		15	P	1	Saturates	7	96
					Aromatics: Bunker C	10	96
Buzzards Bay, Mass.							
Wild Harbor River (Over 1st yr.)	<3	Top 10	P	≅180	Total HC	480-1,630	80,81,177,205
Wild Harbor River (Over 2nd yr.)	<3	Top 10	P		Total HC	230-670	80,81,177,205
Wild Harbor Marsh	0	Surface	P	1	Aromatics	8	190
Wild Harbor	<10	Surface	P	1	Total HC	350[b]	206
Off Silver Beach (Over 1st yr.)	1.5	Top 10	P	≅60	Total HC	550-12,400	80,81,177,205
Off Silver Beach (Over 2nd yr.)	1.5	Top 10	P		Total HC	1,200-3,000	80,81,177,205

a P = polluted; N = non-polluted; N-P = samples ranged from non-polluted to polluted; U = condition unknown. Pollution condition indicated by original author(s) or assessed by the authors of this chapter.

b Wet weight in $\mu g/cm^3$.

c IR - infrared spectrometry

Description of sediment samples					Hydrocarbon type, sub-class, or class	Hydrocarbon content, wet weight (µg/g)	Reference
Location and type (Sampling period)	Water depth (m)	Sample depth (cm)	Condition of area[a]	Number of samples			
Buzzards Bay, Mass.							
Open bay	>10	Surface	P	5	Total HC	66-609[b]	206
Off New Bedford Harbor, Mass.	7.5	Surface	P	1	Total HC	840	207
					Aromatics	63	207
Narragansett Bay, R.I.							
Mouth of Providence River	-	Top 8-10	P	2	Total HC	820-3,560	119
	-	Surface	P	1	Saturates	1,960	208
					Aromatics	30	208
Middle of bay	-	Top 8-10	P	2	Total HC	350-440	119
Long Cove, Maine	-	Surface	P	8	Total HC	53-705	209
South Louisiana, Offshore	10-18	Top 4	P	16	C_{12-33}	31[b]	63
					Total HC	151[b]	63
Santa Barbara Channel, Calif.							
After spill (April, 1969)	50-600	Surface	N-P	35	Total HC	<200-10,800	79
(May, 1969)	50-600	Surface	N-P	30	Total HC	<200-21,500	79
(June, 1969)	20-600	Surface	N-P	26	Total HC	<200-10,600	79
(July, 1969)	20-600	Surface	N-P	23	Total HC	<200-6,700	79
(August, 1969)	20-600	Surface	N-P	33	Total HC	<200-9,100	79
(Feb., 1970)	20-600	Surface	N-P	56	Total HC	<200-8,700	79
(June, 1970)	20-600	Surface	N-P	20	Total HC	<200-14,100	79
Oil seep area	20	0-5	P	6	CCl_4 extracts-IR[c]	285-1,300[b]	210
	20	5-15	P	4	CCl_4 extracts-IR	4,500-4,800[b]	210
	40	0-5	P	6	CCl_4 extracts-IR	4,500-22,500[b]	210
	40	5-15	P	4	CCl_4 extracts-IR	8,100-60,000[b]	210
	60	0-5	P	6	CCl_4 extracts-IR	4,200-9,800[b]	210
	60	5-15	P	2	CCl_4 extracts-IR	6,200-8,000[b]	210
Coal Oil Point, Seep area	-	0-15	U	4		94	211
San Francisco Bay, Calif.	-	2	P	7	Paraffins	110-1,680	212
				7	Aromatics	65-1,020	212
				14	Total HC	1,588	212
Anacortes, Wash.	Intertidal	Surface	P	7	Saturates	0.003-.480 (wet)	213
Port Angeles Harbor, Wash.	Intertidal	0-3	P	13	Saturates	76-1,100	214
					Unsaturates	94-430	214
Washington coast, beach sand	Intertidal	6	P	1	n-C_{14-37}	0.23	84
Gulf of Mexico,							
Buccaneer field	18-19	Surface	P	1	n-Paraffins	306	215
Le Havre, France	Subtidal	Surface	P	2	Total HC	450-920	216

Description of sediment samples							
Location and type (Sampling period)	Water depth (m)	Sample depth (cm)	Condition of area[a]	Number of samples	Hydrocarbon type sub-class, or class	Hydrocarbon content, dry weight (μg/g)	Reference
Lake Maracaibo, Venezuela	5-33	5	N-P	24	Saturates	4.3-3,700	162
					Aromatics	6.5-4,900	162
					Asphaltenes & NSO	67-9,356	162
	-	-	-	8	Hydrocarbons	24-116	217, 218
					Non-hydrocarbons	266-1,600	217, 218
British Isles, Inshore	-	Surface	N-P	22	n-C_{15-33}	0.1-22.0	156
Cape Cod, Mass.							
Marsh	<1	Surface	P	9	Total HC	22-59[b]	206
Falmouth Harbor	<10	Surface	P	1	Total HC	187[b]	206
Eel Pond, Woods Hole	<10	Surface	P	1	Total HC	153[b]	206
Open bay, Station 31	11	10	N	10-20	Total HC	50-70	80, 81
Buzzards Bay, Open water	15	0-2	N	1	n-C_{20-31}	3.0	180
					Unresol. envelope	60	180
	15	70-72	N	1	n-C_{20-31}	2.6	180
					Unresol. envelope	10	180
	<20	Surface	N	13	Total HC	2-8[b]	206
Megansett Harbor	<10	Surface	N	1	Total HC	4[b]	206
Quissett Harbor	<10	Surface	N	1	Total HC	13[b]	206
Buzzards Bay, Open water	15	Surface	N	1	Total HC	100	207
					Aromatics	6.7	207
	16-17.5	Surface	N	2	Aromatics	5	190
Sippewissett Marsh	<1	Surface	N	3	Total HC	9-11[b]	206
	<1	Surface	N	1	Aromatics	19	190
	<1	Surface	N	1	n-C_{20-31}	21.4	180
					Unresol. envelope	6	180
	<1	25-30	N	1	n-C_{20-31}	15.6	180
					Unresol. envelope	28	180
Vineyard Sound	6-9	0-7	N	1	n-C_{14-32}	1.7	150
Pictou, Nova Scotia	5	5	N	2	Aromatics:Bunker C	3-5	165
Long Cove, Maine	-	Surface	N	2	Total HC	7-59	209
Narragansett Bay, R.I.							
Entrance	-	8-10	N	2	Total HC	50-60	119
	3.1	Top 10	N	1	Saturates	147	219
Hudson Trough, 40°N, 73°W	78	5	N	1	Total HC	48.0-58.0	219
Abyssal Plain, 30°N, 60°W	5252	5	N	1	Total HC	1.2-2.9	219
NW Florida, Salt marsh							
Spartina dominant plant	Shallow	0-5	N	1	Aliphatics	510	151
Juncus dominant plant	Shallow	0-5	N	1	Aliphatics	22	151
Mississippi, Coastal bog	0-1.5	-	U	1		350	220

Description of sediment samples					Hydrocarbon type, sub-class, or class	Hydrocarbon content dry weight (μg/g)	Reference
Location and type (Sampling period)	Water depth (m)	Sample depth (cm)	Condition of area[a]	Number of samples			
Louisiana, Coastal marsh	>0	0.8-1.5	U	1	Total HC	9	221
					Paraffin & naphthene	4	221
					Aromatics	5	221
Salt marsh	Shallow	Surface	P	1	Alkanes	180	222
Salt marsh	Shallow	Surface	N	1	Alkanes	100	222
Freshwater	Shallow	Surface	P	1	Alkanes	210	222
Freshwater	Shallow	Surface	N	1	Alkanes	110	222
Louisiana, Coastal mud flat	-	≧0.8	U	1	Total HC	52	221
					Paraffin & naphthene	28	221
					Aromatics	24	221
Mississippi River Delta	0-60	Top 30	U	3	Total HC	32-110	221
					Paraffin & naphthene	22-57	221
					Aromatics	10-68	221
Southern Louisiana, Offshore	10-18	Top 4	N	23	C_{12-33}	1[b]	63
					Total HC	20[b]	63
11 km off Grand Isle	-	1-1.5 m	U	1	Paraffin & naphthene	20	221
					Aromatics	5	221
					NSO	40	221
9 km off Grand Isle	<60	Top 2 m	N	1	Total HC	6	223
46 km off Eugene Is.	<60	Top 2.5 m	N	1	Total HC	8	223
Texas, Laguna Madre, Tide flats							
Algal muck	-	Surface	N	1	Paraffin & naphthene	170	221
					Aromatics	20	221
					NSO	90	221
Algal clay	-	Surface	N	1	Paraffin & naphthene	50	221
					Aromatics	0	221
					NSO	1,200	221
16 km off High Is., Clay	<60	0.3-2 m	N	1	Total HC	7	223
27 km off Corpus Christi	-	Surface	N	1	Paraffin & naphthene	20	221
					Aromatics	8	221
					NSO	30	221
Buccaneer oil field	18-19	Surface	N	33	*n*-Paraffins	0.006-0.530	215
Gulf of Mexico, Open waters	>200	-	N	10	Hydrocarbons	12-63	217,218
					Non-hydrocarbons	113-790	217,218
Gulf of Batabano, Off Cuba	-	-	U	10	Hydrocarbons	15-85	217,218
					Non-hydrocarbons	136-1,023	217,218
Cariaco Trench, Caribbean Sea	>200	-	U	16	Hydrocarbons	56-352	217,218
					Non-hydrocarbons	224-2,600	217,218

Description of sediment samples					Hydrocarbon type, subclass, or class	Hydrocarbon content, dry weight (µg/g)	Reference
Location and type (Sampling period)	Water depth (m)	Sample depth (cm)	Condition of area[a]	Number of samples			
Orinoco Delta, Mud flats	-	Surface	U	2	Paraffin & naphthene	20-30	221
					Aromatics	7-30	221
					NSO	30-140	221
Exposed coast	-	Surface	U	1	Paraffin & naphthene	20	221
					Aromatics	70	221
					NSO	630	221
Orinoco Delta	-	-	U	10	Hydrocarbons	27-110	217,218
					Non-Hydrocarbons	283-1,355	217,218
California, Estuary pond	0.2	Surface	U	1	Total HC	800	221
					Paraffin & naphthene	500	221
					Aromatics	300	221
San Clemente Basin	2,060	0-5	N	1	Saturates/aromatics	43	224
Tanner Basin	1,510	3-38	N	1	Saturates/aromatics	166	224
San Nicolas Basin	-	-	N	1	n-C_{14-36}	2.4	225
Santa Cruz Basin	1,960	30-45	N	1	Paraffin & naphthene	40	221
					Aromatics	60	221
					Asphaltics	270	221
Catalina Basin	1,300	0-30	N	1	Paraffin & naphthene	10	226
					Aromatics	20	226
					Asphaltics	740	226
Santa Barbara Basin	600	0-8	N	1	Paraffin & naphthene	130	226
					Aromatics	250	226
					Asphaltics	1,200	226
	600	0-50	N	1	Saturates/aromatics	208	224
Santa Barbara Channel	50-350	Surface	N	11	CCl_4 extract - IR[c]	2,400-6,800	79
Away from seeps	-	0-15	U	4	—	94	211
Central exposed coast	-	2	N	1	Paraffins	64	212
					Aromatics	21	212
Washington							
Dungeness	Intertidal	0-3	N	12	Saturates	1.5-5.5	214
					Unsaturates	17-37	214
Lummi Island	Intertidal	Surface	N	1	Saturates	(wet) 0.002	213
Port Valdez, Alaska	Subtidal	Surface	N	9	C_{16-28}	(wet) 0.5-2.5	169
Southern Beaufort Sea (1974)	7-191	Surface	N	14	Paraffins	2.6-18.0	157
					Total HC	18.4-164	157
(1975)	14-155	Surface	N	15	Paraffins	6.0-23.6	157
					Total HC	36.9-109	157
					Polycyclic aromatic	0.12-2.89	157

Description of sediment samples							
Location and type (Sampling period)	Water depth (m)	Sample depth (cm)	Condition of area[a]	Number of samples	Hydrocarbon type sub-class, or class	Hydrocarbon content dry weight (μg/g)	Reference
Southern Beaufort Sea							
Nearshore	-	Surface	N	-	Pyrene + benzo[a]pyrene	0.0004-0.020	157
Benthic	-	Surface	N	1	+ perylene + coronene	0.295-0.299	157
River Blyth, U.K.							
Above effluent outfall, Sand		Surface	N	1	Alkanes	36	227
Below effluent outfall, Sand		Surface	P	1	Alkanes	950	227
River at tidal head, Sand		Surface	U-P	1	Alkanes	106	227
Intertidal flats, Mud		Surface	P	1	Alkanes	3,320	227
Firth of Clyde, Scotland	<20	10	N	1	n-C_{18-33}	0.611	148
British Isles, Offshore	>20	Surface	N	34	n-C_{15-33}	0.1-4.7	156
Seine Estuary, France	Subtidal	Surface	N	1	Total HC	33	216
Bay of Les Veys, France	Subtidal	Surface	N	1	Total HC	31	216
Mediterranean Sea, Open waters	>200	-	N	1	Hydrocarbons	29	217
					Non-hydrocarbons	461	217
West Africa							
Continental shelf	110	Top 15-20	N	1	Paraffin & naphthene	7	221
					Aromatics	5	221
					NSO	10	221
Continental slope	275-840	Top 15-20	N	3	Paraffin & naphthene	0.2-8	221
					Aromatics	0.1-4	221
					NSO	0.4-15	221
Norwegian Sea, Lower shelf	360	104 - 285	N	1	Total HC	11	228
Ocean floor	1,200	46 - 188	N	1	Total HC	10	228
Black Sea, Continental flank	800	40 - 107	N	1	Total HC	170	228
Western Pacific Ocean	264	100 - 300	N	2	Total HC	14-16	228

ments from unpolluted coastal areas and deep marginal seas or basins is usually below 70 ppm (Table 10). Sediment from remote deep sea areas generally contain a few ppm total hydrocarbons or less, most of which are thought to be of biogenic origin [2].

Sediment can serve as a long-term sink for water-borne petroleum hydrocarbons. Significant levels of petroleum from spills have remained in the sediment for periods of several years [80,81,93,95,96,209]. Chronic buildup of petroleum hydrocarbons in marine sediments exposed to industrial and sewage pollution has also been reported [118,208,212,285]. Wakeham and Carpenter [285] have determined the aliphatic hydrocarbon concentration at different depths in sediment from Lake Washington, near Seattle, Washington. The hydrocarbon content was about 30 μg/g dry weight at depths below 30 cm in the sediment; the levels began to increase at a depth of 30 cm and rose steadily to a value of about 1,500 μg/g at the surface. The depth at which the hydrocarbon concentration began to increase corresponded to the time when the metropolitan area around the lake began to grow rapidly, after about 1880, and the hydrocarbon profiles with sediment depth showed a progressive increase in hydrocarbon level over the last century closely paralleling the increase in population [15]. Since there have been no major oil spills or recent sewage inflow to Lake Washington, Wakeham and Carpenter [285] concluded the most likely source of the aliphatic hydrocarbons was from urban stormwater runoff containing lubricating oils and pyrolysis products released by motor vehicles. The hydrocarbon pattern and content of recent sediments may provide an indication of the past history of petroleum input to local environments.

SEAWATER

Conflicting results and controversy about the validity of analytical methods often make quantitative data on the composition of petroleum residues in seawater open to question, especially for samples taken in areas remote from known sources of petroleum pollutants. The hydrocarbon content of the surface layer of seawater (upper few meters) in the open ocean appears to be about a few micrograms per liter. Hydrocarbon levels of seawater in arctic areas or from depth (below 100-200 m) in the open ocean are less than levels found in the surface waters of the open ocean (Table 11). Values for seawater from areas suspected of having oil pollution can be as much as 100 times greater than those for the open ocean surface water [2].

Much progress has been made in recent years to expand the global coverage of hydrocarbon analyses and to better define

TABLE 11

Hydrocarbon levels in the water column

Description of water samples			Method of analysis[a]	Hydrocarbon type, subclass, or class	Hydrocarbon content (μg/1)[b]	Reference
Location (Sampling period)	Water depth (m)	Number of samples				
St. Lawrence River, Canada	2	6	GC	n-C_{14-27}	0.034-0.096	229
	50	6	GC	n-C_{14-27}	0.137-0.372	229
	2	6	FS	Crude oil	0.5-1.1	229
	50	6	FS	Crude oil	0.4-0.9	229
	Surface: 1	8	FS	Bunker C	2.1-4.4	230
	Mid: 10-150	6	FS	Bunker C	1.5-3.7	230
	Bottom: 10-300	8	FS	Bunker C	1.9-5.5	230
Gulf of St. Lawrence, Canada	2	9	GC	n-C_{14-37}	0.024-0.209	229
	50	9	GC	n-C_{14-37}	0.076-1.005	229
	2	9	FS	Crude oil	0.6-1.1	229
	50	9	FS	Crude oil	0.5-1.5	229
	Surface: 1	5	FS	Bunker C	1.5-4.2	230
	Mid: 30-75	5	FS	Bunker C	1.3-2.4	230
	Bottom: 75-175	5	FS	Bunker C	1.3-2.9	230

a CA: carbon analyzer
FC: flow calorimeter
FS: fluorescence spectrometry
GC: gas chromatography
GR: gravimetric
IR: infrared spectrometry
MS: mass spectrometry
TLC: thin-layer chromatography

b μg/1, unless otherwise indicated

c Particulate materials (>0.5 μm diameter)

d Dissolved materials (<0.5 μm diameter)

e ($\bar{x}$) = mean value

f Total oil particles collected by 80-100 μm net

Description of water samples			Method of analysis[a]	Hydrocarbon type, subclass, or	Hydrocarbon content (μg/1)[b]	Reference
Location (Sampling period)	Water depth (m)	Number of samples				
Come-by-Chance Bay, Canada	0-10	196	FS	Crude oil	0.6-5.5	231
Labrador Current, off Canada	0	27	FS	Crude oil	0.8-10.8	232
Labrador Shelf, off Canada	0-152	6	FS	Crude oil	13.0-115.4	232
Labrador Shelf, off Canada	7-165	18	FS	Crude oil	0.1-6.2	232
Grand Banks	1-60	9	FS	Crude oil	2.9-38.5	232
Off Sable Island, Canada	1-135	9	FS	Crude oil	5-30	232
Scotia Shelf, Canada	1-175	97	FS	Crude oil	0.2-18.6	232
	2	3	GC	n-C_{14-27}	0.078-0.099	229
	50	3	GC	n-C_{14-27}	0.080-0.154	229
	2	3	FS	Crude oil	0.2-0.8	229
	50	3	FS	Crude oil	0.2-2.0	229
Off eastern N.S.	0-350	19	FS	Bunker C	0.4-2.5	233
Off central N.S.	2-1,000	52	FS	Bunker C	0.2-3.2	233
Off central N.S.	2	11	FS	Bunker C	1.7-13.0	230
Off central N.S.	10	11	FS	Bunker C	1.6-13.5	230
Off western N.S.	10-140	8	FS	Bunker C	0.6-2.0	233
Nova Scotia coast	5	7	FS	Bunker C	0.5-2.6	234
Nova Scotia, unpolluted bays	Surface	≥2	FS	Bunker C	0.3	234
Halifax Harbor, Nova Scotia	-		FS	Bunker C	1.9-71.1	234
Bedford Basin, Nova Scotia	<60	160	FS	Crude oil	1-60	67
	2	27	GC	n-C_{14-27}	0.065-0.590	229
	50	27	GC	n-C_{14-27}	0.025-0.395	229
	2	27	FS	Crude oil	1.6-9.3	229
	50	27	FS	Crude oil	1.4-4.3	229
Chedabucto Bay, Nova Scotia	5-50	37	FS	Bunker C	0-4.0	234
Inner part (3 months)	1-55	5	FS	Bunker C	16-41	230
Outer part (3 months)	1	4	FS	Bunker C	5-16[c]	230
Outer part (3 months)	1	3	FS	Bunker C	15-90[d]	230
Outer part (3 months)	Bottom: 20-140	4	FS	Bunker C	5-11[c]	230
Outer part (3 months)	Bottom: 20-140	4	FS	Bunker C	4-9[d]	230
Upper intertidal	-	≧7	FS	Bunker C	4-170	95
Middle intertidal	-	≧8	FS	Bunker C	1-28	95
Lower intertidal	-	≧4	FS	Bunker C	1	95
Intertidal interstitial water	-	≧8	FS	Bunker C	$(5-269) \times 10^3$	95

Description of water samples			Method of analysis[a]	Hydrocarbon type, sub-class, or class	Hydrocarbon content (μg/l)[b]	Reference
Location (Sampling period)	Water depth (m)	Number of samples				
Boston Harbor, Mass.	0-5 cm	17	IR	CCl_4 extract	190-816	235
	25-35 cm	17	IR	CCl_4 extract	114-234	235
Woods Hole, Mass.	Surface	1	GC	Saturates	11	174
Narragansett Bay, R.I.	0.2	1	GC	Paraffins	5.9	45
Sewage	Surface	8	GC/GR	Total HC	1,000-12,700	236
New York Bight	0	8	IR	Total HC	1-21	237
	0	4	IR	Saturates	1-11	237
	0	4	IR	Aromatics	1-3	237
Atlantic Ocean, Open water	1-10	≧26	IR	Non-vol. HC	1.3-13	238
Off Bermuda	1-30	7	IR	Non-vol. HC	2.3-6.3	239
	50-500	10	IR	Non-vol. HC	0.3-3.0	239
	500-2,500	6	IR	Non-vol. HC	≦0.7	239
	0.2-0.3	5	GC	Paraffins	31-143	46
Azores-Bermuda section	0.2-0.3	7	GC	Paraffins	20-239	46
Halifax-Bermuda section	1	23	FS	Crude oil	$(\overline{0.56})$	240
	5	24	FS	Crude oil	$(\overline{0.37})$	240
	≦25	32	FC	HC	3.7-13.6	241
Bermuda-Rhode Island sec.	0.2-0.3	5	GC	Paraffins	13-131	46
Off Virginia Capes	0	(5)	IR	Saturates	3	237
	0	(5)	IR	Aromatics	1	237
	10	(5)	IR	Saturates	1	237
	10	(5)	IR	Aromatics	<1	237
Western Sargasso Sea	0	7	IR	Total HC	5-22	237
	0	1	IR	Saturates	21	237
	0		IR	Aromatics	1	237
	10	7	IR	Total HC	1-7	237
	10	1	IR	Saturates	2	237
	10	1	IR	Aromatics	1	237
	2	4	GC	n-C_{14-27}	0.02-0.118	229
	50	4	GC	n-C_{14-27}	0.054-0.145	229
North of Bahamas	0	2	IR	Total HC	3-14	237
Florida Strait	1	1	GC	Non-polar HC	52	242
	500	1	GR	Non-polar HC	Trace	242
	1-500	6	GR	Non-polar HC	$(\overline{47})$	242

Description of water samples						
Location (Sampling period)	Water depth (m)	Number of samples	Method of analysis[a]	Hydrocarbon type, sub-class, or class	Hydrocarbon content (μg/l)[b]	Reference
Atlantic Ocean, Open water						
Florida Strait	0	1	IR	Saturates	2	237
	0	1	IR	Aromatics	<1	237
Gulf of Mexico, Open water	0	8	IR	Total HC	2-8	237
	0	2	IR	Saturates	2	237
	0	2	IR	Aromatics	≦1	237
	10	5	IR	Total HC	1-2	237
	10	2	IR	Saturates	2	237
	10	2	IR	Aromatics	≦1	237
	1	1	GR	Non-polar HC	12	242
	500	1	GR	Non-polar HC	13	242
	1-500	5	GR	Non-polar HC	$(\overline{12})$	242
South Louisiana, Offshore	1	1	GC	*n*-Alkanes	0.63	178
East Bay	3	1	GC	*n*-Alkanes	0.2	178
Texas, off Corpus Christi	3	1	GC	*n*-Alkanes	0.1	178
Yucatan Strait	1-500	5	GR	Non-polar HC	$(\overline{12})$	242
Caribbean Sea, Open water	0	3	IR	Total HC	6-50	237
	0	1	IR	Saturates	5	237
	0	1	IR	Aromatics	1	237
	10	3	IR	Total HC	4-14	237
	10	1	IR	Saturates	12	237
	10	1	IR	Aromatics	2	237
East central	200	1	GR	Non-polar HC	8	242
Cariaco Trench	900	1	GR	Non-polar HC	5	242
Lake Maracaibo, Venezuela	1	9	GC	Total HC	30-6,000	162
	1	12	GC	Saturates & olefins	<10-220	162
	1	12	GC	Aromatics	<10	162
	1	12	GC	Polars	<10-5,800	162
North Pacific, Station PAPA	Surface		FS	As chrysene	$(\overline{0.016})$	157
Vancouver Is. - PAPA section	Surface	11	FS	As chrysene	0.011-0.027	243
Santa Barbara Channel, Calif.						
Seep area	1	1	—	—	16	211
	10	1	—	—	0.4	211
	55	1	—	—	1.0	211
Non-seep area	1-400	4	—	—	$(\overline{0.3})$	211

Description of water samples			Method of analysis[a]	Hydrocarbon type, subclass, or class	Hydrocarbon content (μg/l)[b]	Reference
Location (Sampling period)	Water depth (m)	Number of samples				
San Francisco Bay, Calif.	2	8	TLC	Paraffins	14-280	212
	2	8	TLC	Aromatics	<5-59	212
Puget Sound, Wash., Open water	Surface	1	GC	n-C_{14-37}	2.33	168
Polluted harbor water	Surface	1	GC	n-C_{14-37}	7.32	168
Alaska						
Dayville mud flats, Valdez	0	3	GC	Total HC	2.7	244
	10	2	GC	Total HC	1.6	244
Off Old Valdez	0	1	GC	Total HC	14.9	244
	10	3	GC	Total HC	0.95	244
Prince William Sound,						
Siwash Bay	0	1	GC	Total HC	3.1	244
	10	3	GC	Total HC	2.0	244
Squirral Bay	0	3	GC	Total HC	4.6	244
	10	1	GC	Total HC	5.4	244
MacLeod Harbor	0	2	GC	Total HC	7.5	244
	10	4	GC	Total HC	4.0-31.4	244
Off Hinchinbrook Island	0	1	GC	Total HC	5.4	244
Gulf of Alaska, Anchor Cove	0	2	GC	Total HC	8.2	244
	10	2	GC	Total HC	3.7	244
Off Middleton Island	0	1	GC	Total HC	5.5	244
NE Gulf of Alaska, Katella	0	3	GC	Total HC	3.4	244
Cape Yakataga	0	3	GC	Total HC	11.4	244
Lynn Canal, Alaska	Surface	4	GC	Total HC	0.1	245
Cook Inlet, Alaska	180	70	GC	C_{10-25}	$\leqq 0.3$	36
South Beaufort Sea - Nearshore	0		FS	As chrysene	0.013-0.045	157
- Offshore	Deep		FS	As chrysene	0.016	157
Firth of Clyde, Scotland	Surface	1	GC	n-Paraffins	4.91	148
	Mid-depth	1	GC	n-Paraffins	0.21	148
British Isles - Inshore	1	22	GC	n-Paraffins	0.4-5.2	156
- Offshore	1	24	GC	n-Paraffins	0.2-11.8	156
Off Brest, France	Surface	1	GR	Non-polar HC	137	246
Off Roscoff, France	Surface	1	GR	Non-polar HC	46	246
Baltic Sea and approaches						
Open Water	0-5	5	IR	Pol. & nonpolar	<50-100	247
	—	—	FC	—	8.9	241

Description of water samples			Method of analysis[a]	Hydrocarbon type, subclass, or class	Hydrocarbon content (μg/l)[b]	Reference
Location (Sampling period)	Water depth (m)	Number of samples				
Open water	Surface	—	—	Oil products	300-1,000	248
	20	2	CA	Satd. & Monoarom.	0.9-1.0[c]	249
	20	1	CA	Satd. & Monoarom.	48[d]	249
	70	2	CA	Satd. & Monoarom.	1.0[c]	249
	70	1	CA	Satd. & Monoarom.	58[d]	249
	110	2	CA	Satd. & Monoarom.	1.7-2.3[c]	249
	110	1	CA	Satd. & Monoarom.	58[d]	249
	150	2	CA	Satd. & Monoarom.	0.8-1.0[c]	249
	150	1	CA	Satd. & Monoarom.	58[d]	249
	200	2	CA	Satd. & Monoarom.	0.5-0.8[c]	249
	200	1	CA	Satd. & Monoarom.	64[d]	249
	1-200	40	—	Satd. aromatics	0-50	250
Kattegat	0-5	4	IR	Pol. & nonpolar	80-190	247
Gulf of Bothnia	0-5	8	IR	Pol. & nonpolar	<50-130	247
Gulf of Finland	0-5	2	IR	Pol. & nonpolar	110-120	247
Göteborg Harbor (polluted)	0-5	2	IR	Pol. & nonpolar	470-710	247
In islands, 11 km away	0-5	2	IR	Pol. & nonpolar	80-280	247
In islands, 24 km away	0-5	2	IR	Pol. & nonpolar	100	247
Mediterranean Sea						
Off Villefranche, France	50	1	GR	Non-polar HC	75	246
Bay of Brusc (clean)	0.5	1	GC	n-C_{12-40}	0.3[c]	251
	0.5	1	GC	n-C_{12-40}	9.5[d]	251
Etang de Berre (polluted)	0.5	1	GC	n-C_{12-40}	7.2[c]	251
	0.5	1	GC	n-C_{12-40}	2.4[d]	251
Gulf of Sidra	1	1	IR	Non-vol. HC	4	252
Off Sardinia	1	1	IR	Non-vol. HC	195	252
Off Corsica	1	1	IR	Non-vol. HC	170	252
Off Italy	Surface	2	IR	Non-vol. HC	154-195	239
	10	2	IR	Non-vol. HC	6-8	239
Central, Open waters	Surface	15	IR	Non-vol. HC	<1-63	239
	10	13	IR	Non-vol. HC	2-7	239

Description of water samples			Method of analysis[a]	Hydrocarbon type, subclass, or class	Hydrocarbon content (μg/l)[b]	Reference
Location (Sampling period)	Water depth (m)	Number of samples				
Atlantic Ocean, Off West Africa						
Nearshore	50	1	GR	Non-polar HC	95	246
Offshore	50	1	GR	Non-polar HC	43	246
	500	1	GR	Non-polar HC	19	246
	2,000	1	GR	Non-polar HC	10	246
	4,500	1	GR	Non-polar HC	37	246
Lower Chesapeake Bay, Virginia	Surface	3	GC	C_{1-4}	771 nl/l	253
Potomac River, Virginia	Surface	1	GC	C_{1-4}	3,800 nl/l	253
York River, Virginia	Surface	6	GC	C_{1-4}	2,800 nl/l	253
N. Atlantic Ocean, Open water	0	1	GC	C_{1-4}	58.4 nl/l	254
	30	1	GC	C_{1-4}	66.6 nl/l	254
	500	1	GC	C_{1-4}	50.6 nl/l	254
Sargasso Sea	Surface	26	GC	C_{1-4}	46-77 nl/l	253
Trinidad Shelf	Surface	2	GC	C_{1-4}	112 nl/l	253
East of the Antilles	Surface	20	GC	C_{1-4}	34-43 nl/l	253
Bahamas	Surface	10	GC	C_{1-4}	45-53 nl/l	253
Gulf Stream	Surface	10	GC	C_{1-4}	69 nl/l	253
Coastal Carolinas & north	Surface	3	GC	C_{1-4}	42-285 nl/l	253
Coastal Florida	Surface	21	GC	C_{1-4}	39-318 nl/l	253
Miami Harbor, Florida	Surface	4	GC	C_{1-4}	1,380 nl/l	253
Gulf of Mexico, Open water	0	1	GC	C_{1-4}	73.8 nl/l	254
	30	1	GC	C_{1-4}	295 nl/l	254
	500	1	GC	C_{1-4}	26.1 nl/l	254
	Surface	3	GC	C_{1-4}	41-83 nl/l	253
	3	1	GC	C_{1-4}	54 nl/l	255
Gulf of Mexico, Western	Surface	30	GC	C_2H_6	2.3-255 nl/l	256
Florida Shelf	Surface	1	GC	C_{1-4}	121 nl/l	253
Louisiana Shelf	Surface	4	GC	C_{1-4}	450-7,400 nl/l	253
	3	1	GC	C_{1-4}	1,030 nl/l	255
Mississippi River mouth	Surface	4	GC	C_{1-4}	74-8,500 nl/l	253
	3	1	GC	C_{1-4}	1,900 nl/l	257
Texas Shelf	Surface	6	GC	C_{1-4}	140-919 nl/l	253
Galveston Harbor	3	1	GC	CH_4	500 nl/l	257
Puerto Mexico Harbor	Surface	3	GC	C_{1-4}	117-1,400 nl/l	253
East Mexico Shelf	Surface	1	GC	C_{1-4}	63 nl/l	253

Description of water samples			Method of analysis[a]	Hydrocarbon type, subclass, or class	Hydrocarbon content (μg/l)[b]	Reference
Location (Sampling period)	Water depth (m)	Number of samples				
Yucatan Shelf	Surface	5	GC	C_{1-4}	41-62 nl/l	253
Caribbean Sea						
Misteriosa Bank (near tanker)	3	1	GC	C_{1-4}	140 nl/l	255
West of Antilles	Surface	13	GC	C_{1-4}	44-61 nl/l	253
Southeast coastal	Surface	13	GC	C_{1-4}	47-70 nl/l	253
Central, Open water	Surface	5	GC	C_{1-4}	52-60 nl/l	253
Northwestern, Open water	Surface	8	GC	C_{1-4}	40-62 nl/l	253
Scotland, Nearshore	Surface	7	GC	C_{1-4}	70 nl/l	253
Arctic Ocean, Atlantic	Surface	32	GC	C_{1-4}	79-112 nl/l	253
Norwegian Sea	Surface	56	GC	C_{1-4}	65-75 nl/l	253
Mediterranean Sea	Surface	2	GC	C_{1-4}	57-58 nl/l	253
Black Sea	Surface	1	GC	C_{1-4}	91 nl/l	253
Red Sea	Surface	1	GC	C_{1-4}	54 nl/l	253
Arabian Sea	Surface	1	GC	C_{1-4}	58 nl/l	253
Pacific Ocean						
Guayaquil R., Ecuador	2	1	GC	C_{1-3}	137 nl/l	258
Off Central America, Coastal	Surface	4	GC	C_{1-4}	70 nl/l	253
Off Galapagos Islands	2	3	GC	C_{1-3}	60-67 nl/l	258
Eastern Equatorial	Surface	59	GC	C_{1-4}	49-72 nl/l	253
Ecuador to Hawaii	2	13	GC	C_{1-3}	43-70 nl/l	258
Off Hawaiian Islands	Surface	3	GC	C_{1-4}	49 nl/l	253
Pearl Harbor, Hawaii	2	1	GC	C_{1-3}	47 nl/l	258
Hawaii to Tahiti	Surface	40	GC	C_{1-4}	47-52 nl/l	253
Hawaii to Tahiti	2	9	GC	C_{1-3}	43-46 nl/l	258
Tahiti Harbor	2	1	GC	C_{1-3}	42 nl/l	258
Tahiti to Panama	2	14	GC	C_{1-3}	41-61 nl/l	258
Lake Nitinat, Canada	Surface	1	GC	C_{1-4}	860 nl/l	253
Southern Beaufort Sea (1974)	0-5	22	GC	C_{1-4}	68-563 nl/l	157
(1975)	0-5	33	GC	C_{1-4}	43-163 nl/l	157
(1974)	10-35	6	GC	C_{1-4}	125-620 nl/l	157
(1975)	10-35	14	GC	C_{1-4}	69-1,155 nl/l	157
(1974)	48-70	1	GC	C_{1-4}	188 nl/l	157
(1975)	48-70	8	GC	C_{1-4}	78-412 nl/l	157
(1975)	>100	2	GC	C_{1-4}	16-243 nl/l	157

Description of water samples			Method of analysis[a]	Hydrocarbon type, subclass, or class	Hydrocarbon content (μg/l)[b]	Reference
Location (Sampling period)	Water depth (m)	Number of samples				
Central South Pacific	Surface	44	GC	C_{1-4}	45-55 nl/l	253
New Zealand, Coastal	Surface	5	GC	C_{1-4}	80 nl/l	253
Antarctic Convergence	Surface	21	GC	C_{1-4}	61-74 nl/l	253
Antarctic, Shelf	Surface	8	GC	C_{1-4}	51-83 nl/l	253
World Oceans						
Clean Open waters	Surface	452	GC	CH_4	31-80($\overline{49.5}$) nl/l	253
				C_2H_6	0.1-1.7($\overline{0.50}$) nl/l	253
				C_3H_8	0.05-1.4($\overline{0.34}$) nl/l	253
				C_4H_{10}	($\overline{0.05}$) nl/l	253
				C_2H_4	0.7-12.1($\overline{4.8}$) nl/l	253
				C_3H_6	0.1-5.8($\overline{1.4}$) nl/l	253
Contaminated coastal waters	Surface	452	GC	CH_4	103-3,800($\overline{1,250}$) nl/l	253
				C_2H_6	1.4-650($\overline{71}$) nl/l	253
				C_3H_8	1.0-1,100($\overline{95}$) nl/l	253
				C_2H_4	1.9-35($\overline{11.0}$) nl/l	253
				C_3H_6	0.1-16($\overline{2.8}$) nl/l	253
Chedabucto Bay, Nova Scotia	5	31	GR	Tot. oil part.[f]	$0.000-0.020 \times 10^{-3}$[b]	65
	10	11	GR	Tot. oil part.	$0.0006-0.020 \times 10^{-3}$	65
	30	9	GR	Tot. oil part.	$0.001-0.003 \times 10^{-3}$	65
	50-80	5	GR	Tot. oil part.	$0.000-0.002 \times 10^{-3}$	65
Atlantic Ocean						
Off eastern Nova Scotia	5	33	GR	Tot. oil part.	$0.000-0.005 \times 10^{-3}$	65
	10	14	GR	Tot. oil part.	$0.000-0.010 \times 10^{-3}$	65
Off central Nova Scotia	5	64	GR	Tot. oil part.	$0.000-.0025 \times 10^{-3}$	65

the types and sources of petroleum pollution [66,237-239]. Gordon, Keizer, and Dale [259,260] have pointed out that the sampling method for dissolved hydrocarbons in seawater can seriously influence the results and that adsorption of organic substances in the seawater on the inner surfaces of samplers is especially troublesome. Some of the earlier results (prior to 1974) listed in Table 11 may therefore be subject to reevaluation (See references 259 and 260 for discussion of analytical procedures). An extensive collection of data on hydrocarbon content in oceanic waters was prepared by Myers and Gunnerson [286].

TAR BALLS

Recently, several quantitative measurements have been reported on tar balls collected by surface nets [72,73,83,261, 262]. The data for various parts of the oceans are summarized in Table 12. The higher levels of tar balls on the sea surface occurred either along the ocean routes supporting high densities of oil tanker traffic or downstream from these routes [2,83,273,278].

Two distinct types of floating tarry residues have been identified. The first type appears to be from urban and industrial waste products. Tarry residues from New York Harbor, for example, contain substantial amounts of specific components near carbon numbers C_{20} and C_{23} (gas chromatographic analysis), only trace amounts of sulfur, and high levels of oxygenated organic material. The second type consists primarily of petroleum-based materials which have been modified in varying degrees by weathering in the marine environment. Tarry residues collected off the coast of Florida contain high levels of sulfur, suggesting that they originated from discharges of tankers carrying "sour" crude petroleum. In general, high levels of *n*-paraffins in tarry residues are characteristics of materials originating from the discharged washings of oil cargo tanks [278].

The variability of sampling with a towed surface neuston net [263] for pelagic tar is very high and even the results of successive tows may differ by a factor of ten [267,280]. The data given in Table 12 represent mean values.

Several marine organisms may be associated with the tarry lumps in the North Atlantic. An isopod, *Idotea metallica* (10-25 mm length), tended to stay with the tar ball when placed in an aquarium aboard a research vessel [261]. The goose barnacle, *Lepas pectinata*, frequently was found attached to tarry lumps, particularly the firm weathered lumps. At one location, four tar lumps had a total of 150 barnacles (2-8 mm length) whose growth rate was measured at 1 mm/week indicating a two-month minimum age for the tar lumps [261]. Morris [281] found

TABLE 12

Tar residue levels on the sea and beach surfaces

A. Sea surface

Description of tar samples		Amount of tar residue (mg/m^2)			Reference
Type / Geographic area	Time period	Maximum	Geometric mean	Arithmetic mean	
Atlantic Ocean					
Scotia Shelf		2.4	-	0.9	263
Lat. 38° to 42° N, Long. 50° W		9.7	-	2.2	263
Virginia to Cape Cod, Coastal	Winter	4.4	-	1.04	264,265
	Summer		-	0.18	264,265
	-	0.2	-	0.04	124
Offshore	Winter	11	-	0.05	264,265
	Summer		-	0.77	264,265
North Carolina to Florida	Winter	-	-	1.22	264,265
	Summer	-	-	0.23	264,265
	-	20	-	5.5	124
North Antilles & Bahamas	Winter	87	-	4.8	264,265
	Summer		-	3.9	264,265
Lesser Antilles		8.37	-	1.12	266
Ocean Station BRAVO, Labrador Current		0.003	-	0.00	267
Ocean Station CHARLIE, North Atlantic		1.83	-	0.12	267
Ocean Station DELTA, North Atlantic		10.73	-	1.15	267
Ocean Station ECHO, Sargasso Sea		21.62	-	2.64	267
Off Bermuda		-	-	0.6	263
Northeastern North Atlantic		480	-	-	83,261
		14.2	-	4.8	268
		1	-	0.6	124
Barents Sea		3.0	-	0.15	269
Norwegian Shelf		0.4	-	0.04	269
Northern North Sea		0.2	-	0.02	269
Skagerrak		12.1	-	0.32	269
Central eastern North Atlantic		22.6	-	9.8	268
Gulf Stream		0.8	-	0.3	124
		6.7	-	3.8	268

Description of tar samples		Amount of tar residue (mg/m^2)			Reference
Type / Geographic area	Time period	Maximum	Geometric mean	Arithmetic mean	
Atlantic Ocean					
Sargasso Sea		40	—	9.4	73
		1.4	—	0.2	264,265
		6	—	3	124
		90.6	—	25	268
Central Atlantic					
Canary Current		7.69	—	2.02	266
North Equatorial Current		0.27	—	0.16	266
Equatorial Countercurrent		0.04	—	0.02	266
South Equatorial Current		0.57	—	0.11	266
Equatorial Current Region		63.6	—	12.7	270
Caribbean Sea		1.5	—	0.4	271
		0.9	—	0.2	264,265
		4.5	—	0.74	272
		13.4	—	1.62	266
Gulf of Mexico		10.0	—	1.2	272
		3.5	—	1.1	271
		6.0	—	1.12	266
Mediterranean Sea		540	—	20	261,263
Ionian Sea	1969	540	60	130	261,273
	1975	110	5	16.0	273
Alboran Sea	1969	10	—	6.5	261,273
	1975	45	4.4	11.0	273
Tyrrhenian Sea	1969	20	—	1.5	261,273
	1975	15	1.4	3.2	273
Balearic Sea	1969	10	2.2	2.4	261,273
	1975	1	0.4	0.5	273
Central (tarballs)		6.1	—	—	72
Eastern (tarballs)		10.0	—	4.1	72
Central (emulsions)		0.30	—	—	72
Eastern (emulsions)		0.36	—	0.14	72
Northwest Pacific Ocean					
Lat. 35°N, Long. 140°E to 175°W		14	—	3.8	262
Lat. 25° to 40°N, Long. 140° to 160° W		16.3	—	2.1	274

Description of tar samples	Amount of tar residue (mg/m²)			Reference
Type Geographic area Time period	Maximum	Geometric mean	Arithmetic mean	
Northwest Pacific Ocean				
Outside the Kuroshio Current	—	—	0.4	274
Northeast Pacific Ocean				
Lat. 35°N, Long. 175° to 130°W	3	—	0.4	262
Lat. 35°N, Long. 175° to 130°W	—	—	0.03	274
South Pacific Ocean	—	—	0.0003	274

B. Beach surfaces

Description of tar samples: Geographic area / Time period	Amount of tar residue $(g/m^2)^a$	Reference
Southwest Florida coast, 1 km of shoreline	23	276
Bermuda	190 (mean)	83
Barbados, Windward shore	40-62	266
Barbados, Leeward shore	4.5	266
Puerto Rico, North shore	52-112	266
Puerto Rico, West shore	12-20	266
Central Caribbean islands, Windward shore	13-230	266
Central Caribbean islands, Leeward shore	0-2.2	266
Hondurus, Windward shore	90-127	266
Southern California		
Sunset Beach (Long Beach)	0.018-1.35	86
Torrance (Los Angeles)	0.006-0.92	86
Mussel Shoals (Sea Cliff)	0.003-0.60	86
Summerland Beach (Santa Barbara)	0.002-0.38	86
Coal Oil Point	1.22-23.9	86
Gaviota Beach	0.023-1.01	86
Beaufort Sea, Canada 264 km of shoreline (1974)	No tar residues	277
Yukon 16 km of shoreline (1974)	380 g grease	277
N.W.T. 10 km of shoreline (1974)	425 g grease	277
N.W.T. 86 km of shoreline (1975)	157 g grease & asphalt	277
West India coast	4,480 (max.)	278

a In g/m^2 except where otherwise indicated.

that a marine barnacle, *Lepas fascicularis*, residing on tar balls took up hydrocarbons from the tar (about 5% of the solvent-extractable material of the barnacles was non-biogenic); the organisms did not alter the straight-chain hydrocarbon fraction to any extent, but apparently discharged such hydrocarbons, based on gas-liquid chromatographic analyses of the tar lumps and the extracts.

Tar has been found in the stomach of the saury, *Scomberesox saurus*, a fish that feeds on surface-dwelling crustaceans [261]. Because large predaceous fishes feed on saury, the tar lumps thus ingested may provide a direct source of petroleum for ocean food webs [261]; however, evidence obtained so far does not suggest that the magnification of petroleum hydrocarbons occurs in marine food webs [2] (See Chapter 3 of Volume II).

SEA SLICKS

Hydrocarbons tend to accumulate at the air-sea interface (Table 13). Oil slicks and other surface films (e.g., "natural slicks") may act as media for the accumulation of materials such as trace metals, vitamins, amino acids, and some lipophilic pollutants such as DDT residues and PCBs [45,46].

When it occurs, petroleum hydrocarbon contamination of the open ocean appears to be limited to the upper few meters of the water column [45,237]. Most of the hydrocarbons are contained in the surface film. Analyses for the vertical distribution of hydrocarbons in the seawater column have been made for only a few areas of the oceans [240,280]. Surface films (less than 0.5 μm thick) from the Gulf of Mexico contained 50 different hydrocarbon compounds ranging in carbon number from C_{16} to C_{36}, based on gas chromatographic-mass spectrometric analyses. The major fraction by weight (70%) of the hydrocarbons were branched paraffins; the 3-methyl branched substituents accounted for half of this. Cycloparaffins accounted for 13%; only 3% of the surface film hydrocarbons were normal paraffins. The hydrocarbons from samples of seawater collected 10 m below the surface consisted predominately of *n*-paraffins from C_{14} to C_{36} [282]. Tar balls from the Gulf of Mexico varied considerably in chemical composition but contained substantial levels of normal paraffins. The data of Ledet and Laseter [282] suggest that the branched and cyclic paraffins make up the bulk of the paraffins in the sea surface film and that the *n*-paraffins accumulate in the water column.

TABLE 13

Hydrocarbon content in surface slicks

Description of slick samples						
Location	Thickness (μm)	Number of samples	Method of analysis[a]	Hydrocarbon type, subclass, or class	Hydrocarbon content	Reference
North Atlantic						
Bermuda to Halifax	1,000-5,000	53	FS	Aromatics	9.28 μg/l	240
Bermuda to Halifax	1,000-3,000	43	FS	Aromatics	20.4 μg/l	260
					60 $\mu g/m^2$	260
Sargasso Sea						
Azores to Bermuda	100-300	7	GC	Paraffins	14-199 μg/l	46
Near Bermuda	100-300	5	GC	Paraffins	87-559 μg/l	46
Bermuda to Rhode Island	100-300	5	GC	Paraffins	35-343 μg/l	46
Narragansett Bay, R.I.	100-150	1	GC	Paraffins	8.5 μg/l	46
Offshore Louisiana	0.01-0.50	5	GR	Paraffins	700 $\mu g/m^2$	282
Nearshore Louisiana	0.01-0.50	10	GR	Paraffins	210 $\mu g/m^2$	282
Timbalier Bay, Louisiana	0.01-0.50	43	GR	Paraffins	360 $\mu g/m^2$	282
Offshore Florida	0.01-0.50	-	GR	Paraffins	180 $\mu g/m^2$	282
Firth of Clyde, Scotland	-	1	GC	n-C_{18-33}	1.96 $\mu g/m^2$	148
British Isles						
Inshore	-	22	GC	n-C_{15-33}	13.7-145.4 $\mu g/m^2$	156
Offshore	-	29	GC	n-C_{15-33}	4.3-84.2 $\mu g/m^2$	156
Mediterranean Sea						
Central	5-30	1	-	-	162 mg/m^2	72
Eastern	5-30	4	-	-	34-196 mg/m^2	72
Gulf of Trieste, Harbors	-	78	TLC	Total HC	0.33-310 mg/m^2	283
Gulf of Trieste, Inshore	-	621	TLC	Total HC	0-80,000 mg/m^2	283
Gulf of Trieste, Offshore	-	9	TLC	Total HC	2.33-2,666 mg/m^2	283

a GC: gas chromatography; FS: fluorescence spectrometry; GR: gravimetric; TLC: thin-layer chromatography

b Particulate material (>0.5 μm diameter)

c Dissolved material (<0.5 μm diameter)

Description of slick samples			Method of analysis[a]	Hydrocarbon type, subclass, or class	Hydrocarbon content	Reference
Location	Thickness (μm)	Number of samples				
Mediterranean Sea						
Bay of Brusc, Clean water	440	1	GC	n-C_{12-40}	1.9 μg/1[b]	251
	440	1	GC	n-C_{12-40}	4.8 μg/1[c]	251
	50-80	1	GC	n-C_{12-40}	3.8 μg/1[b]	251
	50-80	1	GC	n-C_{12-40}	8.1 μg/1[c]	251
Etang de Berre, Polluted	440	1	GC	n-C_{12-40}	1,214 μg/1[b]	251
	440	1	GC	n-C_{12-40}	15.0 μg/1[c]	251
	50-80	1	GC	n-C_{12-40}	4,911 μg/1[b]	251
Washington coast						
Sea foam, Polluted	-	1	GC	n-C_{14-37}	260 μg/g	84

ARCTIC SAMPLES

Hydrocarbon levels in presumably uncontaminated arctic marine environments have been determined in only a limited number of samples of organisms, seawater, and sediments (See Tables 7-13). Wong and coworkers [157] analyzed one of the largest suites of environmental samples from the Beaufort Sea. They reported data on saturated and aromatic hydrocarbons levels in seawater, fish, plankton, and marine sediments [157] and gave information on the occurrence of tar balls and particulate pollutants on the shore [277]. On the basis of data obtained with neuston-net tows, the southern Beaufort Sea appeared to be free of tar balls.

Only low levels of saturated low molecular weight hydrocarbons were found; however, methane was relatively high in near-bottom waters, attributable to biogenic production in the sediments. The low values indicated to Wong and coworkers [157] an absence of petrogenic inputs. No low-molecular weight aromatic hydrocarbons (e.g., benzene and toluene) were found in the seawater. The deeper shelf waters of the Beaufort Sea had levels of polycyclic aromatic hydrocarbons (calculated as chrysene equivalents) comparable to levels found in uncontaminated waters of the Northeast Pacific Ocean (Table 11). The water near the mouth of the Mackenzie River showed higher, but fluctuating, values for polycyclic aromatic hydrocarbons, possibly due to the fact the river flows from regions with known fossil fuel deposits and natural oil seeps [157].

The few fish samples analyzed showed from 23 to 67% unresolved hydrocarbons (calculated on the basis of percent of total hydrocarbons) compared to 12 to 16% for Northeast Pacific Ocean fish [157] and over 80% for Sargasso Sea fish [154]. Wong and coworkers devised an empirical scale for estimating the level of petroleum pollution in fish based on the non-polar hydrocarbon gas chromatographic fingerprint (See original reference for details). Using this scale the authors considered Northeast Pacific fish to be uncontaminated and the Sargasso Sea fish to be contaminated, and the Beaufort Sea fish to be "marginally" contaminated. However, the authors stressed that their technique is limited because of the small number of samples analyzed to establish the scale, the natural variability in hydrocarbon content of the various species of fish, and the limitations of the analytical methods for estimating unresolved organic components [157].

Wong and coworkers [157] found 9 to 31 μg/kg wet weight of polycyclic aromatic hydrocarbons (defined as the sum of chrysene, benzanthracene, perylene, phenanthrene, and pyrene, and certain isomers) in the flesh of three species of southern Beaufort Sea fish (pomfret, least cisco, and Arctic cisco). Salmon and tuna caught at Ocean Station PAPA (central Northeast Pacific Ocean) waters and considered uncontaminated by the authors contained 20 to 82 μg/kg and 184 μg/kg wet weight of polycyclic aromatic hydrocarbons, respectively.

Mixed zooplankton samples from the southern Beaufort Sea contained *n*-alkanes from C_{16} to C_{38}. The gas chromatographic pattern showed maximum *n*-alkane content at C_{26}, a slight predominance of odd-carbon number alkanes, and a significant envelope of unresolved hydrocarbons. Wong and coworkers [157] were unable to conclude that the plankton samples were contaminated by petroleum residues.

The distribution of *n*-alkanes in the marine sediments of the southern Beaufort Sea showed a pattern characteristic of sediments containing marine and terrestrial biological and detrital materials. The predominance of odd-carbon number *n*-alkanes, characteristic of terrestrial plant material, was evident in most of the sediments of this region suggesting an influx of detritus from the Mackenzie River. The level of polycyclic aromatic hydrocarbons (defined as the sum of pyrene, benzo[a]pyrene, perylene, and coronene) was about 0.4 to 20 μg/kg for nearshore sediments and about 300 μg/kg for benthic sediments on a dry weight basis. The authors attributed the difference in levels between nearshore and offshore sediments to the fact that the nearshore sediments contained much lower levels of organic materials and were exposed to more severe weathering.

Wong and coworkers concluded that the very low levels of hydrocarbons in the southern Beaufort Sea implied an uncontaminated marine environment with respect to petroleum hydrocarbons [157].

PROSPECTUS

We have only superficial knowledge of how the various weathering processes might, singly or in combination, modify petroleum in the marine environment. Additional knowledge is needed to assess and predict the effects of petroleum contamination of the marine environment and to provide a basis for instituting realistic actions to prevent or mitigate damage to the marine system. A systematic classification of the photochemical processes involved in the degradation of petroleum and the combined effects of evaporation, dissolution, and emulsification is essential. Data on the rates of sedimentation, the accumulation of sedimented petroleum, and the fate of such petroleum are generally lacking. Pertinent information is needed on the form, concentration, and distribution of petroleum components and corresponding biogenic hydrocarbons dissolved in the water column or absorbed in and adsorbed on particulate material. More information is needed on the weathering of petroleum in the open sea, including those areas beyond the confines of the original oil slick or the contaminating source. Because of the importance of the air-sea interface, a better understanding of the reactions of hydrocarbons in the atmosphere is needed.

Because of the complicated nature of the problems, the vast area of the marine environment, and the various scientific and engineering disciplines involved, the information required will only be achieved through a coordinated systems approach. This approach involves (1) a thorough analysis of the specific problem areas, (2) an orderly planning of the research programs required, and (3) a coordinated, systematic study of the physical, chemical, and biological aspects of the entire ecosystem and its subsystems.

Although the analytical data tabulated in this chapter cover a wide range of environment samples, very little information is provided for understanding the transport mechanisms of petroleum in the marine environment. For example, systematic, in-depth information is lacking on (1) the identity and quantity of many hydrocarbons in petroleum; (2) the reactions of these components, singly and in combination, under various environmental and geographic conditions; and (3) the utilization and degradation of petroleum by micro- and macroorganisms.

Biogenic hydrocarbons are continuously being formed by terrestrial and marine plants in amounts that appear to be two to three orders of magnitude greater than those arising from petroleum sources. Therefore, an understanding of the nature of the hydrocarbons introduced into the marine environment through biosynthetic pathways and processes is also needed so we can discriminate between the contribution of hydrocarbons from biogenic sources and those from petroleum pollution.

The measurements of hydrocarbons in the various compartments of the marine environment should provide information on the routes and reservoirs involved in the transport and distribution of petroleum contaminants and a basic understanding of the biochemical and geochemical processes which govern the rates of diffusion and fluxes of petroleum hydrocarbons through the marine environment. Since there are a multitude of sources of hydrocarbons in arctic and subarctic marine environments, knowledge of the sources contributing various classes or types of hydrocarbons in a given sample, the relative contribution of each source, and the geochemical and biochemical processes altering the source composition is required. Such studies should be carried out in beach areas, in estuaries, and on the open ocean. These would be particularly pertinent to the arctic and subarctic marine environments where preliminary results indicate that the effects of petroleum contamination on marine organisms and the environment may be more severe and longer lasting than those in more temperate environments.

Special emphasis should be given to long-term and large-scale systematic monitoring of the seawater column, the sea surface, the marine atmosphere, and marine organisms. Although monitoring programs should be conducted over a period of time sufficiently long to gain an understanding of the magnitude of the natural variability of the system, some short-term, temporal studies involving restricted environments should be included to develop reliable models on the behavior of petroleum in the marine system.

Assessing the effects of petroleum on the marine environment necessarily depends upon reliable and accurate analytical procedures. Certainly, improved techniques for sampling water, wastewater, sediments, air, and organisms are needed. Standardized methodology must be developed, adopted, and disseminated to obtain accurate data and to provide a common basis for interpretation of the results.

We now realize and appreciate the nature and complexity of the problems relative to petroleum contamination of the arctic and subarctic marine environment. We also know what we must do to provide a basis for the assessment of the effects of such contamination. The most difficult part lies ahead: the collection and interpretation of the data, the development of the predictive concepts, and the formation of procedures to minimize the impact of petroleum pollution on the marine environment.

REFERENCES

1. Farrington, J.W. and P.A. Meyers (1975). Hydrocarbons in the marine environment. In: Environmental Chemistry (G. Elington, ed.), Vol. 1, p. 109-36. The Chemical Society, London.
2. National Academy of Sciences (1975). Petroleum in the Marine Environment, Washington, D.C., 107 p.
3. International Petroleum Encyclopedia (1975). Petroleum Publishing Co., Tulsa, Okla., 480 p.
4. Ryther, J.H. (1969). Photosynthesis and fish production in the sea. Science 166:72-6.
5. Blumer, M., P.C. Blokker, E.B. Cowell, and D.F. Duckworth (1972). Petroleum. In: A Guide to Marine Pollution (E.D. Goldberg, ed.), p. 19-40. Gordon and Breach, New York.
6. Cox, G.V. (1974). Marine Bioassays Workshop Proceedings 1974. Marine Technology Society, Washington, D.C., 308 p.

7. McAuliffe, C.D. (1976). Surveillance of the marine environment for hydrocarbons. Mar. Sci. Commun. 2(1):13-42.
8. Anderson, J.W., R.C. Clark, Jr., and J.J. Stegeman (1974). Petroleum hydrocarbons. In: Marine Bioassays Workshop Proceedings 1974 (G.V. Cox, ed.), p. 36-75. Marine Technology Society, Washington, D.C.
9. Blumer, M. (1971). Scientific aspects of the oil spill problem. Environ. Aff. 1:54-73.
10. Wardley-Smith, J. (1973). Occurrence, cause and avoidance of the spilling of oil by tankers. In: Proceedings of 1973 Joint Conference on Prevention and Control of Oil Spills, p. 15-20. American Petroleum Institute, Washington, D.C.
11. Nelson-Smith, A. (1973). Oil Pollution and Marine Ecology. Plenum Press, New York, 260 p.
12. Holdsworth, M.P. (1971). Oil pollution at sea. Symposium on Environmental Pollution. University of Lancaster, U.K.
13. Report of the Study of Critical Environmental Probe (SCEP) (1970). Man's Impact on the Global Environment. Assessment and Recommendations for Action. Massachusetts Institute of Technology Press, Cambridge, Mass., 319 p.
14. Washington State Department of Ecology (1973). A report of oil pollution prevention and control, Olympia, Wash., 6 p.
15. Wakeham, S.G. (1976). The geochemistry of hydrocarbons in Lake Washington. Ph.D. Thesis, University of Washington, Seattle, 192 p.
16. Gilmore, G.A., D.D. Smith, A.H. Rice, E.H. Senton, and W.H. Moser (1970). Systems study of oil spill cleanup procedures. Vol. I: Analysis of oil spills and control materials. Am. Petrol. Inst. Publ. 4024, p. 7-51.
17. Keith, V.F. and J.D. Porricelli (1973). An analysis of oil outflows due to tanker accidents. In: Proceedings of 1973 Joint Conference on Prevention and Control of Oil Spills, p. 3-14. American Petroleum Institute, Washington, D.C.
18. Oceanographic Commission of Washington (1974). Offshore petroleum transfer systems for Washington State. Pac. Northwest Sea 7(3/4):3-23.
19. Brummage, K.G. (1973). The sources of oil entering the sea. In: Background Papers: Inputs, Fates, and Effects of Petroleum in the Marine Environment, Workshop, p. 1-6. National Academy of Sciences, Washington, D.C.
20. Weeks, L.G. (1965). World offshore petroleum resources. Bull. Am. Assoc. Pet. Geol. 49:1680-93.

21. Wilson, R.D. (1973). Estimate of annual input of petroleum to the marine environment from natural marine seepage. In: Background Papers: Inputs, Fates, and Effects of Petroleum in the Marine Environment, Workshop, p. 59-96. National Academy of Sciences, Washington, D.C.
22. Blumer, M. (1972). Submarine seeps: Are they a major source of open ocean oil pollution? Science 176:1257-8.
23. Fay, J.A. (1969). The spread of oil slicks on a calm sea. In: Oil on the Sea (D.P. Hoult, ed.), p. 53-63. Plenum Press, New York.
24. McAuliffe, C.D. (1976). Personal communication. Chevron Oil Field Research Co., La Habra, Calif.
25. American Petroleum Institute (1963). Manual on Disposal of Refinery Wastes. 1. Waste Water Containing Oil, 7th Ed. American Petroleum Institute, New York.
26. Jeffery, P.G. (1973). Large-scale experiments on the spreading of oil at sea and its disappearance by natural factors. In: Proceedings of 1973 Joint Conference on Prevention and Control of Oil Spills, p. 469-74. American Petroleum Institute, Washington, D.C.
27. Smith, J.E. (1968). *Torrey Canyon* Pollution and Marine Life. Cambridge University Press, U.K., 196 p.
28. Garrett, W.D. (1973). The surface activity of petroleum and its influence on the behavior of oil at sea. In: Background Papers: Inputs, Fates, and Effects of Petroleum in the Marine Environment, Workshop, p. 451-61. National Academy of Sciences, Washington, D.C.
29. Fay, J.A. (1971). Physical processes in the spread of oil on a water surface. In: Proceedings of 1971 Joint Conference on Prevention and Control of Oil Spills, p. 463-7. American Petroleum Institute, Washington, D.C.
30. Johnson, B.H. and T. Aczel (1967). Analysis of complex mixtures of aromatic compounds by high-resolution mass spectrometry at low-ionizing voltages. Anal. Chem. 39: 682-5.
31. Dudley, G. (1968). The problem of oil pollution in a major oil port. In: The Biological Effects of Oil Pollution on Littoral Communities (J.D. Carthy and D.R. Arthur, eds.), Suppl. to Field Studies, Vol. 2, p. 21-9. Obtainable from E.W. Classey, Ltd., Hampton, Middx., England.
32. Dean, R.A. (1968). The chemistry of crude oils in relation to their spillage on the sea. In: The Biological Effects of Oil Pollution on Littoral Communities (J.D. Carthy and D.R. Arthur, eds.), Suppl. to Field Studies, Vol. 2, p. 1-6. Obtainable from E.W. Classey, Ltd., Hampton, Middx., England.

33. McMinn, T.J. and P. Golden (1973). Behavioral characteristics and cleanup technique of North Slope crude oil in arctic winter environment. In: Proceedings of 1973 Joint Conference on Prevention and Control of Oil Spills, p. 263-76. American Petroleum Institute, Washington, D.C.
34. Kreider, R.E. (1971). Identification of oil leaks and spills. In: Proceedings of 1971 Joint Conference on Prevention and Control of Oil Spills, p. 119-24. American Petroleum Institute, Washington, D.C.
35. Smith, C.L. and W.G. MacIntyre (1971). Initial aging of fuel oil films of sea water. In: Proceedings of 1971 Joint Conference on Prevention and Control of Oil Spills, p. 457-61. American Petroleum Institute, Washington, D.C.
36. Kinney, P.J., D.K. Button, and D.M. Schell (1969). Kinetics of dissipation and biodegradation of crude oil in Alaska's Cook Inlet. In: Proceedings of 1969 Joint Conference on Prevention and Control of Oil Spills, p. 333-40. American Petroleum Institute, Washington, D.C.
37. Sivadier, H.O. and P.G. Mikolaj (1973). Measurement of evaporation rates from oil slicks on the open sea. In: Proceedings of 1973 Joint Conference on Prevention and Control of Oil Spills, p. 475-84. American Petroleum Institute, Washington, D.C.
38. Harrison, W., M.A. Winnik, P.T.Y. Kwong, and D. Mackay (1975). Crude oil spills. Disappearance of aromatic and aliphatic components from small sea-surface slicks. Environ. Sci. Technol. 9:231-4.
39. Duce, R.A., J.G. Quinn, and T.L. Wade (1974). Residence time for non-methane hydrocarbons in the atmosphere. Mar. Pollut. Bull. 5:59-61.
40. Ehrhardt, M. and M. Blumer (1972). The source identification of marine hydrocarbons by gas chromatography. Environ. Pollut. 3:179-94.
41. Doss, M.P. (1943). Physical Constants of the Principal Hydrocarbons. The Texas Co., New York, 215 p.
42. Handbook of Chemistry and Physics (1967). Table of physical properties of organic compounds, p. C53-C601. Chemical Rubber Co., Cleveland, Ohio.
43. Wasik, S.P. and R.N. Boyd (1974). Determination of aromatic hydrocarbons in sea water using an electrolytic stripped cell. In: Marine Pollution Monitoring (Petroleum). Natl. Bur. Stand. Spec. Publ. 409, p. 117-8.
44. Baier, R.E. (1972). Organic films on natural waters: Their retrieval, identification, and modes of elimination. J. Geophys. Res. 77:5062-75.

45. Duce, R.A., J.G. Quinn, C.E. Olney, S.R. Piotrowicz, B.J. Ray, and T.L. Wade (1972). Enrichment of heavy metals and organic compounds in the surface microlayer of Narragansett Bay, Rhode Island. Science 176:161-3.
46. Wade, T.L. and J.G. Quinn (1975). Hydrocarbons in the Sargasso Sea surface microlayer. Mar. Pollut. Bull. 6:54-7.
47. McAuliffe, C. (1966). Solubility in water of paraffin, cycloparaffin, olefin, acetylene, cycloolefin, and aromatic hydrocarbons. J. Phys. Chem. 70:1267-75.
48. McAuliffe, C.D. (1969). Solubility in water of normal C_9 and C_{10} alkane hydrocarbons. Science 163:478-9.
49. Sutton, C. and J.A. Calder (1974). Solubility of higher-molecular-weight *n*-paraffins in distilled water and seawater. Environ. Sci. Technol. 8:654-7.
50. Parker, C.A., M. Freegarde, and C.G. Hatchard (1971). The effect of some chemical and biological factors on the degradation of crude oil at sea. In: Water Pollution by Oil (P. Hepple, ed.), p. 237-44. Institute of Petroleum, London.
51. Eganhouse, R.P. and J.A. Calder (1976). The solubility of medium molecular weight aromatic hydrocarbons and the effects of hydrocarbon co-solutes and salinity. Geochim. Cosmochim. Acta 40:555-61.
52. Boehm, P.D. and J.G. Quinn (1973). Solubilization of hydrocarbons by the dissolved organic matter in sea water. Geochim. Cosmochim. Acta 37:2459-77.
53. Boehm, P.D. and J.G. Quinn (1975). Correspondence to the Editor. Environ. Sci. Technol. 9:365.
54. Sutton, C. and J.A. Calder (1975). Reply to Correspondence to the Editor. Environ. Sci. Technol. 9:365-6.
55. McAuliffe, C.D. (1969). Determination of dissolved hydrocarbons in subsurface brines. Chem. Geol. 4:225-33.
56. Boylan, D.B. and B.W. Tripp (1971). Determination of hydrocarbons in seawater extracts of crude oil and crude oil fractions. Nature 230:44-7.
57. Frankenfeld, J.W. (1973). Factors governing the fate of oil at sea; variations in the amounts and types of dissolved or dispersed materials during the weathering process. In: Proceedings of 1973 Joint Conference on Prevention and Control of Oil Spills, p. 485-95. American Petroleum Institute, Washington, D.C.
58. Lysyj, I. and E.C. Russell (1974). Dissolution of petroleum-derived products in water. Water Res. 8:863-8.
59. Burwood, R. and G.C. Speers (1974). Photo-oxidation as a factor in the environmental dispersal of crude oil. Estuarine Coastal Mar. Sci. 2:117-35.

60. Anderson, J.W., J.M. Neff, B.A. Cox, H.E. Tatem, and G.H. Hightower (1974). Characteristics of dispersions and water-soluble extracts of crude and refined oils and their toxicity to estuarine crustaceans and fish. Mar. Biol. (Berl.) 27:75-88.
61. Roubal, W.T., D.H. Bovee, T.K. Collier, and S.T. Stranahan (1977). Flow through system for chronic exposure of aquatic organisms to seawater-soluble hydrocarbons from crude oil: Construction and applications. In: Proceedings of 1977 Oil Spill Conference. In press. American Petroleum Institute, Washington, D.C.
62. Gordon, D.C., Jr., P.D. Keizer, and N.J. Prouse (1973). Laboratory studies on the accommodation of some crude and residual fuel oils in sea water. J. Fish. Res. Board Can. 30:1611-8.
63. McAuliffe, C.D., A.E. Smalley, R.D. Groover, W.M. Welsh, W.S. Pickle, and G.E. Jones (1975). Chevron Main Pass block 41 oil spill: Chemical and biological investigations. In: Proceedings of 1975 Conference on Prevention and Control of Oil Pollution, p. 555-66. American Petroleum Institute, Washington, D.C.
64. Milne, D. (1950). Character of waste oil emulsions. Sewage Ind. Wastes 22:326-30.
65. Forrester, W.D. (1971). Distribution of suspended oil particles following the grounding of the tanker *Arrow*. J. Mar. Res. 29:151-70.
66. Wicks, M., III. (1969). Fluid dynamics of floating oil containment by mechanical barriers in the presence of water currents. In: Proceedings of 1969 Joint Conference on Prevention and Control of Oil Spills, p. 55-106. American Petroleum Institute, Washington, D.C.
67. Keizer, P.D. and D.C. Gordon, Jr. (1973). Detection of trace amounts of oil in sea water by fluorescence spectroscopy. J. Fish. Res. Board Can. 30:1039-46.
68. Alpine Geophysical Associates, Inc. (1971). Oil pollution incident Platform Charlie, Main Pass block 41 field Louisiana, Proj. 15080 FTU. Water Pollution Control Research Series. Environmental Protection Agency, Washington, D.C., 134 p.
69. Dodd, E.N. (1971). The effects of natural factors on the movement, dispersal, and destruction of oil at sea. Cited in: National Academy of Sciences (1975). Petroleum in the Marine Environment, p. 47. Washington, D.C. Obtainable from National Technical Information Service, U.S. Dep. of Commerce, Springfield, Va., AD 763042.
70. Davis, S.J. and C.F. Gibbs (1975). The effect of weathering on a crude oil residue exposed at sea. Water Res. 9:275-85.

71. Zajic, J.E. and B. Supplisson (1972). Emulsification and degradation of Bunker C fuel oil by microorganisms. Biotechnol. Bioeng. 14:331-43.
72. Morris, R.J. and F. Culkin (1974). Lipid chemistry of eastern Mediterranean surface layers. Nature 250:640-2.
73. Morris, B.F. and J.N. Butler (1973). Petroleum residues in the Sargasso Sea and on Bermuda beaches. In: Proceedings of 1973 Joint Conference on Prevention and Control of Oil Spills, p. 521-9. American Petroleum Institute, Washington, D.C.
74. Stehr, E. (1967). Über Ölverschmutzung durch Tankerunfäller auf hoher See. Gas. Wasserfach. 108:53-4. Cited in: Nelson-Smith, A. (1973), Oil Pollution and Marine Ecology, p. 83. Plenum Press, New York.
75. Dennis, J.V. (1959). Oil pollution survey of the United States Atlantic Coast with special reference to southeast Florida Coast conditions. Am. Petrol. Inst. Publ. 4054, 81 p.
76. Conomos, T.J. (1975). Movement of spilled oil as predicted by estuarine nontidal drift. Limnol. Oceanogr. 20:159-73.
77. Meyers, P.A. and J.G. Quinn (1973). Association of hydrocarbons and mineral particles in saline solutions. Nature 244:23-4.
78. Suess, E. (1968). Calcium carbonate interactions with organic compounds. Ph.D. Thesis, Lehigh University, Bethlehem, Pa.
79. Kolpack, R.L., J.S. Mattson, H.B. Mark, Jr., and T-C. Yu (1971). Hydrocarbon content of Santa Barbara Channel sediments. In: Biological and Oceanographical Survey of the Santa Barbara Channel Oil Spill 1969-1970, Vol. II, p. 276-95. Allan Hancock Foundation, University of Southern California, Los Angeles.
80. Blumer, M. and J. Sass (1972). Oil pollution: Persistence and degradation of spilled fuel oil. Science 176: 1120-2.
81. Blumer, M. and J. Sass (1972). Indigenous and petroleum-derived hydrocarbons in a polluted sediment. Mar. Pollut. Bull. 3:92-3.
82. Shaw, D.G., A.J. Paul, L.M. Cheek, and H.M. Feder (1976). *Macoma balthica*: An indicator of oil pollution. Mar. Pollut. Bull. 7:29-31.
83. Butler, J.N., B.F. Morris, and J. Sass (1973). Pelagic tar from Bermuda and the Sargasso Sea. Bermuda Biol. Stn. Spec. Publ. 10, 346 p.

84. Clark, R.C., Jr., J.S. Finley, B.G. Patten, and E.E. DeNike (1975). Long-term chemical and biological effects of a persistent oil spill following the grounding of the *General M.C. Meigs*. In: Proceedings of 1975 Conference on Prevention and Control of Oil Pollution, p. 479-87. American Petroleum Institute, Washington, D.C.
85. Guard, H.E. and A.B. Cobet (1973). The fate of a bunker fuel in beach sand. In: Proceedings of 1973 Joint Conference on Prevention and Control of Oil Spills, p. 827-34. American Petroleum Institute, Washington, D.C.
86. Ludwig, H.F. and R. Carter (1961). Analytical characteristics of oil-tar materials on southern California beaches. J. Water Pollut. Control Fed. 33:1123-39.
87. Blumer, M., M. Ehrhardt, and J.H. Jones (1973). The environmental fate of stranded crude oil. Deep-Sea Res. 20:239-59.
88. Gebelein, C.D. (1971). Sedimentology and ecology of a carbonate facies mosaic. Ph.D. Thesis, Brown University, Providence, R.I.
89. Rashid, M.A. (1974). Degradation of Bunker C oil under different coastal environments in Chedabucto Bay, Nova Scotia. Estuarine Coastal Mar. Sci. 2:137-44.
90. Spooner, M. (1969). Some ecological effects of marine oil pollution. In: Proceedings of 1969 Joint Conference on Prevention and Control of Oil Spills, p. 313-6. American Petroleum Institute, Washington, D.C.
91. Clark, R.C., Jr., J.S. Finley, B.G. Patten, D.F. Stefani, and E.E. DeNike (1973). Interagency investigations of a persistent oil spill on the Washington Coast: Animal populations studies, hydrocarbon uptake by marine organisms, and algal response following the grounding of the troopship *General M.C. Meigs*. In: Proceedings of 1973 Joint Conference on Prevention and Control of Oil Spills, p. 793-808. American Petroleum Institute, Washington, D.C.
92. Green, D.R., C. Bawden, W.J. Cretney, and C.S. Wong (1974). The Alert Bay oil spill: A one-year study of the recovery of a contaminated bay. Pac. Mar. Sci. Rep. 74-9, Environment Canada, Victoria, B.C., 42 p.
93. Mayo, D.W., D.J. Donovan, and L. Jiang (1974). Long term weathering characteristics of Iranian crude oil: The wreck of the *Northern Gulf*. In: Marine Pollution Monitoring (Petroleum). Natl. Bur. Stand. Spec. Publ. 409, p. 201-8.
94. Kolpack, R.L. (1969). Santa Barbara oil pollution project progress report; marine geology. Mar. Pollut. Bull. 1(18):5-8. (Old Series).

95. Vandermeulen, J.H. and D.C. Gordon, Jr. (1976). Reentry of 5-year-old stranded Bunker C fuel oil from a low-energy beach into the water, sediment, and biota of Chedabucto Bay, Nova Scotia. J. Fish. Res. Board Can. 33:2002-10.
96. Vandermeulen, J.H., P.D. Keizer, and T. Ahern (1976). Compositional changes in beach sediment-bound *Arrow* Bunker C: 1970-1976. Unpublished manuscript. Fisheries Improvement Committee. Int. Counc. Explor. Sea CM 1976/ E:51, 13 p.
97. Ahearn, D.G. and S.P. Meyers (1973). The Microbial Degradation of Oil Pollutants. Publ. No. LSU-SG-73-01. Center for Wetland Resources, Louisiana State University, Baton Rouge, La., 322 p.
98. Cundell, A.M. and R.W. Traxler (1973). Microbial degradation of petroleum at low temperature. Mar. Pollut. Bull. 4:125-7.
99. Walker, J.D., L. Cofone, Jr., and J.J. Cooney (1973). Microbial petroleum degradation: The role of *Cladosporium resinae*. In: Proceedings of 1973 Joint Conference on Prevention and Control of Oil Spills, p. 821-5. American Petroleum Institute, Washington, D.C.
100. Walker, J.D., R.R. Colwell, and L. Petrakis (1975). A study of the biodegradation of a South Louisiana crude oil employing computerized mass spectrometry. In: Proceedings of 1975 Conference on Prevention and Control of Oil Pollution, p. 601-5. American Petroleum Institute, Washington, D.C.
101. ZoBell, C.E. (1973). Bacterial degradation of mineral oils at low temperatures. In: The Microbial Degradation of Oil Pollutants (D.G. Ahearn and S.P. Meyers, eds.), p. 153-61. Publ. No. LSU-SG-73-01. Center for Wetland Resources, Louisiana State University, Baton Rouge, La.
102. Floodgate, G.D. (1973). A threnody concerning the biodegradation of oil in natural waters. In: The Microbial Degradation of Oil Pollutants (D.G. Ahearn and S.P. Meyers, eds.), p. 17-24. Publ. No. LSU-SG-73-01. Center for Wetland Resources, Louisiana State University, Baton Rouge, La.
103. ZoBell, C.E. (1969). Microbial modification of crude oil in the sea. In: Proceedings of 1969 Joint Conference on Prevention and Control of Oil Spills, p. 317-26. American Petroleum Institute, Washington, D.C.
104. Davis, J.B. (1968). Paraffinic hydrocarbons in the sulfate-reducing bacterium *Desulfovibrio desulfuricans*. Chem. Geol. 3:155-60.

105. ZoBell, C.E. (1962). Importance of microorganisms in the sea. In: Proceedings Low Temperature Microbiology Symposium, 1961, p. 107-32. Campbell Soup Co., Camden, N.J.
106. ZoBell, C.E. (1971). Sources and biodegradation of carcinogenic hydrocarbons. In: Proceedings of 1971 Joint Conference on Prevention and Control of Oil Spills, p. 441-51. American Petroleum Institute, Washington, D.C.
107. Jannasch, H.W., K. Eimhjellen, C.O. Wirsen, and A. Farmanfarmaian (1971). Microbial degradation of organic matter in the deep sea. Science 171:672-5.
108. Jannasch, H.W. and C.O. Wirsen (1973). Deep-sea microorganisms: *In situ* response to nutrient enrichment. Science 180:641-3.
109. Scott, G. (1965). Atmospheric Oxidation and Antioxidants. Elsevier Publishing Co., Amsterdam, 528 p.
110. Monaghan, P.H. and C.B. Koons (1973). Petroleum in the marine environment: Gulf of Mexico. Gulf Coast Association of Geological Societies Proceedings, Houston, Texas. Cited in: Petroleum in the Marine Environment (1975), p. 49. National Academy of Sciences, Washington, D.C.
111. Freegarde, M. and C.G. Hatchett (1970). The ultimate fate of crude oil at sea. Interim Report. Admiralty Materials Laboratory, U.K. Cited in: Petroleum in the Marine Environment (1975), p. 48. National Academy of Sciences, Washington, D.C.
112. Brunnock, J.V., D.F. Duckworth, and G.G. Stephens (1968). Analysis of beach pollutants. J. Inst. Petrol. 54:310-25.
113. Klein, A.E. and N. Pilpel (1974). The effects of artificial sunlight upon floating oils. Water Res. 8:79-83.
114. Lacaze, J.C. and O. Villedon de Naïde (1976). Influence of illumination on phytotoxicity of crude oil. Mar. Pollut. Bull. 7:73-6.
115. Conover, R.J. (1971). Some relations between zooplankton and Bunker C oil in Chedabucto Bay following the wreck of the tanker *Arrow*. J. Fish. Res. Board Can. 28:1327-30.
116. Parker, C.A. (1970). The ultimate fate of crude oil at sea: Uptake of oil by zooplankton. Cited in: Water Pollution by Oil (P. Hepple, ed) (1971), p. 242. Institute of Petroleum, London.
117. Alyakrinskaya, I.O. (1966). Behavior and filtering ability of the Black Sea *Mytilus galloprovincialis* on oil polluted water. Zool. Zh. 45:998-1003. Cited in: Biol. Abstr. 48(14):6494; citation 72515.

118. Blumer, M., G. Souza, and J. Sass (1970). Hydrocarbon pollution of edible shellfish by an oil spill. Mar. Biol. (Berl.) 5:195-202.
119. Farrington, J.W. and J.G. Quinn (1973). Petroleum hydrocarbons in Narragansett Bay. I. Survey of hydrocarbons in sediments and clams (*Mercenaria mercenaria*). Estuarine Coastal Mar. Sci. 1:71-9.
120. Clark, R.C., Jr. and J.S. Finley (1974). Paraffin hydrocarbon patterns in petroleum-polluted mussels. Mar. Pollut. Bull. 4:172-6.
121. Clark, R.C., Jr. and J.S. Finley (1975). Uptake and loss of petroleum hydrocarbons by the mussel, *Mytilus edulis*, in laboratory experiments. Fish. Bull. 73:508-15.
122. Ehrhardt, M. (1972). Petroleum hydrocarbons in oysters from Galveston Bay. Environ. Pollut. 3:257-71.
123. Levy, E.M. (1972). Evidence for the recovery of the waters off the east coast of Nova Scotia from the effects of a major oil spill. Water, Air, Soil Pollut. 1:144-8.
124. Attaway, D., J.R. Jadamec, and W. McGowan (1973). Rust in floating petroleum found in the marine environment. Unpublished manuscript, U.S. Coast Guard. Cited in: Figs. 7 and 8 of Butler, J.N., B.F. Morris, and J. Sass (1973). Pelagic tar from Bermuda and the Sargasso Sea. Bermuda Biol. Stn. Spec. Publ. 10, p. 23.
125. Feldman, M.H. and D.E. Cawlfield (1974). Marine environmental monitoring: Trace elements in persistent tar ball oil residues. In: Marine Pollution Monitoring (Petroleum). Natl. Bur. Stand. Spec. Publ. 409, p. 237-41.
126. Allen, A.A., R.S. Schlueter, and P.G. Mikolaj (1970). Natural oil seepage at Coal Oil Point, Santa Barbara, California. Science 170:974-7.
127. Butler, J.N. (1975). Evaporative weathering of petroleum residues: The age of pelagic tar. Mar. Chem. 3:9-21.
128. Snow, N.B. and B.F. Scott (1975). The effect and fate of crude oil spilt on two arctic lakes. In: Proceedings of 1975 Conference on Prevention and Control of Oil Pollution, p. 527-34. American Petroleum Institute, Washington, D.C.
129. Glaeser, J.L. (1971). A discussion of the future oil spill problem in the arctic. In: Proceedings of 1971 Joint Conference on Prevention and Control of Oil Spills, p. 479-84. American Petroleum Institute, Washington, D.C.
130. Deslauriers, P.C. (1975). Oil pollution in ice infested waters (A survey of recent development). Unpublished manuscript. 9104 Red Branch Road, Columbia, MD 21045, 29 p.

131. Campbell, W.J. and S. Martin (1973). Oil and ice in the Arctic Ocean: Possible large-scale interactions. Science 181:56-8.
132. Martin, S. and W.J. Campbell (1974). Oil spills in the Arctic Ocean: Extent of spreading and possibility of large-scale thermal effects (response). Science 186: 845-6.
133. Ayers, R.C., Jr., H.O. Jahns, and J.L. Glaeser (1974). Oil spills in the Arctic Ocean: Extent of spreading and possibility of large-scale thermal effects. Science 186: 843-5.
134. NORCOR Engineering Research Ltd. (1975). The interaction of crude oil with Arctic Sea ice. Beaufort Sea Tech. Rept. 27, Environment Canada, Victoria, B.C., 206 p.
135. Barber, F.G. (1971). An oiled Arctic shore. Arctic 24:229.
136. Barber, F.G. (1971). Oil spilled with ice: Some qualitative aspects. In: Proceedings of 1971 Joint Conference on Prevention and Control of Oil Spills, p. 133-7. American Petroleum Institute, Washington, D.C.
137. Glaeser, J.L. and G.P. Vance (1971). A study of the behavior of oil spills in the arctic. AD717142. National Technical Information Service, U.S. Dep. of Commerce, Springfield, Va., 60 p.
138. Martin, S., P. Kauffman, and P.E. Welander (1976). A laboratory study of the dispersion of crude oil within sea ice grown in a wave field. Unpublished Report. University of Washington, Dep. Oceanogr. Spec. Rep. 69, 34 p.
139. Topham, D.R. (1975). Hydrodynamics of an oil well blowout. Beaufort Sea Tech. Rept. 33, Environment Canada, Victoria, B.C., 52 p.
140. Outer Continental Shelf Environmental Assessment Program (1976). Arctic Project Office, Geophysical Institute, University of Alaska, Fairbanks. Arct. Proj. Bull. 9, p. 32.
141. Keevil, B.E. and R.O. Ramseier (1975). Behavior of oil spilled under floating ice. In: Proceedings of 1975 Joint Conference on Prevention and Control of Oil Pollution, p. 497-501. American Petroleum Institute, Washington, D.C.
142. Barber, F.G. (1970). Report of the Task Force: Operation Oil (Cleanup of *Arrow* Oil Spill in Chedabucto Bay). Vol. 3, p. 35-54. Ministry of Transport, Ottawa. Cited in: Proceedings of 1975 Conference on Prevention and Control of Oil Pollution, p. 500. American Petroleum Institute, Washington, D.C.

143. Wolfe, L.S. and D.P. Hoult (1973). Oil and ice. Tech. Rev. 75(6):45-6.
144. Wolfe, L.S. and D.P. Hoult (1974). Hopeful findings on oil and ice. Tech. Rev. 76(4):73-4.
145. Atlas, R.M. (1973). Fate and effects of oil pollutants in extremely cold marine environments. AD 769895. National Technical Information Service, U.S. Dep. of Commerce, Springfield, Va., p. 13. Cited in: Deslauriers, P.C. (1975). Oil pollution in ice infested waters (A survey of recent development), p. 17. Unpublished manuscript. 9104 Red Branch Road, Columbia, Md.
146. Mackay, D., P.J. Leinonen, J.C.K. Overall, and B.R. Wood (1975). The behaviour of crude oil spilled on snow. Arctic 28:9-20.
147. Von Arx, W.S. (1962). An Introduction to Physical Oceanography. Addison-Wesley Publishing Co., Reading, Mass., 422 p.
148. Mackie, P.R., K.J. Whittle, and R. Hardy (1974). Hydrocarbons in the marine environment. I. *n*-Alkanes in the Firth of Clyde. Estuarine Coastal Mar. Sci. 2:359-74.
149. Oró, J., T.G. Tornabene, D.W. Nooner, and E. Gelpi (1967). Aliphatic hydrocarbons and fatty acids of some marine and freshwater microorganisms. J. Bacteriol. 93:1811-8.
150. Clark, R.C., Jr. and M. Blumer (1967). Distribution of *n*-paraffins in marine organisms and sediment. Limnol. Oceanogr. 12:79-87.
151. Johnson, R.W. and J.A. Calder (1973). Early diagensis of fatty acids and hydrocarbons in a salt marsh environment. Geochim. Cosmochim. Acta 37:1943-55.
152. Clark, R.C., Jr. (1976). Puget Sound Hydrocarbon Studies. Unpublished data. Northwest and Alaska Fisheries Center, NMFS, NOAA, U.S. Dep. of Commerce, Seattle, Wash.
153. Youngblood, W.W., M. Blumer, R.L. Guillard, and F. Fiore (1971). Saturated and unsaturated hydrocarbons in marine benthic algae. Mar. Biol. (Berl.) 8:190-201.
154. Burns, K.A. and J.M. Teal (1973). Hydrocarbons in the pelagic *Sargassum* community. Deep-Sea Res. 20:207-11.
155. Hunter, L., H.E. Guard, and L.H. DiSalvo (1974). Determination of hydrocarbons in marine organisms and sediments by thin layer chromatography. In: Marine Pollution Monitoring (Petroleum). Natl. Bur. Stand. Spec. Publ. 409, p. 213-6.
156. Whittle, K.J., P.R. Mackie, R. Hardy, A.D. McIntyre, and R.A.A. Blackman (1975). UK area hydrocarbon baseline survey: Main findings, preliminary conclusions and implications for future survey and monitoring programmes, Unpublished manuscript. Fisheries Improvement Committee Int. Counc. Explor. Sea CM 1975/E:28, 13 p.

157. Wong, C.S., W.J. Cretney, P. Christensen, and R.W. Macdonald (1976). Hydrocarbon levels in the marine environment of the southern Beaufort Sea. Beaufort Sea Tech. Rept. 38. Unpublished manuscript. Environment Canada, Victoria, B.C., 113 p.
158. Farrington, J.W. (1972). A study program to identify problems related to oceanic environmental quality. Summary of intercalibration measurements and analysis of open ocean organisms for recent biosynthesized hydrocarbons and petroleum hydrocarbons. In: Baseline Studies of Pollutants in the Marine Environment. Background Papers, Workshop, 24-26 May, 1972, p. 583-631. Brookhaven National Laboratories, New York.
159. Bowen, V.T. (1971). A study program to identify problems related to oceanic environmental quality. Progress Report to NSF-IDOE, GX-25334. National Science Foundation, Washington, D.C., 30 p. Cited in: National Academy of Sciences (1975). Petroleum in the Marine Environment, p. 62. Washington, D.C.
160. Tatem, H.E. (1975). The toxicity and physiological effects of oil and petroleum hydrocarbons on estuarine grass shrimp *Palaemonetes pugio* Holthuis. Ph.D. Thesis, Texas A & M University, College Station, Tex.
161. Anderson, J.W. (1974). Unpublished data. Cited in: Table VI of Anderson, J.W., R.C. Clark, Jr. and J.J. Stegeman (1974). Petroleum hydrocarbons. In: Marine Bioassays Workshop, 1974 (G.V. Cox, ed.), p. 51-6. Marine Technology Society, Washington, D.C. as "Cox and Anderson, unpublished results, 1974".
162. Battelle Pacific Northwest Laboratories (1974). Study of effects of oil discharges and domestic and industrial wastewaters on the fisheries of Lake Maracaibo, Venezuela, Vol. II. Fate and effects of oil. Research Report, Contract 212B00899. Creole Petroleum Corp., Richland, Wash., 192 p.
163. Templeton, W.L., E.A. Sutton, R.M. Bean, R.C. Arnett, J.W. Blaylock, R.E. Wildung, and H.J. Moore (1975). Oil pollution studies on Lake Maracaibo, Venezuela. In: Proceedings of 1975 Conference on Prevention and Control of Oil Pollution, p. 489-96. American Petroleum Institute, Washington, D.C.
164. Giam, C.S., H.S. Chan, and G.S. Neff (1976). Distribution of *n*-parrafins [paraffins] in selected marine benthic organisms. Bull. Environ. Contam. Toxicol. 16: 37-43.
165. Scarratt, D.J. and V. Zitko (1972). Bunker C oil in sediments and benthic animals from shallow depths in Chedabucto Bay, N.S. J. Fish. Res. Board Can. 29: 1347-50.

166. Zitko, V. (1971). Determination of residual fuel oil contamination of aquatic animals. Bull. Environ. Contam. Toxicol. 5:559-64.
167. MacLeod, W.D., D.W. Brown, R.G. Jenkins, L.S. Ramos, and V.D. Henry (1976). A pilot study on the design of a petroleum hydrocarbon baseline investigation for Northern Puget Sound and Strait of Juan de Fuca. NOAA Tech. Memo. ERL MESA-8. Unpublished manuscript. National Analytical Facility, NMFS, NOAA, U.S. Dep. of Commerce, Seattle, Wash., 54 p.
168. Clark, R.C., Jr., J.S. Finley, and G.G. Gibson (1974). Acute effects of outboard motor effluent on two marine shellfish. Environ. Sci. Technol. 8:1009-14.
169. Kinney, P.J. (1973). Baseline hydrocarbon concentrations. In: Environmental Studies of Port Valdez (D.W. Hood, W.E. Shiels, and E.J. Kelley, eds.), p. 397-410. University of Alaska Inst. Mar. Sci. Occas. Publ. 3.
170. Fossato, V.U. and E. Siviero (1974). Oil pollution monitoring in the lagoon of Venice using the mussel *Mytilus galloprovincialis*. Mar. Biol. (Berl.) 25:1-6.
171. Dunn, B.P. and H.F. Stich (1975). The use of mussels in estimating benzo[a]pyrene contamination of the marine environment. Proc. Soc. Exp. Biol. Med. 150:49-51.
172. DiSalvo, L.H., H.E. Guard, and L. Hunter (1975). Tissue hydrocarbon burden of mussels as a potential monitor of environmental hydrocarbon insult. Environ. Sci. Technol. 9:247-51.
173. Meiggs, T.O. (1974). Unpublished results. Cited in: Table VI of Anderson, J.W., R.C. Clark, Jr., and J.J. Stegemen (1974). Petroleum hydrocarbons. In: Marine Bioassays Workshop Proceedings 1974 (G.V. Cox, ed.), p. 36-75. Marine Technology Society, Washington, D.C.
174. Stegemen, J.J. and J.M. Teal (1973). Accumulation, release and retention of petroleum hydrocarbons by the oyster *Crassostrea virginica*. Mar. Biol. (Berl.) 22: 37-44.
175. Anderson, R.D. (1973). Effects of petroleum hydrocarbons on the physiology of the American oyster, *Crassostrea virginica* Gmelin. Ph.D. Thesis, Texas A & M University, College Station, Tex., 145 p.
176. Farrington, J.W. and G.C. Medeiros (1975). Evaluation of some methods of analysis for petroleum hydrocarbons in marine organisms. In: Proceedings of 1975 Conference on Prevention and Control of Oil Pollution, p. 115-21. American Petroleum Institute, Washington, D.C.
177. Blumer, M., J. Sass, G. Souza, H.L. Sanders, J.F. Grassle, and G.R. Hampson (1970). The West Falmouth oil spill. Unpublished manuscript. Woods Hole Oceanogr. Inst. Tech. Rep. 70-44, 32 p.

178. Parker, P.L., J.K. Winters, and J. Morgan (1972). A base-line study of petroleum in the Gulf of Mexico. In: Baseline studies of pollutants in the marine environment. Background Papers, Workshop. Brookhaven National Laboratories, New York, p. 555-81.
179. Hardy, R., P.R. Mackie, K.J. Whittle, and A.D. McIntyre (1974). Discrimination in the assimilation of *n*-alkanes in fish. Nature 252:577-8.
180. Teal, J.M. and J.W. Farrington (1976). A comparison of hydrocarbons in animals and their benthic habitats. Int. Counc. Explor. Sea, Rapports et Proces Verbaux, London, In press.
181. Ackman, R.G. and D. Noble (1973). Steam distillation: A simple technique for recovery of petroleum hydrocarbons from tainted fish. J. Fish. Res. Board Can. 30:711-4.
182. Burns, K.A. and J.M. Teal (1971). Hydrocarbon incorporation into the salt marsh ecosystem from the West Falmouth oil spill. Unpublished manuscript. Woods Hole Oceanogr. Inst. Tech. Rep. 71-69, 14 p.
183. Mallet, L. and J. Sardou (1964). Recherche de la Presence de l'Hydrocarbure Polybenzenique Benzo-3,4 Pyrene dans le Milieu Planctonique de la Region de la Baie de Villefranche (Alpes-Maritimes). Comp. Rend. 258:5264-67.
184. Zitko, V. (1975). Aromatic hydrocarbons in aquatic fauna. Bull. Environ. Contam. Toxicol. 14:621-31.
185. Ehrhardt, M. and J. Heinemann (1974). Hydrocarbons in blue mussels from the Kiel Bight. In: Marine Pollution Monitoring (Petroleum). Natl. Bur. Stand. Spec. Publ. 409, p. 221-5.
186. Stegeman, J.J. (1974). Hydrocarbons in shellfish chronically exposed to low levels of fuel oil. In: Pollution and Physiology of Marine Organisms (F.J. Vernberg and W.B. Vernberg, eds.), p. 329-47. Academic Press, New York.
187. Cahnmann, H.J. and M. Kuratsune (1957). Determination of polycyclic aromatic hydrocarbons in oysters collected in polluted water. Anal. Chem. 29:1312-7.
188. Shipton, J., J.H. Last, K.E. Murray, and G.L. Vale (1970). Studies on a kerosene-like taint in mullet (*Mugil celphalus*). II. Chemical nature of the volatile constituents. J. Sci. Food Agric. 21:433-6.
189. Lee, R.F., R. Sauerheber, and G.H. Dobbs (1972). Uptake, metabolism and discharge of polycyclic aromatic hydrocarbons by marine fish. Mar. Biol. (Berl.) 17:201-8.
190. Youngblood, W.W. and M. Blumer (1975). Polycyclic aromatic hydrocarbons in the environment: Homologous series in soils and recent marine sediments. Geochim. Cosmochim. Acta 39:1303-14.

191. Harvey, G.R. and J.M. Teal (1970). PCB and hydrocarbon contamination of plankton by nets. Bull. Environ. Contam. Toxicol. 9:287-90.
192. Lee, R.F. (1975). Fate of petroleum hydrocarbons in marine zooplankton. In: Proceedings of 1975 Conference on Prevention and Control of Oil Pollution, p. 549-53. American Petroleum Institute, Washington, D.C.
193. Rossi, S.S., J.W. Anderson, and G.S. Ward (1976). Toxicity of water-soluble fractions of four test oils for the polychaetous annelids, *Neanthes arenaceodentata* and *Capitella capitata*. Environ. Pollut. 10:9-18.
194. Anderson, J.W. (1976). Personal communication. Battelle Pacific Northwest Laboratories, Sequim, Wash.
195. Anderson, J.W. (1975). Laboratory studies on the effects of oil on marine organisms: An overview. Am. Petrol. Inst. Publ. 4249, 70 p.
196. Cox, B.A., J.W. Anderson, and J.C. Parker (1975). An experimental oil spill: The distribution of aromatic hydrocarbons in the water, sediment, and animal tissues within a shrimp pond. In: Proceedings of 1975 Conference on Prevention and Control of Oil Pollution, p. 607-12. American Petroleum Institute, Washington, D.C.
197. Anderson, J.W. and J.M. Neff (1974). Accumulation and release of petroleum hydrocarbons by edible marine animals. In: Proceedings Recent Advances in the Assessment of the Health Effects of Environmental Pollutants, Vol. III, p. 1461-7. Commission of the European Communities, U.S. Environmental Protection Agency, and World Health Organization, Paris.
198. Lee, R.F., R. Sauerheber, and A.A. Benson (1972). Petroleum hydrocarbons: Uptake and discharge by the marine mussel *Mytilus edulis*. Science 177:344-6.
199. Vaughan, B.E. (1973). Effects of oil and chemically dispersed oil on selected marine biota - A laboratory study. Am. Petrol. Inst. Publ. 4191, 105 p.
200. Anderson, J.W. (1973). Uptake and depuration of specific hydrocarbons from oil by the bivalves *Rangia cuneata* and *Crassostrea virginica*. In: Background Papers: Inputs, Fates, and Effects of Petroleum in the Marine Environment, Workshop, p. 689-708. National Academy of Sciences, Washington, D.C.
201. Anderson, R.D. (1975). Petroleum hydrocarbons and oyster resources of Galveston Bay, Texas. In: Proceedings of 1975 Conference on Prevention and Control of Oil Pollution, p. 541-8. American Petroleum Institute, Washington, D.C.

202. Neff, J.M. and J.W. Anderson (1975). An ultraviolet spectrophotometric method for the determination of naphthalene and alkylnaphthalenes in the tissues of oil-contaminated marine animals. Bull. Environ. Contam. Toxicol. 14:122-8.
203. Neff, J.M. and J.W. Anderson (1975). Accumulation, release, and distribution of benzo[a]pyrene-C^{14} in the clam *Rangia cuneata*. In: Proceedings of 1975 Conference on Prevention and Control of Oil Pollution, p. 469-71. American Petroleum Institute, Washington, D.C.
204. Anderson, J.W., J.M. Neff, B.A. Cox, H.E. Tatem, and G.W. Hightower (1974). The effects of oil on estuarine animals: Toxicity, uptake and depuration, respiration. In: Pollution and Physiology of Marine Organisms (F.J. Vernberg and W.B. Vernberg, eds.), p. 285-310. Academic Press, New York.
205. Blumer, M. and J. Sass (1972). The West Falmouth oil spill: Data available in November, 1971. II. Chemistry. Unpublished manuscript. Woods Hole Oceanogr. Inst. Tech. Rep. 72-19, 57 p.
206. Michael, A.D., C.R. Van Raalte, and L.S. Brown (1975). Long-term effects of an oil spill at West Falmouth, Massachusetts. In: Proceedings of 1975 Conference on Prevention and Control of Oil Pollution, p. 573-82. American Petroleum Institute, Washington, D.C.
207. Giger, W. and M. Blumer (1974). Polycyclic aromatic hydrocarbons in the environment: Isolation and characterization by chromatography, visible, ultraviolet, and mass spectrometry. Anal. Chem. 46:1663-71.
208. Zafiriou, O.C. (1973). Petroleum hydrocarbons in Narragansett Bay. II. Chemical and isotopic analysis. Estuarine Coastal Mar. Sci. 1:81-7.
209. Mayo, D.W., C.G. Cogger, D.J. Donovan, R.A. Gambardella, L.C. Jiang, and J. Quan (1975). The ecological, chemical and histopathological evaluation of an oil spill site. Part II. Chemical studies. Mar. Pollut. Bull. 6:166-71.
210. Straughan, D. (1974). Field sampling methods and techniques for marine organisms and sediments. In: Marine Pollution Monitoring (Petroleum). Natl. Bur. Stand. Spec. Publ. 409, p. 183-7.
211. Koons, C.B. and D.E. Brandon (1975). Hydrocarbons in water and sediment samples from Coal Oil Point area, offshore California. In: Proceedings 1975 Offshore Technical Conference III, p. 513-21. Offshore Technology Conference, Dallas, Tex. Cited in: McAuliffe, C.D. (1976). Surveillance of the marine environment for hydrocarbons. Mar. Sci. Commun. 2(1):29.

212. DiSalvo, L.H. and H.E. Guard (1975). Hydrocarbons associated with suspended particulate matter in San Francisco Bay waters. In: Proceedings of 1975 Conference on Prevention and Control of Oil Pollution, p. 169-73. American Petroleum Institute, Washington, D.C.
213. Watson, J.A., J.P. Smith, L.C. Ehrsam, R.H. Parker, W.G. Blanton, D.E. Solomon, and C.J. Blanton (1971). Biological Assessment of Diesel Oil Spill, Anacortes, Washington, May 1971. Final report to the Environmental Protection Agency (68-01-0017). Texas Instruments Inc., Dallas, Tex., 168 p.
214. MacLeod, W.D., Jr., D.W. Brown, R.G. Jenkins, and L.S. Ramos (1977). Intertidal hydrocarbon levels at two sites on the Strait of Juan de Fuca. In: Proceedings of the Symposium on Fate and Effects of Petroleum Hydrocarbons in Marine Ecosystems and Organisms. In press. Pergamon Press, New York.
215. Giam, C.S. (1976). A preliminary environmental assessment of the Buccaneer oil/gas field. IV. Levels of petroleum hydrocarbons in sediments, Buccaneer Oil field. Report to NMFS (03-6-042-35110). Texas A & M University, Galveston, Tex., p. 44-51.
216. Tissier, M. and J.L. Oudin (1973). Characteristics of naturally occurring and pollutant hydrocarbons in marine sediments. In: Proceedings of 1973 Joint Conference on Prevention and Control of Oil Spills, p. 205-14. American Petroleum Institute, Washington, D.C.
217. Meinschein, W.G. (1969). Hydrocarbons - saturated, unsaturated and aromatic. In: Organic Geochemistry: Methods and Results (G. Eglinton and M.T.J. Murphy, eds), p. 330-56. Springer-Verlag, New York.
218. Hunt, J.M. (1961). Distribution of hydrocarbons in sedimentary rocks. Geochim. Cosmochim. Acta 22:37-49.
219. Farrington, J.W. and B.W. Tripp (1975). A comparison of analysis methods for hydrocarbons in surface sediments. In: Marine Chemistry in the Coastal Environment. Am. Chem. Soc. Symp. Ser. 18, p. 267-84.
220. Sever, J.R., T.F. Lytle, and P. Haug (1972). Lipid geochemistry of a Mississippi coastal bog environment. Contrib. Mar. Sci. 16:149-61. Cited in: Table 3-5 of National Academy of Sciences (1975) Petroleum in the Marine Environment, Washington, D.C., p. 57.
221. Smith, P.V., Jr. (1954). Studies on origin of petroleum: Occurrence of hydrocarbons in recent sediments. Am. Assoc. Pet. Geol. Bull. 38:377-404.
222. Whelan, T., III, J.T. Ishmael, and W.S. Bishop (1976). Long-term chemical effects of petroleum in South Louisiana wetlands. I. Organic carbon in sediments and water. Mar. Pollut. Bull. 7:150-5.

223. Stevens, N.P., E.E. Bray, and E.D. Evans (1956). Hydrocarbons in sediments of Gulf of Mexico. Am. Assoc. Pet. Geol. Bull. 40:975-83.
224. Orr, W.L. and J.R. Grady (1967). Perylene in basin sediments off southern California. Geochim. Cosmochim. Acta 31:1201-9.
225. Kvenvolden, K.A. (1966). Molecular distributions of normal fatty acids and paraffins in some Lower Cretaceous sediments. Nature 209:573-7.
226. Orr, W.L. and K.O. Emery (1956). Composition of organic matter in marine sediments: Preliminary data on hydrocarbon distribution in basins off Southern California. Bull. Geol. Soc. Am. 67:1247-58.
227. Cooper, B.S., R.C. Harris, and S. Thompson (1974). Land-derived pollutant hydrocarbons. Mar. Pollut. Bull. 4: 15-6.
228. Shishenina, Ye. P., N.P. Popova, T.G. Chernova, M.S. Telkova, and R.M. Morozova (1974). Geochemistry of hydrocarbons from organic matter in recent sediments. Geokhimiya 1974(8):1212-9. Translation in Geochem. Int. 11(4):831-8.
229. Keizer, P.D., D.C. Gordon, Jr., and J. Dale (1975). *n*-Alkanes in eastern Canadian marine waters. Unpublished manuscript. Fisheries Improvement Committee Int. Counc. Explor. Sea CM 1975/E:22, 17 p. To be published in J. Fish. Res. Board Can. 34:(1977).
230. Levy, E.M. (1971). The presence of petroleum residues off the east coast of Nova Scotia, in the Gulf of St. Lawrence, and the St. Lawrence River. Water Res. 5: 723-33.
231. Gordon, D.C., Jr., P.D. Keizer, and P.S. Chamut (1974). Estimation of hydrocarbon concentrations in the water column of Come-by-Chance Bay, 1971-1973. Fish. Res. Board Can. Tech. Rep. 442, 15 p.
232. Gordon, D.C., Jr. and P.D. Keizer (1974). Hydrocarbon concentrations detected by fluorescence spectroscopy in seawater over the Continental Shelf of Atlantic Canada - Background levels and possible effects of oil exploration activity. Fish. Res. Board Can. Tech. Rep. 448, 24 p.
233. Levy, E.M. (1972). Evidence for the recovery of the waters off the east coast of Nova Scotia from the effects of a major oil spill. Water, Air, Soil Pollut. 1:144-8.
234. Gordon, D.C., Jr. and P.A. Michalik (1971). Concentration of Bunker C fuel oil in the waters of Chedabucto Bay, April 1971. J. Fish. Res. Board Can. 28:1912-4.

235. Ahmed, A.M., M.D. Beasley, A.C. Efromson, and R.A. Hites (1974). Sampling errors in the quantitation of petroleum in Boston Harbor water. In: Marine Pollution Monitoring (Petroleum). Natl. Bur. Stand. Spec. Publ. 409, p. 109-11.
236. Farrington, J.W. and J.G. Quinn (1973). Petroleum hydrocarbons and fatty acids in wastewater effluents. J. Water Pollut. Control Fed. 45:704-12.
237. Brown, R.A., T.D. Searl, J.J. Elliott, B.G. Phillips, D.E. Brandon, and P.H. Monaghan (1973). Distribution of heavy hydrocarbons in some Atlantic Ocean waters. In: Proceedings of 1973 Joint Conference on Prevention and Control of Oil Spills, p. 505-19. American Petroleum Institute, Washington, D.C.
238. Brown, R.A. and H.L. Huffman, Jr. (1976). Hydrocarbons in open ocean waters. Science 191:847-9.
239. Brown, R.A., J.J. Elliott, J.M. Kelliher, and T.D. Searl (1975). Sampling and analysis of nonvolatile hydrocarbons in ocean water. In: Analytical Methods in Oceanography. Am. Chem. Soc. Adv. Chem. Ser. 147, p. 172-87.
240. Gordon, D.C., Jr. and P.D. Keizer (1974). Hydrocarbon concentrations in seawater along the Halifax-Bermuda section: Lessons learned regarding sampling and some results. In: Marine Pollution Monitoring (Petroleum). Natl. Bur. Stand. Spec. Publ. 409, p. 113-5.
241. Zsolnay, A. (1974). Hydrocarbon content and chlorophyll correlation in the waters between Nova Scotia and the Gulf Stream. In: Marine Pollution Monitoring (Petroleum). Natl. Bur. Stand. Spec. Publ. 409, p. 255-6.
242. Iliffe, T.M. and J.A. Calder (1974). Dissolved hydrocarbons in the eastern Gulf of Mexico Loop Current and the Caribbean Sea. Deep-Sea Res. 21:481-8.
243. Cretney, W.J. and C.S. Wong (1974). Fluorescence monitoring study at ocean weather station "P". In: Marine Pollution Monitoring (Petroleum). Natl. Bur. Stand. Spec. Publ. 409, p. 175-7.
244. Chesler, S.N., B.H. Gump, H.S. Hertz, W.E. May, S.M. Dyszel, and D.P. Enagonio (1976). Trace hydrocarbon analysis: The National Bureau of Standards Prince William Sound/Northeastern Gulf of Alaska baseline study. Natl. Bur. Stand. Tech. Note 889, 73 p.
245. Shaw, D.G. (1976). Procedures and quality control for hydrocarbons: Natural distributions and dynamics of the Alaskan Outer Continental Shelf. Unpublished manuscript. University of Alaska, Fairbanks, 8 p.
246. Barbier, M., D. Joly, A. Saliot, and D. Tourres (1973). Hydrocarbons from sea water. Deep-Sea Res. 20:305-14.

247. Carlberg, S.R. and C.B. Skarstedt (1972). Determination of small amounts of non-polar hydrocarbons (oil) in sea water. J. Cons. Cons. Int. Explor. Mer 34:506-15.

248. Simonov, A. and A. Justchak (1970). The effect on the chemical content of sea water with a limited exchange of water from a large ocean of polluting discharges of chemicals (with the Baltic Sea as an example). In: Advances in Water Pollution Research, Proceedings 5th International Conference, San Francisco and Hawaii (W.H. Jenkins, ed.), Vol. 2. Pergamon Press, London. Cited in: Table 3-4 of Petroleum in the Marine Environment (1975), p. 56. National Academy of Sciences, Washington, D.C.

249. Zsolnay, A. (1972). Preliminary study of the dissolved hydrocarbons and hydrocarbons on particulate material in the Götland Deep of the Baltic. Kiel. Merresforsch. 27: 129-34.

250. Zsolany, A. (1973). Personal communication. Bermuda Biological Station, St. George's West, Bermuda. Cited in: Table 3-4 of Petroleum in the Marine Environment (1975), p. 56. National Academy of Sciences, Washington, D.C.

251. Daumas, R.A., P.L. Laborde, J.C. Marty, and A. Saliot (1976). Influence of sampling method on the chemical composition of water surface film. Limnol. Oceanogr. 21:319-26.

252. Monaghan, P.H., J.H. Seelinger, and R.A. Brown (1973). The persistent hydrocarbon content of the sea along certain tanker routes. A preliminary report. American Petroleum Institute Tanker Conference, Hilton Head, S.C., 7-9 May. Cited in: Table 3-4 of Petroleum in the Marine Environment (1975), p. 56. National Academy of Sciences, Washington, D.C.

253. Swinnerton, J.W. and R.A. Lamontagne (1974). Oceanic distribution of low-molecular-weight hydrocarbons. Baseline measurements. Environ. Sci. Technol. 8:657-63.

254. Swinnerton, J.W. and V.J. Linnenbom (1967). Gaseous hydrocarbons in sea water: Determination. Science 156: 1119-20.

255. Brooks, J.M., A.D. Fredericks, W.M. Sackett, and J.W. Swinnerton (1973). Baseline concentrations of light hydrocarbons in Gulf of Mexico. Environ. Sci. Technol. 7:639-42.

256. Sackett, W.M. and J.M. Brooks (1974). Use of low molecular-weight-hydrocarbon concentrations as indicators of marine pollution. In: Marine Pollution Monitoring (Petroleum). Natl. Bur. Stand. Spec. Publ. 409, p. 171-3.

257. Brooks, J.M. and W.M. Sackett (1973). Sources, sinks, and concentrations of light hydrocarbons in the Gulf of Mexico. J. Geophys. Res. 78:5248-58.
258. Lamontagne, R.A., W.D. Smith, and J.W. Swinnerton (1975). C_1-C_3 hydrocarbons and chlorophyll A concentrations in the Equatorial Pacific Ocean. In: Analytical Methods in Oceanography. Am. Chem. Soc. Adv. Chem. Ser. 147, p. 163-71.
259. Gordon, D.C., Jr. and P.D. Keizer (1974). Estimation of petroleum hydrocarbons in seawater by fluorescence spectroscopy: Improved sampling and analytical methods. Fish. Res. Board Can. Tech. Rep. 481, 28 p.
260. Gordon, D.C., Jr., P.D. Keizer, and J. Dale (1974). Estimates using fluorescence spectroscopy of the present state of petroleum hydrocarbon contamination in the water column of the Northwest Atlantic Ocean. Mar. Chem. 2: 251-61.
261. Horn, M.H., J.M. Teal, and R.H. Backus (1970). Petroleum lumps on the surface of the sea. Science 168:245-6.
262. Wong, C.S., D.R. Green, and W.J. Cretney (1974). Quantitative tar and plastic waste distributions in the Pacific Ocean. Nature 247:30-2.
263. Morris, B.F. (1971). Petroleum: Tar quantities floating in the northwestern Atlantic taken with a new quantitative neuston net. Science 173:430-2.
264. Sherman, K., J.B. Colton, R.L. Dryfoos, and B.S. Kinnear (1973). Oil and plastics contamination and fish larvae in surface waters of the Northwest Atlantic. MARMAP Operational Test Survey Report: July-August 1972, January-March 1973. Unpublished report. NMFS, MARMAP Field Office, Narragansett, R.I. Cited in: Table 1 of Butler, J.N., B.F. Morris, and J. Sass (1973). Pelagic tar from Bermuda and the Sargasso Sea, p. 20. Bermuda Biol. Stn. Spec. Publ. 10.
265. Sherman, K., J.B. Colton, R.L. Dryfoos, K.D. Knapp, and B.S. Kinnear (1974). Distribution of tar balls and neuston sampling in the Gulf Stream system. In: Marine Pollution Monitoring (Petroleum). Natl. Bur. Stand. Spec. Publ. 409, p. 243-4.
266. Sleeter, T.D., B.F. Morris, and J.N. Butler (1976). Pelagic tar in the Caribbean and equatorial Atlantic, 1974. Deep-Sea Res. 23:467-74.
267. McGowan, W.E., W.A. Saner, and G.L. Hufford (1974). Tar ball sampling in the western North Atlantic. In: Marine Pollution Monitoring (Petroleum). Natl. Bur. Stand. Spec. Publ. 409, p. 83-4.
268. Sleeter, T.D., B.F. Morris, and J.N. Butler (1974). Quantitative sampling of pelagic tar in the North Atlantic, 1973. Deep-Sea Res. 21:773-5.

269. Smith, G.B. (1976). Pelagic tar in the Norwegian Coastal Current. Mar. Pollut. Bull. 7:70-2.
270. Polikarpov, G.G., N. Yegorov, V.N. Ivanov, A.V. Tokareva, and I.A. Feleppov (1971). Oil areas as an ecological niche. Priroda No. 11 (translated from Russian by Precoda, N.) Pollut. Abstr. 3:72-5TC-0451. Cited in: Sleeter, T.D., B.F. Morris, and J.N. Butler (1976). Pelagic tar in the Caribbean and equatorial Atlantic, 1974. Deep-Sea Res. 23:467-74.
271. Jeffrey, L.M. (1973). Preliminary report on floating tar balls in the Gulf of Mexico and Caribbean Sea. Unpublished report. Sea Grant Project 53399, Texas A & M University, College Station, Tex. Cited in: Table 3-2 of Petroleum in the Marine Environment (1975), p. 53. National Academy of Sciences, Washington, D.C. and in Fig. 5 of Butler, J.N., B.F. Morris, and J. Sass (1973). Pelagic tar from Bermuda and the Sargasso Sea, p. 18. Bermuda Biol. Stn. Spec. Publ. 10.
272. Jeffrey, L.M., W.E. Pequegnat, E.A. Kennedy, A. Vos, and B.M. James (1974). Pelagic tar in the Gulf of Mexico and Caribbean Sea. In: Marine Pollution Monitoring (Petroleum). Natl. Bur. Stand. Spec. Publ. 409, p. 233-5.
273. Morris, B.F., J.N. Butler, and A. Zsolany (1975). Pelagic tar in the Mediterranean Sea, 1974-75. Environ. Conserv. 2:275-81.
274. Wong, C.S., D.R. Green, and W.J. Cretney (1976). Distribution and source of tar on the Pacific Ocean. Mar. Pollut. Bull. 7:102-6.
275. Lee, R.F. (1973). Private communication to J.N. Butler. Scripps Institution of Oceanography, La Jolla, Calif. Cited in: Table 3-2 of Petroleum in the Marine Environment (1975), p. 53. National Academy of Sciences, Washington, D.C. and in Footnote 32 of Butler, J.N., B.K. Morris, and J. Sass (1973). Pelagic tar from Bermuda and the Sargasso Sea, p. 24. Bermuda Biol. Stn. Spec. Publ. 10.
276. Saner, W.A. and M. Curtis (1974). Tar ball loadings on Golden Beach, Florida. In: Marine Pollution Monitoring (Petroleum). Natl. Bur. Stand. Spec. Publ. 409, p. 79-81.
277. Wong, C.S., D. Macdonald, and W.J. Cretney (1976). Tar and particulate pollutants on the Beaufort Sea coast. Beaufort Sea Tech. Rept. 13, Environment Canada, Victoria, B.C., 96 p.
278. Dwivedi, S.N. and A.H. Parulekar (1974). Oil pollution along the Indian coastline. In: Marine Pollution Monitoring (Petroleum). Natl. Bur. Stand. Spec. Publ. 409, p. 101-5.

279. Mommessin, P.R. and J.C. Raia (1975). Chemical and physical characterization of tar samples from the marine environment. In: Proceedings of 1975 Conference on Prevention and Control of Oil Pollution, p. 155-67. American Petroleum Institute, Washington, D.C.
280. Butler, J.N. and B.F. Morris (1974). Quantitative monitoring and variability of pelagic tar in the North Atlantic. In: Marine Pollution Monitoring (Petroleum). Natl. Bur. Stand. Spec. Publ. 409, p. 75-8.
281. Morris, R.J. (1973). Uptake and discharge of petroleum hydrocarbons by barnacles. Mar. Pollut. Bull. 4:107-9.
282. Ledet, E.J. and J.L. Laseter (1973). Alkanes at the air-sea interface from offshore Louisiana and Florida. Science 186:261-3.
283. Majori, L., F. Petronio, G. Nedoclan, and A. Barbieri (1973). Marine pollution by hydrocarbons in the Northern Adriatic Sea. Rev. Int. Oceanogr. Med. 31-32:137-69.
284. Kator, H. (1973). Utilization of crude oil hydrocarbons by mixed cultures of marine bacteria. In: The Microbial Degradation of Oil Pollutants (D.G. Ahearn and S.P. Meyers, eds.), p. 47-65. Publ. No. LSU-SG-73-01. Center for Wetland Resources, Louisiana State University, Baton Rouge, La.
285. Wakeham, S.G. and R. Carpenter (1976). Aliphatic hydrocarbons in sediments of Lake Washington. Limnol. Oceanogr. 21:711-23.
286. Myers, E.P. and C.G. Gunnerson (1976). Hydrocarbons in the ocean. MESA Special Report. National Oceanic and Atmospheric Administration, U.S. Dep. of Commerce, Boulder, Colo., 44 p.

Chapter 3

ALTERATIONS IN PETROLEUM RESULTING FROM PHYSICO-CHEMICAL AND MICROBIOLOGICAL FACTORS

NEVA L. KARRICK
Environmental Conservation Division

Northwest and Alaska Fisheries Center
National Marine Fisheries Service
National Oceanic and Atmospheric Administration
U.S. Department of Commerce
Seattle, Washington 98112

INTRODUCTION

Petroleum in the environment is subjected to physical, chemical, and biological influences. The resulting reactions occur simultaneously but also are interdependent. The changes are continuous and the reactions create a dynamic state with an almost infinite number of variables. Furthermore, the nature and extent of the changes that occur in petroleum depend upon the combined effects of these interdependent influences. Eventually, the action of these combined forces serve to degrade or remove oil contamination from the marine environment. We do not know all the reactions that occur; nor do we understand many of the factors that initiate reactions, affect the routes of the degradative processes, or impinge upon the mechanisms of the changes. Although a tremendous amount of information has been generated from laboratory and field studies, the complicated, interdependent, and multivariable nature of the influences preclude development of concepts that can define the exact sequence of events that can occur in the degradation of petroleum. Furthermore, the effects of a given set of influences will also depend upon the state of degradation of the petroleum and upon the physical location of the petroleum product (such as on the surface of seawater, on deep or shallow sediments, or on the beach). Thus, the nature of the changes that take place in petroleum will be determined by the particular set of physical, chemical, and biological conditions existing at the moment. Nevertheless, some very general patterns for the degradation of petroleum are beginning to emerge as a result of recent research studies. In this chapter we will discuss the specific nature of the changes that take place in petroleum in the marine environment as a result of physical, chemical, and microbiological forces and then, from this information, we will attempt to estimate the general pattern or trends that might be expected after addition of petroleum into the marine environment with particular reference to the arctic and subarctic areas.

Physical and chemical characteristics of petroleum products that enter the environment have a major influence on the nature and rate of the degradation that takes place. Crude oils from different sources often vary widely in both physical and chemical characteristics, such as viscosity, density, specific gravity, and chemical composition. Refined products may have even wider variations depending upon the fractionation processes used in their manufacture. Petroleum starts to change (or weather), immediately upon its introduction into the marine environment. Degradation patterns and the resultant products will depend upon the original chemical composition of the petroleum as influenced by the physical nature and reactivity of the components. In addition, degradation patterns will vary with the rate of input of the petroleum into the environment, e.g., introduced as a large oil spill or slowly and continuously in industrial discharges. The rates of weathering, too, will be influenced by the nature of the products produced during the course of degradation of the petroleum.

PHYSICAL FACTORS

ENVIRONMENTAL CONDITIONS

Environmental conditions will influence the changes in petroleum or its products as well as the rates at which these changes occur. The areas of greatest physical and chemical activity are the seawater-air and sediment-seawater interfaces.

Water

The movement of the seawater has a major impact on disposal and degradation of the oil by diluting dissolved petroleum compounds and reaction products, by replenishing nutrients for biological reactions, and by furnishing oxygen for both chemical and biological reactions. In addition, action of waves tends to break up oil slicks and to form emulsions, thus making the oil more available for chemical and biological reactions. Salinity levels in estuarine areas will also affect the biota and their reactions.

Temperature

Temperature plays a major role in rates of reactions and in some cases may determine what reactions occur. The temperature of 90% of ocean waters (by volume) is 5°C or less [1]. Changes in petroleum will take place at these low temperatures. Temperatures will often be at 20° to 25°C, however, in many surface waters and in the shallow waters of bays and estuaries in temperate regions, where many of the studies on oil degradation have been done. If temperature is the only variable, the rates of abiotic reactions in general will double with every 10° rise in temperature.

Wind

The effects from wind may be as great as those from water movement in evaporation of volatile compounds, in breakup of oil slicks, and in emulsification of the oil and seawater. However, it is impossible to separate the effects of wind and water, since wind is often a major driving force for surface water movement.

Oxygen

Oxygen is consumed by petroleum in abiotic reactions much as by any other organic material. This oxygen requirement is measured in terms of chemical oxygen demand (COD). Microbiological degradation reactions are primarily aerobic, although some studies with sediments have indicated that some anaerobic organisms also utilize petroleum [2,3]. Many of the biological enzymatic reactions that require oxygen use so little (ca. 10^{-7} M) that enough oxygen is ordinarily available even though it is below the sensitivity of contemporary measuring instruments [4]. In general, however, aerobic reaction rates are faster when oxygen levels are higher.

A possible environmental change that can arise from the introduction of petroleum is a decrease in oxygen concentration, particularly in waters with little current or tidal movement. This decrease results from influences such as utilization of oxygen by the abiotic reactions of petroleum, by biological activity increased in the presence of oil, and by the decomposition of dead cells. Reduction of oxygen levels in turn affects the subsequent degradation rates and certain other reactions that occur.

PHYSICAL PROCESSES

As soon as petroleum is released into the marine environment, various physical forces immediately interact with the petroleum, beginning the degradation process known as weathering. For ease of discussion, the various physical forces involved (spreading, evaporation, dissolution, emulsification, and sedimentation) are discussed separately; however, it should be remembered that these physical forces act interdependently as well as simultaneously with other chemical and microbiological forces in their effects on petroleum in the marine environment.

Spreading

Crude oil spilled on the surface of the ocean spreads rapidly and may result in the formation of the familiar oil slick. The amount of spreading is related to the volume of oil and to the physical characteristics of the oil such as viscosity, density, surface tension, pour point, as well as to the wind speed, water currents, and temperature. The amount of spreading helps to determine the thickness of the slick, and in turn affects rates at which the other physical, chemical, and microbiological factors interact with the crude oil. Berridge et al. [5] presented formulae to determine spreading coefficients, but felt that competing degradative processes are of greater significance in controlling the spreading of an oil than the surface tension properties of spilled oil.

Both small experimental spills and large accidental spills in Cook Inlet formed thin oil slicks that had a half-life of less than one day [6]. Crude oil slicks in Arctic seawater were from 0.1 to 1.0 cm thick [7]. Petroleum also spreads under ice [8-10]. The spread of oil on ice has also been investigated [9-12].

Evaporation

Evaporation is important when the oil is exposed on the surface of the water. The rate of evaporation depends initially on ambient conditions such as wind, temperature, and waves. Evaporation of the light (low-boiling) fractions of the crude oil may, in minutes, leave a residue having appreciably different physical properties from that of the original material. The rate of loss of the volatile compounds decreases exponentially with time [13]. As a rough approximation, the loss after one hour is equivalent to the fraction in the crude oil that boils below 150°C, the loss within one day is equivalent to the fraction boiling to 250°C. The residue remaining after evaporative losses from a crude oil slick results in the petroleum residue showing higher values for physical properties such as relative density, viscosity, flash point, and pour point; for contents of wax, asphaltenes, carbon residue, sulfur, and metals; and for temperature ranges of fractional distillates.

Harrison et al. [14] suggested that two rate-loss curves for evaporation be developed for sea-state and roughness: one for extensive white cap formation and the other for little or no whitecapping. Evaporation of crude oil fractions varied with wind speed, degree of wave action, and thickness of slicks. They found that the majority of the compounds boiling below 220°C disappeared from a slick of South Louisiana crude oil within three to eight hours after a spill, with most of the compounds having evaporated within three hours. These studies were carried out at Grand Bahama Island, where the water temperature was 23.6°C, air temperature 20.5° to 27.1°C, and the wind calm to 18 mph with gusts to 22 mph. These results are comparable to those obtained in studies of natural oil seeps in the Santa Barbara Channel, where certain volatile fractions evaporated in two hours under moderate wind and sea conditions [15].

Kinney et al. [6] determined evaporation losses from a small experimental spill in Cook Inlet, Alaska. The wind was 9 to 12 knots at the time of the spill. Levels of hydrocarbon compounds below C_{12} were significantly decreased within eight hours.

The evaporation and abiotic weathering of petroleum under winter conditions in the Arctic has been described [8-10]. When oil was frozen in the ice very little evaporation occurred. However, petroleum on the surface of ice and covered with snow showed as much as 25% evaporation loss over a period of one month at subzero ambient temperature.

Dissolution

The solubility of many petroleum compounds in water is low, but the volume of seawater is so great that a significant amount of oil can be dissolved. The solubilities of pure compounds have been reported [16-18] and the solubilities of the lower straight chain aliphatics are roughly proportional to their vapor pressure [19]. McAuliffe [17] suggested that for each homologous series of hydrocarbons, the logarithm of the solubility in water at room temperature of pure hydrocarbons is a linear function of the hydrocarbon structure. Branching increased water solubility for paraffins, olefins, and acetylene hydrocarbons but not for cycloparaffins, cycloolefins, and aromatic hydrocarbons. For a given carbon number, unsaturation and ring formation increased water solubility. The solubility of pure hydrocarbons in water however will not be the same as that in a competitive two-phase, water-oil system. Furthermore, as oil composition changes during exposure, many of the oxidized derivatives will be more soluble than the parent hydrocarbons; for example, oxygenated compounds such as hydroxy and dihydroxy compounds, aldehydes, and ketones are more soluble than their parent hydrocarbons.

The identification and amounts of hydrocarbons in water-soluble fractions from each crude oil and refined products have been reported from several laboratories [20-22]. The amount of individual compounds and total water-soluble materials will vary depending on the type of crude oil and the method used to prepare the water-soluble fraction. Lower molecular weight compounds are the more soluble, but they also are more volatile and thus evaporate faster. Wasik and Brown [23] suggested that the solubility of aromatic compounds is probably the most important of the various mechanisms that affect the impact of oil on marine organisms.

Harrison et al. [14] analyzed oil slicks and determined that the dissolution rate was much less than half the evaporation rate. The model that they proposed indicated that dissolution rates may be as low as one percent of the evaporation rate.

Burwood and Speers [24] found that an approximate state of equilibrium was reached in less than 48 hours between Middle East crude oil and seawater. The total concentration of aromatic hydrocarbons, expressed in terms of milligrams per liter (mg/l) of benzene, was 23 mg/l in the water. Benzene and toluene were each about 4 mg/l. Trace quantities of cyclohexane, 1,1-dimethylcyclopentane, 3-methylhexane, *n*-heptane and methylcyclohexane, alkylbenzenes with C_3, C_4, or C_5-chains, indone and alkyl-substituted indones, naphthalene and the C_1, C_2, and C_3 alkyl-substituted naphthalenes, and C_1 to C_4 alkyl-substituted phenols were also found. A significant fraction of the soluble compounds was not separated by the gas chromatographic analyses. Polar compounds were in a broad band called "unresolved envelope." The phenolic compounds in this polar fraction represented 0.4 mg/l of crude oil and 15% of the unresolved envelope. The less volatile aromatic compounds represented 7% of this unseparated group of compounds. The bulk of these were saturated compounds and contained 13.4% sulfur. The sulfur compounds were tentatively identified as a complex mixture of thialkane oxides (sulfoxides).

Lysyj and Russell [22] found that when different petroleum products were in contact with distilled water as films, equilibrium occurred in less than 24 hours. This equilibrium state was followed by a period of essentially constant solubility levels of organic material. This stable period varied for different products, i.e., from four days for No. 2 diesel fuel and Navy distillate to four weeks for No. 5 aviation jet fuel. After these stable periods, the amount of oil transferred to water increased, probably due to chemical or microbiological transformation of the hydrocarbons into more soluble compounds.

The NORCOR report on the 1975 Beaufort Sea Project [8] described a method to discriminate between evaporation and solubility in the changes in petroleum that occurred after exposure to the Arctic marine environment. The amounts of benzene, methylcyclopentane, and *n*-heptane in the crude oil were determined and the relative ratios of their content in the oil were calculated. Changes in this ratio during exposure of the crude oil indicated losses from dissolution or from evaporation depending upon the specific changes in the ratio. Using this method, it appeared that the amount of the components of an oil slick lost by evaporation was below the level of detection by laboratory analysis.

Emulsification

Emulsification of oil plays an important role in dispersion of the oil, and is an important factor relative to the impact of the oil on the physical and biological environment and to the eventual degradation of the oil. Emulsification of the oil facilitates dispersion and distribution of small oil droplets in the marine environment. The formation of these droplets increases the exposed surface area of the petroleum, thus facilitating biodegradation.

Emulsions are formed by wind and wave action. Emulsification can be aided by surface-active compounds generated by microorganisms [25,26]. The emulsions can be oil-in-water or water-in-oil. The oil-in-water emulsions disperse rapidly in seawater and thus are carried by existing currents [27]. The water-in-oil emulsions ("chocolate mousse") are relatively stable and may contain up to 80% water. The viscosity of water-in-oil emulsions is higher than that of the component oil; these emulsions sometimes have the appearance of solids or semi-solids. Davis and Gibbs [28] reported that large masses of water-in-oil emulsion weather slowly with no net loss of material during a two-year period. They concluded that this rate of change is limited by the diffusion of oxygen and metals into the mousse. The oil-water interface within the mousse may be very large, with the aqueous phase not continuous with the external seawater. The metabolic activity of any microorganisms in the mousse can be severely limited by lack of oxygen and available nutrients. Berridge et al. [29] carried out laboratory studies on the formation and stability of emulsions of water and oil. They found that under the conditions of their tests, bacteria were not important in the development of a mousse. They concluded that neither bacteria nor other natural phenomena, such as oxidation, had any significant effect on the removal of mousse during the three-month period of their experiment.

Finnerty et al. [25] studied oil-in-water emulsions using scanning electron microscopy. An *Acinetobacter* sp. grew in a broth medium containing hexadecane; the hydrocarbon was dispersed into the culture medium as an emulsion. Microscopic examination showed that the bacteria had uniformly covered the microdroplets of hexadecane. An emulsion did not form when hexadecane was added to uninoculated medium, nor when bacteria were unable to utilize the hydrocarbon, nor when bacteria were not pre-conditioned to growth on hydrocarbons.

Zajic [26] showed that a pseudomonad growing on No. 6 fuel oil (a Bunker C oil) or on aliphatic hydrocarbons formed an extracellular emulsifying agent that appeared to be a high molecular weight polysaccharide. Emulsification of the oil by the extracellular product was not affected by temperatures as low as 6°C. When 3% sodium chloride was added to the mixture an emulsion did not form; instead a patch of surface oil was produced along with oil pellets from 1 to 2 mm in diameter.

When *Pseudomonas aeruginosa* and two yeasts, *Candida petrophilus* and *C. tropicalis*, were grown with hexadecane as the sole source of carbon, the organisms formed extracellular emulsifying agents [30,31]. The products from the yeasts were peptide in nature. The product from the pseudomonad apparently was partially carbohydrate but one which also could be partially destroyed by pancreatic lipase.

The indigenous microflora from Prudhoe Bay and Cook Inlet generated effective emulsifying agents and caused extensive emulsification of both Cook Inlet and Prudhoe Bay petroleum [6,32,33].

Sedimentation

The process of sedimentation has several effects: significant amounts of petroleum can be removed from the water column to the sediments, the petroleum may be dispersed depending on density of suspended particles, and surface area of the petroleum can be increased. Theoretically, sedimentation also can occur when the weathering process results in increases in the density of the petroleum to the point where it can sink. In actual practice most sedimentation occurs from the sorption of petroleum (or the weathered product) onto particulate matter in the water. The petroleum movement then is determined by the fate of this particle.

Hartung and Klinger [34] found under laboratory conditions that the highest sedimentation rate with a diatomaceous earth occurred at 0% sodium chloride and the lowest rate at 3% sodium chloride. A definite difference in sedimentation was detected between oil types, especially at low levels of sodium chloride. Sulfurized lubricating oil sedimented to a much greater degree than did a technical white oil having a medium viscosity and containing only a few additives. They suggested that this difference between the sulfurized and unsulfurized oil may be related to hydrogen bonding on silica surfaces.

Poirier and Thiel [35] studied the effects of eleven kinds of sediments, of the size of suspended particles, and of the amount of oil on the sedimentation process. At high concentrations of sediment, interference between particles formed vertical currents which tended to carry oil toward the surface of the water. Furthermore, a high silt concentration altered the viscosity and density of the petroleum so that the rate of settling was retarded in muddy waters. Fine-grained sediments of less than 1/8 mm carried down more petroleum than did coarser sediments. Microscopic examinations of oil and sediment mixtures showed that the oil was carried down in two somewhat different ways. In most sediments petroleum was in the form of microscopic globules or spheres. In sediments with organic particles, however, petroleum occurred in irregular flaky and stringy forms. This difference was believed to be due to variations in surface or interfacial tension. The spheres were characteristic of an ordinary oil-in-water emulsion where surface tension was sufficient to cause the drops to assume the least possible surface area. The flaky forms were indicative of a tendency for the oil to spread out. An explanation of this difference in surface tension may be found in a fundamental principle of colloids that organic acids lower surface tension and that the higher members of an homologous series of organic compounds have the greatest effect in lowering the surface tension.

Kinney [33] found that suspended silt was not very important in dissipation of oil slicks in Cook Inlet. This situation may have been due to the fact that Cook Inlet waters were turbulent at the time of the observation and the half-life of the slicks was less than one day during these studies. The conditions and the short period that the slicks existed may have precluded an important or observable role for the silt in the disappearance of the slicks.

A detailed discussion of the physical factors that influence the changes that occur in petroleum is presented in Chapter 2.

CHEMICAL FACTORS

The processes for chemical degradation of petroleum are incompletely understood. Furthermore, studies of chemical oxidation are difficult to delineate in complex systems because the products of oxidation are similar whether the reactions are catalyzed by light, metals, or enzymes from microorganisms.

Chemical degradation of petroleum involves primarily photocatalytic oxidations and polymerizations. The oxidations are also catalyzed by metals in the petroleum, e.g., vanadium, but are inhibited by sulfur. Thus, crude petroleum with a high content of paraffin compounds and a low content of sulfur would probably oxidize more rapidly, especially in bright sunlight, than would crude petroleum with a high content of aromatic compounds and a high content of sulfur [5,36].

Chemical oxidation of the hydrocarbons involves a typical free radical chain reaction [5,19]; after formation of free radicals, hydroperoxides are produced, decomposed to hydroxy compounds, and more free radicals are formed. Compounds (e.g., sulfur compounds) that terminate the chain reactions act as inhibitors of oxidation. The hydroxy compounds may be oxidized and dehydrogenated to aldehydes and ketones and then to carboxylic acids of lower molecular weight. These degradation products are usually more soluble than the parent hydrocarbons. Polymerization can occur from radical-radical combinations, from condensation of the aldehydes or ketones with phenols, or from esterification between alcohols and carboxylic acids. The hydroperoxides are involved in the oxidation of sulfur compounds such as oxidation of thioethers to their corresponding sulfoxides [24].

Photocatalyzed oxidations are considered the most likely chemical reactions to occur. Parker et al. [19] reported that the most effective wavelengths for the reactions were from 300 to 350 nm. Klein and Pilpel [37] reported that photo-oxidation initially increased spreading of the oil. The presence of 1-naphthol acted as a photo-sensitizer for crude oil although it was less effective after viscosity of the oil had increased from various degradation processes.

Chemical oxidations are generally slower than physical or microbiological degradative processes. Kinney et al. [6] estimated that ultraviolet-catalyzed oxidation was a relatively insignificant process in the removal of oil from Cook Inlet compared to flushing and biodegradation. They suggested that this may have been the result of the low input of radiant energy in the Cook Inlet area and to rapid dissipation of the oil slicks (the half-life of the slicks was less than one day).

A detailed discussion of the chemical factors that influence the changes that occur in petroleum is presented in Chapter 2.

MICROBIOLOGICAL FACTORS

Microbial degradation of oil is undoubtedly the most important process involved in weathering and eventual disappearance of petroleum from the marine environment. Bacteria, yeasts, and molds attack gaseous, liquid, and solid hydrocarbons, transforming them into more soluble and usually more reactive compounds that in turn are broken down by microorganisms into simpler compounds and, hopefully, eventually to carbon dioxide and water.

The process of biodegradation is more complex than that of physical or chemical degradation. Mere determination of the numbers and types of microorganisms that can utilize petroleum is inadequate to estimate the effects of these organisms on petroleum. Certain microorganisms can degrade hydrocarbons but cannot utilize the degraded products in their metabolic processes. Furthermore, the organisms need not be growing, or even alive, to oxidize the petroleum. The role of microbiological communities is more important in the degradation of petroleum than is the role of any organism operating singly. The ability of certain microorganisms to degrade petroleum seems to be an adaptive process and is governed by environmental conditions. The presence of petroleum may also affect the community through selection of species. In any event organisms from an area where petroleum is present will degrade petroleum more rapidly than will organisms from areas normally free of petroleum components.

DISTRIBUTION OF MICROORGANISMS

Geographical distribution of specific hydrocarbon-utilizing microorganisms is relatively unimportant in view of what appears to be a generalized ability of microorganisms to utilize hydrocarbons [38-42]. Whenever a careful effort has been made to find hydrocarbon-utilizing microorganisms, some are present, although the numbers are greater in locations where oil has been added to an area [3,32,43-47]. Psychrophilic pseudomonads are ubiquitous and generally a dominant species in the marine environment. Bushnell and Haas [39] indicated that the ability to utilize hydrocarbons is a common characteristic of this genus, *Pseudomonas*. ZoBell

[48] showed that psychrophilic organisms from the Arctic degraded petroleum at temperatures as low as -1.1°C. Morita [49] has listed the studies in which microorganisms were isolated and identified from the Arctic and Antarctic. Yeasts [50] as well as aerobic and anaerobic bacteria are found in these areas and obviously have adapted to growth and reproduction at low temperatures.

Kriss [51] described the bacterial flora of the North Beaufort Sea in 1955-1956. Seasonal samples taken from floating ice stations led to the conclusion that bacterial activity occurred only during the summer months; but the techniques that he used did not isolate psychrophilic bacteria. This oversight in methodology is common to a surprising amount of work on organisms from the marine environment, including a number of studies on degradation of petroleum by marine microorganisms. Bunch [38,52] in more recent studies carefully prevented samples from undergoing a rise in temperature. The results from two years' work showed that the abundance and distribution of heterotrophic bacteria were similar to the numbers found in more temperate waters.

Greater numbers and varieties of microorganisms are present in sediments than are present in the water column [53]. Sediments exposed to petroleum contain higher numbers of hydrocarbon-utilizing organisms than those sediments not exposed to petroleum [43]. However, total communities of microorganisms have not been fully studied. Therefore, it is not possible to say whether the presence of petroleum has resulted in major shifts in microbial species or whether differences in microflora have any ecological significance.

The intestinal microflora of marine animals may play a role in biodegradation of petroleum. Strong activity was found in the intestinal microflora of certain deep-sea amphipods collected in the Aleutian Trench [54,55]. The responses of these intestinal bacteria to various substrates and conditions were markedly different from the responses of bacteria isolated from the water and sediment taken from the same area. Petroleum hydrocarbons were not included in the substrates for these studies, and research is needed on this aspect of potential petroleum degradation in deep ocean waters.

YEASTS AND FUNGI

Yeasts and fungi are widely distributed in the ocean; many are capable of assimilating and degrading petroleum hydrocarbons. Much of the work with yeasts has dealt with the synthesis of protein from petroleum fractions. However, work has also been done on both yeasts and fungi in relation to their possible role in the degradation of petroleum in the marine environment. The vegetative cells of yeasts are more resistant than those of bacteria to stress conditions, including exposure to ultraviolet rays and to alterations of osmotic pressure and salinity.

Stokes [56] reviewed the influence of temperature on the growth and metabolism of the yeasts. A special group of yeasts are obligate psychrophilic organisms; these grow most rapidly at about 20°C but may also grow at 0°C or lower. Yeasts have been found in polar regions, and cultures have survived after freezing and storage at temperatures as low as -193°C.

Ahearn, Meyers, and Standard [57] examined yeasts from various marine, fresh water, and terrestrial environments for their capacity to assimilate the hydrocarbons from South Louisiana crude oil or its fractions. The crude oil or a specific fraction at a 4% level was used as the sole source of carbon with seawater or distilled water as diluents. The most rapid utilization of hydrocarbons was obtained with strains of yeasts that had been isolated from oil-polluted habitats. Generally, strains of the same species from non-polluted areas showed a relatively poor capability for utilization of hydrocarbons even after two or three transfers on hydrocarbon media. Several species of yeasts that occurred commonly in oil-polluted environments failed to grow at 4% concentration of hydrocarbons, although good growth occurred at hydrocarbon concentrations below 4%. In other instances, certain compounds such as thiophene (which is an unsatisfactory sole source of carbon) added to the medium along with a source of utilizable carbon, stimulated growth and oxygen uptake of the yeast when added along with a utilizable hydrocarbon. Certain mixed cultures grew in the presence of a substrate that failed to support growth of the individual strains cultured separately.

Most of the yeasts covered the surface of the oil globules. The surface tension of the oil was affected and small concavities were formed on the globule surface. As the yeast crop increased, the surface of the globule was disrupted; the globule divided and subdivided into smaller droplets and finally became completely emulsified in the broth. A few isolates of the yeasts, particularly certain marine strains, developed mycelia on contact with the oil globule. The cells of these strains penetrated the oil globule, filling its interior with mycelia. Consequently, the oil droplets consisted of a small fungal colony with numerous small entrapped oil globules.

Some yeasts grow readily on various straight-chain alkanes and alkenes with 9 to 18 carbons [58-60]. Low concentrations of cyclic compounds, such as benzene and thiophene-2-carboxylic acid, stimulated oxygen uptake with certain yeasts in the presence of 1% tetradecane. However, concentrations of some cyclic compounds as low as 0.005% failed to support growth. Non-starved yeast cells grew in the presence of various aromatics but transfer of the cells to fresh media gave negative results.

The various stimulatory and inhibitory actions of aromatic hydrocarbons on yeast growth were most evident in plate tests. More organisms grew from exposure solely to hydrocarbon vapors than from direct exposure to filter paper saturated with the hydrocarbon compound. The yeast cultures varied in their response to a number of the hydrocarbons. Several cultures grew when exposed to vapors from each of a great number of compounds. Heaviest growth was on plates exposed to vapors of acetophenone or benzaldehyde; the various aromatic compounds inhibited growth. Vapors of aniline, pyridine, and naphthol were lethal to certain yeasts after exposure for 24 hours.

Komagata et al. [61] did a preliminary screening of cultured yeasts for their capacity to assimilate hydrocarbons. They tested 498 different yeasts belonging to 26 genera for their ability to grow on kerosine at 30°C. They found that 56 strains, of which 54 were *Candida*, utilized kerosine. The *n*-paraffins over C_9 were utilized by these yeasts.

Cook, Massey, and Ahearn [62] found that addition of yeast to aquaria that contained oil caused immediate, but partial, emulsification of the oil. However, the degree of emulsification increased during the 30-day test period with an obvious disruption of the oil slick and a reduction of the amount of oil adhering to the sides of the aquaria. The surface slick of a low-asphalt Louisiana crude oil was affected to a greater extent than the surface slick of a high-asphalt Mississippi crude oil. The emulsified oil droplets were completely covered with yeast cells; some droplets eventually had been penetrated by yeast cells and were somewhat lighter in color. In aerated aquaria, the resulting turbulence kept a part of the oil droplets in suspension. In non-aerated tanks containing oil and yeast, the surface slick was considerably disrupted; those tanks without yeast had lost significant amounts of water through evaporation. In the still tanks the yeast growing in the oily surface patches occurred mainly in the stage that had mycelia.

The yeasts maintained a more or less stable population in the aquaria for over 30 days. Higher populations of yeast were found in aquaria containing the Louisiana crude oil than in aquaria containing the Mississippi crude oil. Guppies, *Lesbistes reticulatus*, survived an average of 11 days in aquaria containing Louisiana crude oil; in aquaria that also contained yeast, the guppies survived an average of 12 days. In aquaria containing Mississippi crude oil, guppies survived an average of 25 days; in aquaria that also contained yeast, survival was reduced to an average of 13 days. In aquaria that contained neither oil nor yeast, guppies survived an average of 25 days; in aquaria that contained yeast, guppies survived an average of 28 days. These laboratory experiments suggested that some crude oils and their metabolites or by-products generated during decomposition and emulsification by yeasts may be harmful to some fish.

The authors also examined microscopically the water from the tanks containing yeast and crude oil. Numerous amoebae, flagellates, ciliates, daphnia, and other microinvertebrates were observed feeding on the yeast and ingesting small oil globules. In aquaria with oil, but lacking yeast, approximately one microscopic field in 50 contained invertebrates. The tendency for invertebrate predators to concentrate in the natural environment in areas enriched with yeast had been previously noted. Turner and Ahearn [60] studied a microbial bloom after an oil spill and found that ciliates fed on the yeast and bacterial cells present on the surface of oil globules.

A filamentous fungus, *Cladosporium resinae*, has been shown to degrade petroleum [63,64]. Growth on media containing hydrocarbons was slow; however, degradation of the oil was not slow. The enzyme system involved did not require an induction period and was active in a cell-free preparation. The same fungus has been found in Chesapeake Bay [65] and presumably can play a role in degradation of petroleum in the marine environment.

MICROBIOLOGICAL PROCESSES

Role of Temperature in Microbial Degradation

The range of temperature in much of the ocean is narrow. About 90% of ocean waters show temperatures below 5°C [1]; these are the deep-ocean areas. The remaining 10% above 5°C are the surface waters of parts of the open ocean and the waters of shallow estuaries. The temperature of the shallow areas of the Arctic and Subarctic waters are rarely above 15°C.

These temperatures are critical parameters in the evaluation of microbiological degradation of petroleum. They automatically define and separate the type of microbes and define microbiological conditions that are important in biodegradation in different geographical areas. Obviously the microorganisms in these areas have adapted to the environmental conditions of their habitat. Consequently, evaluation of the existing information on microbial degradation requires comparison of growth and metabolism among organisms that live at different temperatures.

Microorganisms are often classified in accordance with the temperature range at which they are viable. Although many psychrophilic microorganisms are said to grow best at 15° to 20°C, some grow well at 0° to 5°C and may have minimum growth temperatures below 0°C. Mesophilic organisms grow best at temperatures over 20°C, but grow slowly at 15° to 18°C and have minimum growth temperatures at about 10°C. The psychrophilic organisms thus are the more important in the degradation of petroleum in most of the marine environment. The mesophilic organisms, however, are important in surface and shallow water areas of the marine environment. The existence of different microflora in marine areas at different temperatures has been demonstrated many times [44,66-70].

In abiotic reactions, temperature directly affects the rates of the reactions; however, the effects of temperature on biological reactions are more complex. Psychrophilic bacteria, yeasts, and molds possess mechanisms and metabolic pathways that are different from mesophilic microorganisms. Important differences have been found in cellular control mechanisms, transfer of available information from genes to ribosomes, substrate uptake [56], metabolic pathways [71], composition and integrity of semipermeable cell membranes, and enzyme synthesis and activity [72]. Readers who are particularly interested in the reactions of psychophilic microorganisms should consult reviews on the subject [49,56,73-78].

Bacteria that have optimum temperatures for growth at 5°, 10°, 13° and 15°C have been isolated [49]. Organisms have been shown to grow at temperatures as low as -5°C, although growth is usually slow. Morita and Albright [79], however, refuted the common concept that low cell yields and slow growth rates are manifestations of low incubation temperatures. They used a strain of *Vibrio marinus* and found 1.3×10^{12} cells/ml at 15°C in 24 hours and 9×10^{9} cells/ml at 3°C in 24 hours. Morita [47] observed that, in his laboratory, growth of psychrophilic bacteria is excellent when cultures are incubated overnight at 0° to 3°C. They grow many psychrophiles routinely at -2.5°C but have never determined the minimum temperature for growth. Nevertheless, growth did occur at -5.5°C, the lowest temperature tested. Psychrophiles function well in the environment at low temperatures. Morita and Burton [80] found that a psychrophile can outgrow psychrotrophic organisms at temperatures near 0°C. Geesy and Morita [81] found that heat inactivation of cellular processes in psychrophiles could occur at temperatures as low as 13°C. This fact may explain some of the apparently contradictory results in studies on the hydrocarbon-degrading capability of organisms from marine waters and sediments; the temperatures used in a surprising number of these studies have been at 25° and 30°C during some stage of the project, with the result that the organisms selected for study were mesophiles and not psychrophiles.

Role of Other Environmental Factors in Microbial Degradation

The main environmental factors other than temperature that affect the ability of psychrophiles to function are degree of salinity, availability of nutrients, level of hydrostatic pressure, and oxygen content. The effect of salinity and the various cations making up the salinity of the ocean on marine bacteria was reviewed by MacLeod [82,83]. Membrane permeability is also salt-dependent in marine bacteria. In ecological situations such as in estuaries where the salinity and temperature can change due to fresh water intrusions, both evaporation and the temperature-salinity regime may be important factors in the numbers of microbial flora present [84].

Organisms function in the deep-sea environment [85-88]. ZoBell and Kim [89] studied pressure effects on the activity of microbial enzymes in cell-free systems and found that many of the enzymes were not inhibited by pressures up to 1,000 atm, but that other enzymes showed decreased activities at higher pressures. Factors that determined pressure tolerance of enzymes included molecular size and the absence of unstable groups, such as thiol, in the enzymes. Schwarz et al. [90] found that the rate of utilization of *n*-hexadecane was much slower under deep-ocean conditions (500 atm and 4°C) than under near-surface conditions (1 atm and 4°C).

In laboratory experiments, the levels of nutrients had a major effect on the rates and extent of microbial degradation of petroleum; however, availability of nutrients probably is not a problem in highly-active environmental areas where nutrients are continually replenished. In the marine environment, the availability of nutrients may become important when there is an extremely low rate of water movement and exchange or when competition for nutrients may exist, such as during or immediately after a plankton bloom.

Oxygen levels affect microbial degradation of petroleum primarily at the sediment-water interface. Anaerobic and aerobic reactions are discussed in the subsequent section on sediments.

Because of the many combinations of conditions possible with the variables of temperature, salinity, pressure, oxygen levels, and nutrient content, it is impossible to simulate in the laboratory the situations that may occur in the environment. In addition, the effects of these variables on petroleum degradation in the marine environment are influenced by combinations of many other variables. Thus, generalizations about rates and total time required for microbial degradation of petroleum in the marine environment are not possible. Furthermore, any attempt to generalize about rates of degradation of petroleum must also be tempered by recognition that the importance of particular factors will differ depending on the environmental site; the factors that are important at the sea surface or in shallow water areas may not be important at the sediment-water interface in deeper waters.

Microbial Activity

Many marine microorganisms degrade numerous petroleum compounds, but no one species will utilize all the components of petroleum. The degradative interactions are complex and it is misleading to classify organisms as hydrocarbon-utilizing or as non-hydrocarbon-utilizing for several reasons. First, whether or not a microorganism utilizes hydrocarbons often appears to be determined by the methods used during isolation, enrichment, and testing of the isolates [41,91]. Second, degradation of petroleum involves progressive or sequential reactions. In such reactions, the initial attack on the petroleum may be by certain organisms whose activity produces intermediate compounds, which subsequently are attacked by a different group of organisms that contribute to additional degradation [3]. The latter groups of organisms are not hydrocarbon-utilizing by strict definition, but are just as important as the first group in the eventual removal of petroleum from the environment. Third, a time element is important because some organisms require varying periods of exposure to induce activity of the enzymes that oxidize hydrocarbons. These enzymes may show extracellular as well as intracellular activity. An understanding of these concepts is essential to evaluate the role of microorganisms in environmental degradation of oil.

Sequence of Organisms and Degradation of Substrates

Floodgate [92] attributed the succession of dominant organisms to the removal of the most easily degraded compounds and to the formation of products from degradation. These intermediate products are in turn further degraded by microorganisms. Robertson et al. [93] suggested that this succession also may affect the existing microbiological communities after addition of oil to the system: growth of the microflora that oxidize hydrocarbons or the degraded products may be favored, thus increasing the total population in the community but possibly decreasing the number of dominant species. Floodgate [92] stated that we do not have a good picture of the succession of microorganisms.

Cooxidation

Cooxidation, sometimes called cometabolism, is the process by which organisms can degrade a compound, even though the organisms do not utilize the compound for growth.

Leadbetter and Foster [94] showed that *P. methanica* when grown on methane would oxidize ethane, propane, and butane. This phenomenon also occurred with more complex petroleum compounds such as cycloparaffinic and aromatic compounds [95, 96]. Gibson et al. [97,98] used the cooxidation process to isolate and identify early intermediates in degradation of aromatic compounds. Cooxidation has several important ramifications in environmental biodegradation. Horvath [99] reviewed microbial cooxidation and the degradation of organic compounds in nature; the investigations indicated that degradation of organic compounds occurred as cooxidation reactions of natural populations. The environmental studies were with herbicides and a surfactant rather than with petroleum compounds, but they demonstrated that the process of cooxidation can be involved in degradation of complex organic compounds under environmental conditions. Furthermore, increased oxidative degradation occurred both by repeated applications of the contaminant [100] and by simultaneous application of a contaminant and a biodegradable analog of the compound [101].

Cooxidation may also occur in marine waters after addition of petroleum. Various petroleum compounds, including paraffins, cycloparaffins, aromatic, and substituted aromatic compounds, have undergone cooxidation reactions under laboratory conditions [96,102]. Non-hydrocarbons can act as growth substrate for organisms that can oxidize some aromatic hydrocarbons. An important basis for cooxidation is that the enzyme(s) that utilize the growth substrate also attack the non-growth-supporting analogs of the substrate [99].

Extracellular Activity

Extracellular degradation of hydrocarbons can occur from enzymes excreted by microbial cells or from intracellular material released after lysis of the cells. The activity of extracellular enzymes is one of the reasons that the amount of cell growth often does not show a correlation with the actual measured amount of petroleum degradation.

Studies with cell-free extracts have demonstrated that degradative enzymes may remain active after their release into the environment. Thus it appears that extracellular activity of enzymes may account for much of the degradation that occurs in the deep sea where the growth rate of microorganisms is slow [89].

The emulsifying agents produced by the yeast *C. petrophilum* and by the bacterium *P. aeruginosa* were extracellular compounds [30,103]. The factor(s) from yeast was peptide in nature. The factor(s) from the pseudomonad was different in that it was partially destroyed by pancreatic lipase and was anthrone-positive suggesting that it is at least part carbohydrate.

Kim and ZoBell [104] found cell-free enzymes in sediments and seawater from different marine areas. The amount of enzyme activity did not correlate with the total microbial biomass in marine sediments; however, the density of the dominant species present influenced the types and concentrations of the cell-free enzymes found in the samples. Seasonal variations in the abundance and activities of the cell-free enzymes occurred. Their studies of the effects of pressure and temperature on cell-free enzymes indicated that stability of the enzymes varied but that a number of enzymes could be active in the deep-sea environments and could degrade organic compounds.

Microbial Oxidation of Hydrocarbons

The literature on microbial degradation of petroleum hydrocarbons, the metabolic pathways involved, and the products formed is voluminous, and a complete review is beyond the scope of this discussion. Other reviews are available on the various aspects of the degradation of hydrocarbons [87,91,105-112]. Bacteria have been used in most of the metabolic studies, but sufficient information on yeasts [59] and filamentous fungi [63] has been obtained to indicate that their pathways for degradation of petroleum are similar to those of bacteria.

The final products from metabolic degradation (catabolism) of hydrocarbons are carbon dioxide and water; however, during the conversion to carbon dioxide and water, a number of intermediate compounds are formed. Some of these intermediate compounds will be excreted and some will be incorporated into the microbial cell. The two situations need not be differentiated in consideration of potential environmental impact because the intracellular material will be released upon lysis of the cells. The types of intermediate compounds formed and their possible impact on both micro- and macro-organisms are important.

The initial stages of hydrocarbon oxidation and the products formed in the early stages are of primary interest from the standpoint of potential environmental impact. At later stages, the sequence of oxidation steps form intermediate products that enter the usual pathways of cellular catabolism. In this way the hydrocarbons are either degraded to carbon dioxide and water or are used to produce fatty acids that are incorporated into cellular lipids [113].

The initial reactions require the presence of oxygenases or mixed-function oxidases, enzymes that promote the incorporation of oxygen into molecules such as hydrocarbons. Oxygenases require the presence of only a slight amount of oxygen to initiate the oxidation reactions, much less than the instrumentation used to measure oxygen in the marine environment can detect. The enzyme systems can be induced in many organisms. This capability to produce these enzymes that specify (and regulate) the degradative pathways for both octane and naphthalene is genetic in nature [114,115]. The genes that code for these enzymes are extra-chromosomal and are borne on "degradative plasmids" that are essential for the organisms to convert complex organic molecules to simple metabolites [116,117].

Aliphatics

Van Eyk and Bartels [118] studied the induction of the enzyme system for catabolism of aliphatic hydrocarbon compounds as a mechanism of aliphatic adaptation in *P. aeruginosa*. They found that *n*-alkanes induced the synthesis of specific oxidizing enzymes in the bacteria. Some of the compounds that induced the enzymes supported growth (e.g., *n*-heptane) of the organisms; other compounds did not support growth (e.g., *n*-butane). Induction of the enzymes by the *n*-alkane series started with *n*-butane. It is interesting to note that the nitrogen- or sulfur-containing analogs of the alkane series did not induce activity of the enzyme system. However, enzyme activity was induced by other non-hydrocarbon organic compounds.

King and Perry [113] studied the fatty acid pattern in a mycobacterium after growth on *n*-alkanes (C_{14} to C_{18}), 1-alkenes (C_{14} to C_{18}), 2- or 3-methyloctadecane, or 8-heptadecene. Radioactive experiments demonstrated that long-chain saturated fatty acids were converted to the unsaturated fatty acids and that chain elongation did not occur. During growth on 1-alkenes the methyl group was attacked and the bacterium incorporated the ω-unsaturated fatty acids and their β-oxidation products. No direct evidence was found for saturation of the double bond during growth on 1-alkenes, but such activity was proposed as a possible explanation for the increase of the saturated fatty acids homologous to the substrate in cells growing on 1-pentadecene and 1-heptadecene. Cellular fatty acid composition (C_{13} to C_{19}) of the mycobacterium was affected by substrate chain length and additionally was modified by cellular control mechanisms.

Growth of the mycobacterium on C_{14} to C_{18} straight-chain 1-alkenes always resulted in incorporation of ω-unsaturated fatty acids into the cells. Fatty acids extracted from cells grown on 1-pentadecene and 1-heptadecene contained significant amounts of both saturated and unsaturated C_{15} and C_{17} fatty acids.

Branched-chain alkanes are oxidized by bacteria, although rates [91,119] and pathways [110,120] are not fully defined, and contradictory reports are found in the literature. ZoBell [111] stated that mixed cultures of both soil and marine bacteria had oxidized *iso*-octane faster than octane but that some organisms had oxidized octane faster than *iso*-octane. Although branched-chain alkanes are considered relatively inert, sufficient evidence is available to indicate that they are oxidized by microorganisms [53,91,110,111,120, 121]. The oxidation of pristane is of particular interest because values for the ratio of pristane to phytane in environmental samples has been suggested as an indication of the presence of petroleum. McKenna and Kallio [122] proposed a metabolic pathway for the degradation of pristane by a *Corynebacterium*. Pirnik et al. [120] used a strain of *Brevibacterium erythrogenes* that had been isolated from seawater to study the degradative pathways of pristane and other branched-chain alkanes. Nine metabolic products were identified from the oxidation of pristane. Pristane-derived intermediates accumulated as a series of dicarboxylic acids. When both pristane and *n*-hexadecane were present, the hexadecane was preferentially attacked, resulting in delayed onset of the oxidation of pristane. Westlake et al. [123] found that temperature and the quality of the oil used affected the ability of the bacteria to utilize pristane and phytane. At 4°C pristane was sometimes degraded but the results were inconsistent.

The catabolic enzyme system of *P. oleovorans* catalyzed both epoxidation of olefins and hydroxylation of octane [124], even though the two reactions are mutually competitive [125]. Apparently, 1,7-octadiene was more reactive than either 1-octene or octane. This enzyme system did not catalyze oxidation of *n*-hexadecane [42,124]. The epoxide, 1,2-epoxyoctane, was metabolized by the cells, provided that 1-octene was absent. The cells did not use the epoxide for growth [126]. A variant of the strain used for the above studies was isolated. The variant produced about five times more epoxide from 1-octene in one hour at 30°C than the parent culture produced [127]. The variant strain did not grow at 5°C and the induced enzymes appeared to be inactivated during storage at 5°C. The author suggested that the greater production of epoxide by the variant strain may be due to increased stability of the cell from a cell wall or membrane modification rather than to a modification of the enzyme system. This suggestion adds another possible variable to an already complex situation.

Schwartz and McCoy [128] described the kinetics of the enzymatic formation of 7,8-epoxy-1-octene, 1,2-7,8-diepoxyoctane, and 1,2-epoxy-octane. The epoxidation of octane occurred concurrently with exponential growth of the organism, *P. oleovorans*. Epoxidation of 1-octene also proceeded concurrently with growth; but epoxidation of the octadiene lagged behind cell growth suggesting that it was the least reactive substrate. Competition for the enzyme system by various substrates was evident. Furthermore, monoepoxide was toxic to the cells with a resultant decrease in cell viability. This decrease occurred at concentrations of the monoepoxide where it effectively competed with the octadiene for the enzyme system. The monoepoxide was further metabolized to diepoxide, which reacted with or was sequestered by the cellular material. Presumably, the diepoxide was accumulated by viable cells, although this was not proved because bound epoxides cannot be recovered by available methods.

Non-Benzoid Cyclic Hydrocarbons

Ooyama and Foster [129] studied the oxidation of cycloalkanes and cycloolefins by a soil "pseudomonad-like" bacterium. The organism did not grow on the C_3 to C_8 membered rings. All of the rings were oxidized, but only cyclopropane was cleaved.

Monoaromatic Compounds

The degradation of aromatic compounds through incorporation of oxygen by microorganisms and the intermediary pathways of the oxygenated compounds are well known. Both topics have been thoroughly covered in a number of reviews [91,105-111, 130]. Until the late 1960's, studies dealt with the pathways and mechanisms of oxygenated aromatic derivatives. Aromatic compounds are converted to catechol (1,2-dihydroxybenzene) [41]. Bacterial enzymes cleave the catechol ring to form *cis, cis*-muconic acid. This dicarboxylic acid is further oxidized to 3-oxoadipic acid. The degradation of this keto-acid is dependent on coenzyme A. At this step it apparently enters the terminal respiration chain.

The above studies indicated that once the aromatic compounds are degraded to catechol and the ring is cleaved, the products enter the tricarboxylic system and are degraded. This left questions about the initial reactions: What products are formed and what effects might these products have, particularly if they accumulated in the environment? The available information on initial reactions and pathways has been determined using soil microorganisms, but it is reasonable to expect similar mechanisms in marine microorganisms. However, microbial pathways are different from mammalian pathways. The two systems were compared by Gibson [106]. Many bacteria incorporate two oxygen atoms into the aromatic substrate, and the *cis*-dihydrodiols are the first products found in the media. Dehydrogenases catalyze the oxidiation of *cis*-dihydrodiols from benzene [131], benzoic acid [132], and naphthalene [133,134]. Patel and Gibson [135] purified a (+)-*cis*-naphthalene dihydrodiol dehydrogenase; the enzyme oxidized other *cis*-dihydrodiol substrates, including the *cis*-dihydrodiols of anthracene, phenanthrene, biphenyl, ethylbenzene, toluene, and benzene. Bacterial enzymes apparently do not oxidize the *trans*-dihydrodiols, which are the degradation products formed by mammals.

Several catabolic pathways are available for the initial reactions by which microorganisms degrade benzene. Gibson and his coworkers [97,136-139] showed that *cis*-1,2-dihydro-1,2-dihydroxybenzene was an intermediate compound in the degradation of benzene to catechol. They also showed that two atoms of oxygen were incorporated into the aromatic ring and suggested the possibility that a cyclic peroxide was involved in benzene oxidation. Ribbons and Senior [140,141] studied an oxygenase induced in a strain of *P. fluorescens* and found that a 3,4 fission of the benzenoid nucleus occurred. Sparnins and Dagley [142] have shown alternative pathways, namely, enzymes for both *ortho*- and *meta*-fission reactions.

Toluene can be degraded either by oxidation of the methyl group or by incorporation of oxygen in the aromatic ring forming *cis*-2,3-dihydro-2,3-dihydroxytoluene as an intermediate product [143]. Para- and meta-xylene was also degraded by oxidation, involving either the methyl group [144] or the aromatic ring [145].

Naphthalene Compounds

Although naphthalene compounds are considered among the most acutely toxic of the aromatic compounds, microorganisms both utilize and degrade the naphthalenes. Jerina et al. [134] and Catterall et al. [146] identified *cis*-1,2-dihydroxy-1,2-dihydronaphthalene as a product from oxidation by whole cells or by cell-free extracts of pseudomonads. The product is metabolized through several steps via salicylaldehyde, which is reduced to salicylic acid, which in turn is decarboxylated to catechol, which is cleaved to give α-hydroxy muconic semialdehyde [147]. The activities of the enzymes that catalyze three of the five reactions converting naphthalene to salicylate have been measured [148,149]. Naphthalene oxygenase, 1,2-dihydroxy naphthalene oxygenase, and salicylaldehyde dehydrogenase were induced under similar conditions; possibly salicylate or 2-hydroxybenzyl alcohol may induce the enzymes for this series of reactions. Pseudomonads have different pathways [133] for metabolism of naphthalene. Barnsley [133] suggested that enzymes for a long metabolic sequence not only share a single inductive mechanism but may be regulated as a unit.

Raymond [150,151] studied the degradation of naphthalene and its derivatives by six strains of a marine pseudomonad. When the organisms were grown on naphthalene, 1-ethylnaphthalene, or 2-ethylnaphthalene, the corresponding salicylic acids were formed. None of the organisms grew on 1-methylnaphthalene, but all of them oxidized it to 3-methyl salicylic acid. Dimethylnaphthalenes were metabolized by two pathways, namely, oxidation of the ring to form dimethyl salicylic acids and oxidation of one methyl group to form a methylnaphthoic acid (See section on substituted aromatic compounds).

Polyaromatic Compounds

A number of high molecular weight polynuclear aromatic hydrocarbons are considered to be potential carcinogens. Several of the more potent compounds, e.g., benzo[a]pyrene, 1,2-benzanthracene, and 1,2,5,6-dibenzanthracene, are found at low levels in oil. The sources and biodegradation of carcinogenic hydrocarbons were reviewed by ZoBell [152]. Under anaerobic conditions mixed cultures of bacteria synthesized benzo[a]pyrene, but under aerobic conditions the cultures destroyed benzo[a]pyrene. Microbial synthesis of benzo[a]-pyrene under natural conditions in seawater and bottom deposits has been confirmed. Poglazova et al. [153] investigated both the accumulation and transformation of benzo[a]pyrene by soil bacteria at 28°C. They showed by fluorescence microscopic techniques that benzo[a]pyrene accumulated in the fatty inclusions and lipoprotein of cytoplasm. Transformation of benzo[a]pyrene by the bacteria was shown by fluorescence spectroscopy. The amount of benzo[a]pyrene degraded depended on treatment of the bacteria prior to the tests. For example, one strain tested immediately after isolation degraded 66% of the benzo[a]pyrene in eight days; after this same strain was grown on a benzo[a]pyrene-enriched medium, its activity was enhanced and 82% benzo[a]pyrene was destroyed in eight days.

Barnsley [154] determined the degradation of fluoranthene and benzo[a]pyrene by a number of bacterial species. The loss of these polyaromatic hydrocarbons was measured by gas chromatography. The loss of these compounds was detected during the growth stage of the organisms but the losses increased markedly during the stationary phase. The bacterial species that degraded the polyaromatic hydrocarbons also grew on naphthalene; conversely, the species that did not grow on naphthalene did not degrade the polyaromatic hydrocarbons. The rates of metabolism of the polyaromatic hydrocarbons were determined and compared with rates found by other authors. Such rates determined under laboratory conditions, however, cannot be applied to conditions that might occur in the marine environment. The degradation is an aerobic reaction, although some degradation does occur slowly under anaerobic conditions.

The first report on the structural identification of compounds formed by bacterial degradation of benzo[a]pyrene and benzanthracene was recently published by Gibson [155]. The organism used was a mutant strain of a species *Beijerinckia*, which had originally been isolated from a polluted stream and was capable of utilizing biphenyl as a sole source of carbon and energy. Cells of the mutant were grown on succinate in the presence of biphenyl for four hours and were then grown on benzo[a]pyrene or benzanthracene at 25°C. The products produced were mixtures of vicinal dihydrodiols. The major intermediate identified from the degradation of benzo[a]pyrene was *cis*-9,10-dihydroxy-9,10-dihydro-benzo-[a]pyrene. The major intermediate from benzanthracene was *cis*-1,2-dihydroxy-1,2-dihydro-benzanthracene. Other dihydrodiols were also isolated. Gibson [155] stated that the structure of these metabolites suggested that they were formed from arene oxide precursors. He further postulated that microorganisms with a nucleus have a monooxygenase enzyme system to oxidize aromatic hydrocarbons but that non-nucleated microorganisms utilize a dioxygenase system.

Strains of certain species of marine pseudomonads metabolized anthracene and phenanthrene by pathways similar to naphthalene metabolism [150]. Tetraline and fluorene were oxidized to tetralone and fluorenone, respectively.

Substituted Aromatic Compounds

Microorganisms can attack substituted aromatic compounds either at the side chain or by fission of the aromatic ring [138]. When the microorganisms attack the side chain the degradation mechanisms are similar to those for aliphatic compounds. The β-oxidation process results in the production of benzoic acid from side chains with an uneven number of carbons or the production of phenylacetic acid from side chains with an even number of carbons. Gibson [136] stated that metabolism of toluene and isopropylbenzene by *P. putida* converted them to *ortho*-dihydroxy compounds with the side chain intact.

Raymond [150,151] used six strains of marine pseudomonads to study the effects of metabolic products from substituted naphthalenes. The naphthoic acids and metabolic products from 1,3- and 2,3-dimethylnaphthalene were not oxidized. Thus, unless other organisms were present to oxidize these compounds, these oxidation products could be toxic to the bacterial population.

Thiophenes

Three types of sulfur compounds are found in petroleum: thiols, sulfides, and thiophenes. Condensed thiophenes occur in the "heavy" fractions. In general the organic sulfur compounds have received very little attention. Nevertheless, dibenzothiophene (DBT), which is insoluble in water and is toxic to microorganisms, has been utilized by bacteria under aerobic conditions as a sole carbon source [156]. The metabolites were found in the culture broth and were identified as oxygenated derivatives [157,158]. Recently dihydrodiol derivatives have been identified [159]. Hou and Laskin [160] chose DBT as a model to study microbial metabolism of petrosulfur compounds. They found several bacterial strains that would grow in media containing *n*-paraffin saturated with DBT. *Pseudomonas aeruginosa* produced three compounds: 4[2-(3-hydroxy)-thianaphthenyl]-2-hydroxy-3-butenoic acid (I); a tetradecane ester of (I); and a hydroxylated hydrocarbon moiety and the chromophore of (I). The enzyme system of *P. aeruginosa* apparently did not attack the sulfur moiety directly, but opened the benzene ring of DBT.

Mathematical Modeling of Microbial Processes

Mathematical models that attempt to depict the flow of energy and the perturbations in the environment resulting from added petroleum must also consider the dynamics of the physical, biological, and microbiological processes that are involved, if such models are to be used to forecast the impact of petroleum hydrocarbons on aquatic systems. McCarty [161] published an excellent discussion on the energetics of degradation of organic matter. He emphasized that equilibrium models cannot give a good indication of concentrations of materials present. Calculations of the equilibrium concentrations of organic matter that should exist at any electron activity likely to be encountered in natural waters indicated concentrations many orders of magnitude less than those normally found.

Some reactions that are thermodynamically possible do not occur as microbiologically-mediated reactions. Thermodynamic concepts cannot be used to explain all the reactions nor to explain why some reactions do not occur. For example, temperature, oxygen levels, formation of toxic or unusable intermediates, and capability of organisms to produce proper degradative enzyme systems are among the biological factors that determine not only rates but also what reactions occur.

McCarty [162,163] described a relationship between free energy of reaction and maximum cell yield that may apply to both heterotrophic and chemosynthetic autotrophic bacteria. The comparison between energetics of reaction and growth yield was considered good, but many factors resulted in growth yields different from calculated levels. Less than maximum efficiency can result when bacteria are growing under adverse conditions. Lower efficiency might result from what has been termed "nutrient limitation" [164], a condition sometimes occurring during rapid rates of growth when a high percentage of substrate carbon is converted to cellular material. Such a condition may result in (1) accumulation of polymeric products, either in storage form or as unusable wastes, (2) dissipation as heat by "ATPase mechanisms" and (3) activation of shunt mechanisms bypassing energy-yielding reactions. In addition, organisms living under adverse conditions, such as in the presence of inhibiting materials, unbalanced ionic concentrations, or other than optimum pH, may have fewer numbers of cells because of higher energy expenditure to maintain favorable balances within the cell.

Fasoli and Numann [165] reported preliminary attempts to model mathematically the biodegradation of emulsions containing 10 to 20 ppm of crude oil in fresh water at 22°C. Experimental results showed that maximum bacterial concentrations were much greater than had been calculated from nutrient concentrations or from initial hydrocarbon concentrations. The amounts of hydrocarbons were measured by an infrared method that was in common use when these studies were carried out. However, the method was not detailed enough to evaluate degradation rates for different fractions of the crude oil. The authors stated that their results cannot be used to create formulae to forecast natural hydrocarbon pollution phenomena. However, they had attempted what modelers of aquatic systems usually ignore, namely, depiction of the role of microbiological organisms in degradation of organic material.

The redox potential (pE) of an equilibrium system indicates the nature of the reactions likely to occur and the availability of free energy from introduction of organic matter into that system [161]. Aerobic systems are characterized by high pE values and high energy availability following the introduction of organic matter, while anaerobic systems are characterized by low pE and low energy availability. Measurements of redox potential in natural aqueous systems usually represent mixed potentials that are not amenable to quantitative interpretation. However, conceptually defined redox potentials can be calculated from analytical information of a few important components such as the electron acceptors for biological reactions and their reduction products. These data can be used for interpretations of the nature of the system with respect to redox reactions and energy potentials.

Electron transfer rate is a function of temperature, but appears to equal about 1 to 2 electron moles per gram per day at 25°C for a variety of heterotrophic and autotrophic bacteria [162]. Comparison of values in the literature for temperatures between 6° and 35°C showed that the spread was surprisingly low. The close agreement between widely varying energy reactions suggests that electron transport is the rate-limiting step in most reactions mediated by bacteria. Values for electron transfer rate increased with increases in temperature as would be expected. When the Arrhenius equation was used, values for electron transfer were determined from the observed rates in the temperature range from 10° to 40°C. These values varied between about 9 and 18 Kcal/M. This large spread perhaps reflects the limitations of using the Arrhenius equation for biological data, especially over a wide temperature range. Bacteria have strict temperature limitations and inclusion of data for temperature conditions outside of their normal range of growth can lead to unrealistic calculated values for electron transfer rates.

McCarty [161] presented equations for steady-state substrate concentration. An essential assumption was that an individual organic substrate, such as acetate or glucose, was consumed by a single species of bacteria under steady-state conditions. In their equations, the substrate concentration at zero dilution rate was defined as the concentration which permits a sufficient rate of diffusion of energy-yielding substrate into the cell so that the growth rate from substrate consumption just balances the rate of decay of the organisms. In other words, it is the concentration which permits a balance between energy-yielding and energy-consuming reactions. A significant value in the equations is the constant for the saturation coefficient. Vaccaro and Jannasch [166] indicated that this value varied considerably with different organisms. Values as low as 4×10^{-8} were found for certain marine organisms growing on glucose, well below typical reported values of 10^{-5} to 10^{-3} M/l. The marine organisms apparently had adapted to the low concentration of substrates present in the oceans.

McCarty [161] commented that the mechanisms of organic material turnover are no doubt considerably more complicated than steady-state calculations that assume single substrate-single organism interactions. Nevertheless, such calculations can serve as first approximations in ascertaining some of the factors that influence the nature and concentration of organic materials found in natural waters. The use of these reactions as a base to estimate impact on the environment from additions of organic materials, such as petroleum, need additional study. Mill and and Hendry [167] have recently suggested that:

> "knowledge of the physical properties and chemical reactivity of a compound toward specific oxidizing species present in the environment coupled with the recognition that the lifetime (half-life) is often determined by a single, dominant reaction path provides a means of estimating both lifetime and primary products for many classes of pollutants in the environment."

They presented data on the half-life of a number of hydrocarbons to use for models of sunlight-oxidized compounds in aqueous systems.

Many people who are attempting to model perturbations from contaminants in the environment have not been including the microbiological input. Yet, any model that attempts to quantitate the overall results of processes cannot be valid if it ignores the impact on the system from not only microbiological but also other biological processes.

Microbial Degradation of Petroleum in Sediments

Studies on petroleum deposited in sediments have been carried out primarily to follow changes that take place after an oil spill and to determine and explain the processes that occur. Few studies have been reported on petroleum in Arctic sediments; however, some work has been done in other geographical areas. Some of the work involved test conditions at 10°C or lower, and some studies were carried out through several seasons, providing some insight into the influence of temperature on degradation of oil in the sediments. Post-spill information is not the only pertinent investigation that is needed, because a buildup of concentration of petroleum in sediments is characteristic of areas that receive regular exposure to smaller amounts of petroleum.

Johnston [168] investigated, under laboratory conditions, the rate of oil decomposition in sand columns containing natural populations of bacteria, diatoms, protozoa, and interstitial fauna. Beach conditions were simulated by providing a continuous constant downward flow of seawater through the sand column. The tests were done at 10°C. The median grain size in the sand columns was 250 microns. A weathered Kuwait crude oil was added to the columns at two levels: lightly oiled (0.03%) or heavily oiled (2.2%). The reaction sequence in the heavily oiled column was as follows. Soon after the oil was added, the dissolved oxygen content of interstitial water beneath the oiled layer decreased sharply. The deepest level of the sand column soon became anaerobic, and, successively, the shallower levels approached zero oxygen. The effluent water started to smell strongly of hydrogen sulfide; the odor lasted for several weeks. Recovery started at the surface just below the oil layer and gradually spread downward. Recovery was slow and still incomplete at the end of the four-month experiment. The mean long-term (four months) rate of oxygen uptake for a square meter of sand surface at 10°C was 0.45 g m^{-2} day^{-1} and the equivalent rate of oil destruction was 0.09 g oil m^{-2} day^{-1}.

The sequence in the lightly oiled (0.03%) sand column was as follows. The dissolved oxygen content of interstitial water beneath the oil layer decreased for a few days. The amount of oxygen continued to decrease at the deepest level up until the tenth day. The column never became anaerobic and hydrogen sulfide was not formed. A gradual recovery to the original oxygen concentrations occurred at a fairly uniform rate. This recovery varied slightly with depth, but occurred within 26 to 31 days. The oxygen consumption rates of the lightly oiled sand column was about one-half of the corresponding rates for the heavily oiled column. The initial oxygen concentrations were restored while almost 90% of the oil remained in the sand column. This result indicated that the remaining fractions either were resistant to decay or decayed at immeasurably slow rates.

Crude oil added to both columns stimulated the growth of the microorganisms, increased the oxygen demand of the water, and decreased the concentrations of oxygen in the water. After the dissolved oxygen had been consumed, all of the nitrate in the inflowing water was also gone. It was unclear whether the disappearance of nitrate was the result of altered physical-chemical conditions or of microbial demands. Most of the oil remained in the 0 to 5 cm layer of the heavily oiled sand column. Quantitative analysis showed that 1.15% oil remained in the 0 to 5 cm layer and 0.84% oil remained in the 5 to 10 cm layer. These two layers contained nearly all of the oil (2.2%) added at the start of the experiment. Data obtained from gas chromatographic analyses indicated that oil in the 0 to 5 cm layer had been evenly weathered, with disappearance of *n*-alkanes up to about C_{16} and a reduction in those from C_{17} to C_{20}. Oil from the mixed deeper layer (5 to 10 cm) showed disappearance of *n*-alkanes up to C_{20} with some loss of the higher *n*-alkanes. Analyses of the residue from the lightly oiled sand column showed that the 0 to 5 cm layer lost *n*-alkanes up to C_{16} and was depleted of the higher *n*-alkanes. Non-hydrocarbon residues increased. The 5 to 10 cm layer also showed an overall loss of prominent *n*-alkanes. About 10% of the added oil was lost.

Hunt et al. [169] reported a preliminary evaluation of the potential microbial degradation of oil deposited in sediments of the United States continental shelf region. They found that the properties of continental shelf sediments vary considerably with their location. The amount of dissolved oxygen in sediment and water decreases because of the microbial and chemical oxidation of organic matter. This decrease in oxygen content is in part reflected by lower values for the oxidation-reduction potential (E_h). Replenishment of oxygen in the sediments depended on a variety of factors such as grain size, the nature of bottom currents, and the abundance of burrowing organisms. The maximum content of dissolved oxygen in seawater is about 15 ppm; however, in many deeper areas of the continental shelf it is much lower, about 2 to 3 ppm. Such low levels can be depleted rapidly if oxygen is not replenished during degradation of organic material. In any event, anaerobic conditions exist not far below the surface of the sediments. Thus, after the deposited petroleum is covered with other material, it will be exposed to anaerobic conditions. Also, anaerobic conditions often occur in localized areas around oil-loading docks and near the bottom of bays or estuaries subject to continuous or excessive oil pollution [170], due to oxidation of the petroleum. Oxidation of petroleum under anaerobic conditions is at a significantly slower rate than under aerobic conditions [2,169-172].

ZoBell and Prokop [173] found that many of the mud samples collected around Barataria Bay, Louisiana, contained bacteria which anaerobically oxidized USP mineral oil. Floodgate [172] discussed possible anaerobic pathways and reviewed the evidence that anaerobic bacteria such as *Desulphovibrio* can oxidize hydrocarbons using sulfate as a source of oxygen and final hydrogen acceptor.

Hunt et al. [169] demonstrated experimentally the degradation of hydrocarbons under anaerobic conditions in sediments by detection of by-products from ^{14}C-labeled hydrocarbons at atmospheric pressure and 150 psi (the pressure that occurs at a depth of 100 meters in the ocean). Crude oil from Prudhoe Bay fields was mixed at a level of 15% by volume with one liter of sediment, one liter of seawater, and 50 ml of fresh inoculum-sediment, plus 5 μCi each of ^{14}C-hexadecane and ^{14}C-heptadecane. The hydrocarbons were degraded and organic intermediates were formed. Although rates were not determined quantitatively, the rates appeared to increase with longer exposure time. They concluded that degradation of petroleum waste products would occur in continental shelf sediments, but the rate of degradation would be extremely slow under anaerobic conditions.

Shelton and Hunter [2] compared aerobic and anaerobic microbial decomposition by observing the changes in natural sediments containing petroleum that took place in an experimental system over a period of 30 weeks. Their data indicated a greater loss of total organic carbon from sediments under anaerobic than from sediments under aerobic conditions. A slight but measurable drop in pH occurred in the overlying water. Hydrogen sulfide was continually produced by the anaerobic sulfate-reducing organisms. A steady decline in chemical oxygen demand values of the sediment indicated a progressive decrease in decomposition of organic matter, probably as a result of increased solubilization and diffusion of the oxidized products. Release of colored organic substances from the natural bottom sediments occurred during anaerobic decomposition. Volatile solids in the sediments decreased over the 30-week period by 85% under anaerobic conditions and 5.1% under aerobic conditions. The amount of the hexane-extractable fraction (primarily aliphatic compounds) in the sediments held under anaerobic conditions decreased by 11.3% during the 30-week experimental period. The total loss of hexane-extractable fraction in the sediments held under aerobic conditions was 4.45% after the 30-week period. Thus, a greater loss of hexane-extractable compounds occurred under anaerobic conditions. Petroleum in a natural bottom sediment under aerobic conditions persisted longer than other organic material (measured by determination of the amount of volatile solids present). Under anaerobic conditions in the presence of sulfates, however, the reverse was true; the aliphatic fraction of petroleum was lost more readily than other organic material. Shelton and Hunter [2] suggested that possibly the biochemical intermediates formed under anaerobic conditions were more soluble in water than the original aliphatic hydrocarbons. This explanation might account for the greater loss of total organic carbon to the overlying water under anaerobic conditions, i.e., 13.3 mg carbon per week compared with 9.1 mg carbon per week under aerobic conditions. Shelton and Hunter [2] concluded that the more rapid loss of aliphatic hydrocarbons under anaerobic conditions could not be explained by microbial degradation, because the amount of hydrogen sulfide produced was much less than would be available if organic carbon had been lost from the system via biological activity. The percentage of benzene-extractable compounds (aromatic fraction) showed no detectable change during the 30-week period. This result was consistent with the solubility data of the aromatic fraction. The level of oxygen compounds in the sediments was 19.9% at time zero; the level in the anaerobic sediment was 7.54% at week 30. The natural bottom sediment originally contained a higher percentage of oxidized hydrocarbons (10.9%) than was present in South Louisiana crude oil (3.52%). The above two pieces of data served as the basis for the conclusion that oxy-compounds (aldehydes, ketones, carboxylic acids) were lost first from the bottom sediment when the overlying water was anaerobic. Functional group analysis by infrared spectrophotometry of aliphatic, aromatic, and oxy-compound fractions of petroleum indicated an increased loss of the oxy-compounds under anaerobic conditions.

A large amount of Bunker C fuel oil was spilled in Chedabucto Bay, Nova Scotia, in February 1970 (the *Arrow* spill). The areas affected by the spill were studied during subsequent years to determine the changes that occurred with time in the Bunker C fuel oil [174-177]. One aspect of the studies involved the assessment of the persistence or degradation of the oil in beach and shallow sediment areas. Laboratory studies were also carried out for reference purposes or for control experiments to permit extension of the conclusions from the field studies.

Mulkins-Phillips and Stewart [174] collected beach sediments and waters from three areas about nine months (November 1970) after the spill. By the time of this sampling a certain amount of natural cleanup had taken place. The water temperature was 7°C when the samples were taken. Their objective was to determine the capability of bacteria from the beach and water samples to degrade compounds from petroleum, and the effects of temperature on the bacterial action.

All beach and water samples yielded populations of bacteria capable of degrading Bunker C fuel oil [174]. The amount of Bunker C fuel oil degradation varied depending on the area from which the sample was taken, whether it was a beach or water sample, and on the temperature at which the sample was incubated. The amount of Bunker C fuel oil degraded by the microorganisms seemed to be correlated with the visible presence of oil on the beach areas from which the samples were taken. The tests were run at 5°, 10°, and 15°C. At incubation temperatures of 0° and 15°C the beach samples from each area showed greater degradative activity than water samples from the same area. The amounts of degradation by mixed cultures after seven days of incubation were comparable at the two temperatures. Incubation at 5°C for 14 days produced different values for all cultures. The values for degradation at 5°C for 14 days were 21 to 70% less than values obtained from incubation at 10°C for seven days.

The mixed cultures were tested in the laboratory for their capability to degrade 1% by volume concentrations not only of the Bunker C fuel oil but also of a Venezuelan crude oil and a mixture made up of representative compounds from different classes of hydrocarbons. The hydrocarbon mixture contained naphthalene, anthracene, dibenzothiophene, decalin, hexadecane, hexadecene-1, octadecane, dodecane, and *iso*-octane. The cultures grew on the different substrates: aromatic compounds, straight-chain alkanes and alkenes, and cycloalkanes, as well as Bunker C fuel oil and Venezuelan crude oil. They found no apparent correlation between growth on aromatic compounds or cycloalkanes and the amount of degradation of the Bunker C fuel oil. Maximum degradation of Bunker C fuel oil coincided with the presence of maximum numbers of hydrocarbon-utilizing bacteria. The bacterial population increased 10^6 times after four days of incubation at 28°C, indicating that Bunker C fuel oil provided a good source of carbon for growth and that intermediate products toxic to the bacteria apparently were not formed. After maximum degradation of Bunker C fuel oil took place, the hydrocarbon-utilizing bacteria declined in number. When this experiment was repeated at 15°C, extended lag phases were observed for both the growth curve and the course of Bunker C fuel oil degradation; in addition there was a decrease in total growth as compared with the growth obtained at 28°C.

A *Nocardia* sp. was isolated from one of the mixed cultures; it grew at similar rates on hexadecane, the hydrocarbon mixture, glucose, and Venezuelan crude oil [174]. The time required to double the number of cells, or the generation time, was slightly greater when the organism was grown on Bunker C fuel oil, but it grew only slowly on naphthalene. Decreasing the temperature of incubation from 15° to 5°C on the average doubled the generation time. This result agrees with that obtained by ZoBell [178] who estimated a two- to three-fold increase in generation time for each 10°C decrease in temperature. After 14 days incubation at 15°C, the *Nocardia* sp. had utilized 94% of the *n*-alkane fractions in Arabian crude oil and 77% of the *n*-alkane fractions in Venezuelan crude oil. Hexadecane at concentrations of 1 to 10% by volume were tested; no significant differences in generation time were found; in addition, the 10% hexadecane did not inhibit the growth of the organism. During the growth period the pH decreased and addition of a buffering agent decreased the generation time. This could be important in the marine environment. Under conditions of the test, approximately 0.05 mg of elemental nitrogen was calculated as necessary to bring about the disappearance of 1 mg of hexadecane. Phosphorus concentration affected both the rate of growth of the *Nocardia* sp. and the maximum number of cells. Decreased temperatures during growth of *Nocardia* sp. on Bunker C fuel oil resulted in extended lag time for both the growth curve of the organism and the degradation curve of the Bunker C fuel oil. This same effect on lag phase was reported by Atlas and Bartha [179] during their studies on the degradation of Sweden crude oil.

Scarratt and Zitko [175] determined the residual Bunker C fuel oil in sediments of Chedabucto Bay 26 months after the *Arrow* spill. Although the oil in the water column had disappeared long before this time [176], significant amounts remained in the sediments. The surface layer of shallow soft sediments showed little evidence of decrease of oil residue during the 26 months, with 300 μg/g present as measured by fluorescence spectrometry. In general, the amount of oil residue present at deeper sites in subtidal areas had decreased. Samples of gravel-type sediments showed only slightly more residue than samples from areas that had not been contaminated by the spill.

Rashid [177] summarized the chemical and physical changes in Bunker C fuel oil 3.5 years after the spill at Chedabucto Bay and assessed the degree and differences in degradation of the oil in different coastal environments and under laboratory conditions. His results indicated that the extent of degradation of spilled fuel oil depended in large measure upon the environmental conditions of the coastal areas. The degradation was rapid in high-energy environments but was relatively slow in quiet protected areas. Bacterial as well as oxidative processes altered the composition of the fuel oil.

Bunker C fuel oil exposed to high-energy shoreline environments lost *n*-alkanes, probably due to bacterial degradation. Measurement of the changes in the oil from exposure determined that saturated and aromatic hydrocarbons decreased, accompanied by a corresponding increase in the non-hydrocarbon compounds, particularly in resins and in nitrogen-, sulfur-, and oxygen-containing organic compounds. These changes were prominent in the oils exposed in high-energy environments. The rates of degradation of saturated and of aromatic fractions appeared to be the same. Specific gravity and viscosity values of the fuel oil increased markedly during the weathering process. The increased viscosity of the oil residues in the high-energy environment reduced its mobility.

The data suggested that residual fuel oils present in varying amounts on contaminated beaches in protected areas with low- and moderate-energy environments will persist for several years. In high-energy environments, the residual fuel oils are substantially altered due to the loss of *n*-alkanes and a parallel increase in resin and in nitrogen-, sulfur-, and oxygen-containing organic compounds. The resulting residues are highly viscous and adhere firmly to sand and pebble substrates. Biodegradation and oxidation during 3.5 years storage of the sediments with Bunker C fuel oil under laboratory conditions, reduced the oil level in the sediment from 73.1 to 51%. The non-hydrocarbon content of the sediment sample increased from 26.9 to 49%. The loss in saturated and aromatic hydrocarbons from the oils in the low- and moderate-energy environments was less than 5%. The changes in the oil on beaches under the high-energy conditions indicated that the saturated and aromatic hydrocarbons of the oil were reduced to 34% compared with reduction to 51% in the laboratory-stored reference Bunker C fuel oil. A marked difference was also noticed in the non-hydrocarbon fraction, which increased from 49 to 66% (Table 1).

The ratio of hydrocarbons to non-hydrocarbons have been suggested as an index of the degree of changes in petroleum due to weathering. In the original Bunker C fuel oil the ratio was 2.72; in the same oil stored under laboratory conditions for 3.5 years the ratio was 1.04. The ratios in the oil samples extracted from the low-, moderate-, and high-energy environments were 0.96, 0.88, and 0.52, respectively.

Changes occurred in non-hydrocarbons such as hydroxyl, carbonyl, and carboxyl compounds. A 20% reduction was found in the oil from sediments collected in the moderate-energy environment, and 60% reduction in the oil from the high-

TABLE 1

Chemical and physical characteristics of original and residual Bunker C oils extracted from sediments collected in Chedabucto Bay

Characteristics	Bunker C oil Original[a]	Stored sample	Low energy coast	Moderate energy coast	High energy coast
Hydrocarbons (%)					
Saturated	---	26	25	23	18
Aromatic	---	25	24	24	16
Total hydrocarbons	73.1	51	49	47	34
Ratio of saturate to aromatic	---	1.04	1.04	0.96	1.12
Non-hydrocarbons (%)					
Asphaltenes	16.3	20	22	23	22
Resins and NSOs[b]	10.6	29	29	30	44
Total of non-hydrocarbons	26.9	49	51	53	66
Hydrocarbons/ non-hydrocarbons	2.72	1.04	0.96	0.88	0.52
Functional groups (mequiv/g of OM)					
Carbonyl	---	0.55	0.18	0.44	0.22
Physical properties					
Specific gravity	0.950	0.963	0.9953	0.9765	0.9823
Viscosity (cP)	---	19 584	28 600	1210 000	3640 000

[a] Task Force Operation Oil Report, 1970.
[b] NSO: nitrogen, sulfur, oxygen compounds.
Reproduced by permission of Academic Press from Estuarine and Coastal Marine Science, 2:137-144, 1974 [177].
From Rashid [177].

energy environment. The greatest reduction in carbonyl content was in the oil sample extracted from sediments of the low-energy lagoon, where the water temperature was about 5.5° to 8.0°C warmer than the waters of other areas. The high temperature of the shallow water was probably conducive to the oxidation of the carbonyl group. Carboxyl groups were present, particularly in the more weathered samples. Gas chromatographic spectra of the oil from the low-energy environment and from the reference sample did not show any carboxyl peaks. Hydroxyl functional groups were present in all samples except the sample from the high-energy area.

Changes in bacterial populations and activity and in chemical composition of a No. 6 fuel oil were followed for a one-year period after a spill of fuel oil in Narragansett Bay, Rhode Island [53,67,70,180,181]. The total number of heterotrophic bacteria present in beach sediments showed no definite trend, although fluctuations occurred during the year. However, the proportion of hydrocarbon-utilizing bacteria in the total population increased. Within 16 days after the spill, almost all of the organisms grew on hydrocarbons. This fact was true throughout the year of the study. Bacteria were isolated that utilized straight-chain, branched, and cyclic alkanes, and aromatics. The organisms that utilized aromatic compounds also utilized aliphatics, but some of the organisms that utilized aliphatics did not attack the aromatics. Alkylbenzenes, such as hexylbenzene, may have been attacked at the side chain rather than at the aromatic nucleus.

Temperature ranged from 6° to 25.5°C and influenced the species composition of the bacterial population in the sediments. The bacterial population of Narragansett Bay changes from predominantly psychrophilic to predominantly mesophilic at about 12°C [69]. During the winter months, psychrophilic bacteria that utilized hydrocarbons at 5°C were present in the sediment; however, they were absent from the sediments during the summer months. Mesophilic hydrocarbon-degrading bacteria were present in the sediments during both summer and winter.

When the ambient temperature was below 15°C, the biodegradation of the fuel oil by mesophilic bacteria was reduced, and the psychrophilic bacteria emerged. This seasonal selection of psychrophilic and mesophilic bacteria within the sediment led to a temporary reduction in the percentage of hydrocarbon-degrading bacteria in the total bacterial population when the ambient temperature was about 12°C. Organisms isolated from the winter sediments from 2° to 5°C showed good growth on both dodecane and naphthalene at 8° and 0°C [53,70]. Furthermore, a greater number of the isolates utilized naphthalene at 16° than at 24°C. During the winter the total hydrocarbon content of the mid-tide sediment steadily decreased from more than 20 to 3 μg/g of dry sediment, a rate of less than 1 μg of hydrocarbon per gram of dry sediment per day. The *n*-alkane fraction in the low-tide sediment was relatively constant (1.8 to 3.2 μg/g) suggesting that alkane biodegradation equaled the alkane influx into the sediment.

The organic carbon content was lowest in the coarse-grained, well-sorted, mid-tide sediments and highest in fine-grained, low-tide sediments. Heavy pollution and high temperatures led to anaerobic conditions in low-tide sediments during the summer. After one year, compounds from the original fuel oil were still present in the low-tide sediment, but interpretation of the chemical data was complicated by the constant influx of hydrocarbons from other sources. The aromatics and possibly the *n*-alkanes from the fuel oil appeared to be significantly reduced after the year, but the branched and cyclic hydrocarbons persisted [180]. Reports from other areas have also indicated that petroleum persisted in sediments [2,169,175,178,182,183].

Potential Changes in Petroleum Under Arctic and Subarctic Conditions

Transformation of Petroleum Compounds

Microbial oxidation produces extensive physical and chemical changes in petroleum in the marine environment. Microbial action is particularly important in the degradation of weathered petroleum. Because of the many complex mechanisms involved in the degradation of petroleum in the marine environment, it is exceedingly difficult to devise field and laboratory experiments that will provide an understanding of the changes that are taking place [3,121,123,184].

Atlas [32] suggested that the number of petroleum-degrading microorganisms present could be used as an index of low levels of petroleum in an area. Walker and Colwell [44,45] could not directly correlate the number of petroleum-degrading bacteria with the amount of petroleum present in either sediments or water from which the microorganisms had been isolated. A positive correlation was evident, however, between the percentage of petroleum-degrading bacteria in the total viable population and the amount of petroleum.

Walker and Colwell [45] used ^{14}C-labeled hydrocarbons to estimate the hydrocarbon-degrading potential of bacteria from estuarine and marine environments. Oxidation of the hydrocarbons to carbon dioxide and water was found to be related to activity of the bacterial populations during *in situ* incubation. Rates of uptake and degradation to carbon dioxide were greater in samples collected from an oil-polluted harbor as compared with samples from a relatively unpolluted shellfish-harvesting area. The rate of bacterial growth and cellular uptake of hexadecane increased as the temperature increased from 5° to 25°C. The reverse temperature relationship occurred, however, in the complete oxidation of the hexadecane in that the percentage of hexadecane broken down to carbon dioxide was highest at 5°C and decreased with increasing temperature [185].

Our views have changed concerning the relative rates of attack by microorganisms on different chemical classes and compounds in petroleum. The early literature indicated that *n*-alkanes were preferentially attacked, with aromatic compounds utilized after the *n*-alkanes were gone. However, information gained using improved techniques for chemical analyses and changes in microbiological methodology have indicated that this concept is a simplification of the reactions that occur.

In 1973, ZoBell [48] suggested that it was necessary to use mixed microbial strains known to oxidize the different classes of compounds in order to study the microbiological decomposition of crude oil. Since then assessments have been made on the relative rates of degradation of the various classes of compounds in a number of different crude oils using mixed microbial populations from different sources. Results have indicated considerable variation in specific rates and amounts of degradation of the classes of petroleum compounds. The following important generalizations and agreements have been reached among scientists.

During microbial degradation of petroleum, the non-hydrocarbon organic fractions increase [28,121,123,171,177,184]. This fact is usually attributed to the introduction of heteroatoms, such as oxygen, into the hydrocarbons during microbial metabolism. On the other hand, Walker et al. [184] found that these heterocyclic compounds often increased during degradation of fuel oil but decreased during degradation of crude oils; at the same time, the amount of aliphatic and aromatic hydrocarbons decreased. The alkanes, cycloalkanes, and aromatics were all metabolized. The lighter, lower molecular weight compounds in each series were preferentially removed, except possibly for one instance of the rapid removal of the 6-ring cycloalkane [183]. Soli [186] suggested that presence of aliphatic side chains affected the degradation of cycloalkanes. The lower molecular weight cycloalkanes and aromatics were attacked at the same time as the lower molecular weight *n*-alkanes and before the heavier molecular weight *n*-alkanes. The isoprenoids such as pristane and phytane were degraded [123,169,184,185,187] and were consumed after the heavier molecular weight *n*-alkanes and after partial loss of cyclic hydrocarbons. Westlake et al. [123] reported that pristane and phytane were more resistant to bacterial attack under temperature conditions (4°C) favorable to growth of psychrophilic organisms.

Walker et al. [184] characterized the degradation of 0.1% v/v South Louisiana crude oil by a mixed culture of estuarine bacteria. After three weeks growth at 20°C, the petroleum fractions designated as asphaltenes and resins had each increased by 28%; saturated and aromatic hydrocarbons had decreased by 83.4 and 70.5%, respectively. Analysis of the saturated fractions showed that 96.4% of the normal and branched alkanes were removed, including heptadecane, pristane, octadecane, and phytane. The amount of alkanes from C_{28} to C_{32} increased and apparently were produced during microbial growth. Saturated cyclic compounds were less susceptible to attack than were the alkanes. A surprising result was that the 6-ring saturates disappeared completely (Table 2). Loss in aromatic compounds ranged from 25% (perylene) to 85% (alkyl-benzene) (Table 3). Both the number of rings and ring configuration influenced the degree of transformation.

Lee and Ryan [188] used ^{14}C-labeled compounds to measure activity of microorganisms in water containing hydrocarbons.

TABLE 2

Degradation of saturated hydrocarbons present in South Louisiana crude oil by microorganisms from the Colgate Creek sediment sample

	Amount remaining in:			
	weathered sample		degraded sample[a]	
Hydrocarbon	mg	% of weathered sample	mg	% of weathered sample
Alkanes	5.5	100	0.2	3.6
1-Ring	5.1	100	0.35	6.9
2-Ring	3.7	100	0.45	12.1
3-Ring	2.6	100	0.6	23.0
4-Ring	2.9	100	1.4	48.3
5-Ring	2.1	100	0.6	35.0
6-Ring	1.2	100	0.0	0.0

[a] Corrected for weathering.
Reproduced by permission of the National Research Council of Canada from the Canadian Journal of Microbiology, 21:1760-1767, 1975 [184].
From Walker et al. [184].

TABLE 3

Degradation of aromatic hydrocarbons in South Louisiana crude oil by microorganisms from the Colgate Creek sediment sample

	Amount remaining in:			
	weathered sample		degraded sample[a]	
Hydrocarbon	mg	% of weathered sample	mg	% of weathered sample
Monoaromatics	9.2	100	1.95	21.2
Alkylbenzenes	2.8	100	0.40	14.3
Naphthene benzenes	2.8	100	0.55	19.6
Dinaphthene benzenes	3.6	100	1.00	27.8
Diaromatics	7.3	100	2.10	28.8
Naphthalenes	2.3	100	0.45	19.6
Acenaphthenes	2.4	100	0.70	29.2
Fluorenes	2.6	100	0.95	36.5
Triaromatics	2.9	100	1.05	36.2
Phenanthrenes	2.1	100	0.65	31.0
Naphthene phenanthrenes	0.8	100	0.40	50.0
Tetraaromatics	1.1	100	0.55	50.0
Pyrenes	0.8	100	0.40	50.0
Chrysenes	0.3	100	0.15	50.0
Pentaaromatics	0.3	100	0.20	66.0
Perylenes	0.2	100	0.15	75.0
Dibenzanthracenes	0.1	100	0.05	50.0
Sulfur aromatics	1.1	100	0.45	40.9
Benzothiophenes	0.5	100	0.20	40.0
Dibenzothiophenes	0.5	100	0.20	40.0
Naphthobenzothiophenes	0.1	100	0.05	50.0
Unidentified aromatics	1.1	100	0.55	50.0

[a] Corrected for weathering.
Reproduced by permission of the National Research Council of Canada from the Canadian Journal of Microbiology, 21:1760-1767, 1975 [184].
From Walker et al. [184].

The non-volatile hydrocarbons were degraded in decreasing rates: naphthalene, methylnaphthalene, heptadecane, hexadecane, octadecane, fluorene, and benzpyrene. Degradation rates ranged from 4.4 μg/liter/day for naphthalene at a concentration of 130 ppb to 0.002 μg/liter/day for benzpyrenes at 5 ppb. Increased concentration of the hydrocarbons did not increase turnover time (incubation time divided by fraction of labeled hydrocarbon degraded to carbon dioxide) except for benzene and toluene. The addition of a water-soluble extract of No. 2 fuel oil did not significantly change the time required for loss of the pure compounds, although the rate loss of ^{14}C-naphthalene may have been somewhat faster.

Studies in Alaskan Coastal Waters

Early theories that questioned possible degradation of petroleum in Alaska were based on the surmise that data from studies under tropical and subtropical conditions could be applied to Alaskan conditions. Assumptions were made that microbial reaction rates in the Arctic would be similar to those involving abiotic reactions and would decrease by one-half for every 10° decrease in temperature. Another assumption was that organisms would not be active near 0°C and that little or no biodegradation would occur at 0°C or below. At the time it was not known whether psychrophilic organisms could oxidize petroleum hydrocarbons [178]. The field and laboratory studies described in this section have disproved these earlier assumptions.

ZoBell [48] found that psychrophilic oil-oxidizing bacteria were present in 13 of 16 samples collected from oil-polluted waters, oil-soaked soil, and tundra muck of the North Alaska Slope near Prudhoe Bay and oil seeps along the Colville River. Degradation of petroleum occurred at 8°, 4°, and -1.1°C. Reproduction rates were twice as fast at 8° as at -1.1°C; however, the rates at 4° and -1.1°C were almost the same. Direct microscopic observations indicated that the tendency for the lower temperature to retard reproduction was offset in part by the beneficial effects of the solid surfaces provided by the slush ice in the medium at -1.1°C. Colony counts showed that reproduction occurred within two or three days at 4°C. At this temperature, however, 28 days were required for appreciable turbidity from cell growth and for emulsification or disappearance of petroleum.

These mixed cultures also degraded crude oils that had been treated to remove the fraction that was volatile below 60°C. The percentage of the crude oils degraded after 10 weeks are reported in Table 4 [48]. Degradation of the crude oils at 8°C ranged from 38 to 91% in 10 weeks, while degradation of a mineral oil consisting of C_{12} to C_{20} alkanes was virtually complete in 10 weeks. ZoBell [48] suggested that

TABLE 4

Percent of various crude oils degraded by psychrophilic bacteria in 30 ml of mineral salts medium during ten weeks' incubation

Sources of crude oil	Oil in sterile control (mg)	% of oil degraded at: -1.1°C	4°C	8°C
North Alaska, Richfield	214	58	64	74
Prudhoe Bay, BP, Alaska	234	61	68	82
Cymric Field, CA, Associated	198	24	28	46
Abgaio, Arabian, ARAMCO	207	40	43	51
Avoyelles Parish, LA, Amerada	218	59	76	91
Barataria Bay, LA, Texaco	172	48	54	73
Santa Fe Springs, CA, Union	188	29	35	38
Lost Hills, CA, Gen. Petroleum	223	27	30	43
La Guillas, Venezuela, Esso	195	32	37	48
Average	205.4	37.8	43.5	54.6

Reprinted by permission of the author [48].
From ZoBell [48].

the differences may be due partially to refractory materials in the crude oils and partially to the kinds of bacteria, since selection of species had undoubtedly occurred because the enrichment of cultures had been made on a mineral oil medium.

A number of additional investigations have confirmed that microflora oxidize petroleum in the waters surrounding Alaska. Descriptions of these investigations are arranged herein by the geographical location of the study.

Beaufort Sea. Bunch and Harland [38] found that samples from the Beaufort Sea during the open water season in 1974 and 1975 had total viable counts of colony-forming units that ranged from 1.0×10^6 to 3.0×10^7 per liter of seawater. This heterotrophic flora was predominantly psychrophilic. Microorganisms that could utilize petroleum appeared to be ubiquitous to the waters and nearshore (but not offshore) sediments of this region. They were relatively abundant when compared to the levels present in ocean bodies at more southerly latitudes. Optimum rates of degradation by the mixed cultures occurred at 20°C or lower.

In an arctic marine ecosystem, where low temperatures are prevalent throughout the year, the response of an indigenous flora of heterotrophic bacteria to the presence of petroleum would be dependent on the abundance and diversity of the flora and on the amount of metabolic activity at the normally low temperatures. Bunch and Harland [38] determined the petroleum biodegradation potential of the heterotrophic flora of the south Beaufort Sea at low temperatures. Counts obtained on media that contained artificial seawater were generally an order of magnitude higher than counts obtained on ordinary plate count agar. Results suggested that the predominant flora of heterotrophic bacteria after breakup of the ice cover were psychrophiles.

All enriched cultures degraded petroleum at 5°C. Four of the mixed cultures degraded petroleum at 0° and -1°C. Two of the four mixed cultures preferred low temperatures for growth and multiplication on a petroleum substrate. The detection of psychrophilic bacteria in these cultures demonstrated that the process of biodegradation in the Beaufort Sea would result from the activity of a portion of the indigenous flora that would degrade petroleum preferentially at temperatures below 20°C.

The petroleum-oxidizing microflora were found at a depth of one meter at all of the sites used to collect samples. The population was quantitated at three of these sites. The range of values from 2.3×10^2 to 9.3×10^4 cells per liter of seawater corresponds closely to a range of 9.3×10^2 to 4.6×10^5 cells per liter determined in the North Sea [204] and a range of 6.6×10^3 to 7.3×10^4 cells per liter taken along a Halifax-Bermuda transect in the Atlantic Ocean [68]. Low values of 7.9×10^2 and 0.6×10^2 cells per liter were reported for samples taken during the winter off the coast of New Jersey when water temperatures were 4.7° to 4.9°C [189]. The low numbers of petroleum oxidizers in the latter study may have resulted from the quantitation procedure which involved an incubation temperature of 28°C and consequent loss of psychrophilic organisms.

The percent of petroleum oxidizers in a mixed population isolated from waters of the Beaufort Sea was independent of the total number of heterotrophs. The number of petroleum-oxidizing bacteria, however, may indicate the amount of hydrocarbons in the waters sampled. Although data on hydrocarbon concentrations in the Beaufort Sea were not available, the surface waters from the two sites that contained the largest number of petroleum-oxidizers were directly influenced by discharge from the McKenzie River. The river run-off probably contained varying levels of petroleum and petroleum by-products from natural seepage at Norman Wells in the Northwest Territories and from commercial traffic on the river.

The petroleum-oxidizing microflora in the Beaufort Sea together with their demonstrated ability to attack crude oil at the low, natural temperatures of the water assures a degree of petroleum degradation in open waters of the Beaufort Sea. Chemical analyses of crude oil inoculated with microflora from the Beaufort Sea did not reveal any change in the amounts of aliphatic compounds during the first four weeks incubation at 5°C. Complete degradation of the aliphatics occurred in 11 to 12 weeks. Results of experiments to determine the nutrient requirements for *in situ* degradation were inconclusive.

Preliminary experiments suggested that batch cultures required 0.5 to 1.0 g/l of nitrogen and less than 0.05 g/l of phosphorus to degrade the aliphatic fraction in 200 mg of Norman Wells crude oil. These concentrations are one thousand times more than those found at one meter depth in the Beaufort Sea; however, a closed culture system of microflora with petroleum requires an initially high level of nutrients to allow for their utilization during growth cycles. On the other hand nutrients are constantly replenished in the open ocean and thus lower concentrations should be required for utilization and consequently for degradation of the oil.

The effects of crude oil upon the heterotrophic activity of the predominant flora of non-petroleum oxidizers were also studied [38]. The effect of the oil on turnover of dissolved organic material by bacteria in the Beaufort Sea was determined. In all cases but one, neither weathered nor unweathered crude oil appeared to significantly inhibit the rate of degradation of glutamic acid by the flora. Glutamic acid degradation was enhanced when a large amount of hydrocarbon substrate (0.1%) was added to the reaction vessel. The results suggested that the *in situ* activity of heterotrophs in the cycling of organic material would not be adversely affected by added petroleum. However, added petroleum would contribute over a period of time to the natural carbon cycle. The studies did not show whether temperature or nutrient supply would be a limiting factor for *in situ* biodegradation at the observed environmental temperatures. The nutrient supply in surface waters apparently was rapidly depleted by the summer blooms of phytoplankton [190]. The rate of degradation of the petroleum from a spill that occurred immediately after a phytoplankton bloom thus would probably be limited by the available nutrients and would increase only after the nutrients are replenished. This replenishment of nutrients in this area probably would not take place until the following spring. In the event of a spill or a release of ice-entrained oil before the annual bloom of phytoplankton, petroleum-oxidizers would presumably compete with the phytoplankton for nutrients, providing the bloom would occur under a layer of oil.

Cook Inlet. Kinney et al. [6,33] conducted studies to quantitatively define the potential magnitude of oil pollution problems in Alaska's Cook Inlet. Physical dissipation and biodegradation rates of oils were determined and the data were integrated with estimates of hydrocarbon input rates to assess the fate of oil in Cook Inlet. The question of accumulation of crude oil components within the Inlet was approached from the above studies and by direct analysis. Hydrocarbon accumulation was less than their analytical limits of detection. The results further showed that microflora in Cook Inlet water oxidized Cook Inlet crude oil and that biodegradation of the petroleum was essentially complete in a few months. Since complete flushing of Cook Inlet requires ten months, biodegradation would be more important than physical flushing in removing hydrocarbons from the Inlet.

Several of the microorganisms isolated from Cook Inlet water were able to utilize the hydrocarbons of kerosine as the sole source of carbon. Chromatographic analyses indicated that the various species selected different hydrocarbon components as the sources for their carbon and energy requirements. The most probable number determinations showed approximately 10^4 organisms per liter of water from Cook Inlet, with approximately 10% of them capable of utilizing crude oil. Significant numbers of the microorganisms were found in the oil phase, rather than in the aqueous phase where they would have worked only on the surface of the oil droplets. Particles of water-in-oil emulsion were in the water phase. The droplets of oil in these emulsions were oxidized by the microorganisms existing in both the oil and the water phases. These emulsified particles were said to be more dense than seawater due to the presence of cellular material and oxidized hydrocarbon intermediates; but moderate turbulence kept these particles in suspension. Cook Inlet water supplemented with autoclaved Cook Inlet crude oil placed in sterile bottles and incubated at 10°C contained 10^9 organisms per liter after five days. At this rate of growth it was estimated that a thousandfold increase in the microbial population would require 5.7 days. A 1 mg drop of oil inoculated with a single bacterium would be degraded in 17 days.

Ultraviolet-induced oxidation of crude oils in Cook Inlet appears to be negligible as compared to biodegradation and flushing because of the low radiant energy input in Alaska and the short half-life of a thin surface film of oil. However, little information is available on photochemically catalyzed reactions of petroleum in the environment so that firm conclusions cannot be made. The half-life of an oil slick on the surface of Cook Inlet was less than one day; tide-driven turbulence was the chief factor in breaking up the slicks. The hydrocarbon-oxidizing organisms were present at a concentration of about 10^3 cells per liter, and at this level seemed to serve as an adequate inoculum to degrade oil slicks. Biodegradation of Cook Inlet crude oil in Cook Inlet was essentially complete in one to two months. Hydrocarbon components of less than about C_{12} evaporated within eight hours. Accumulated petroleum hydrocarbons in the range of C_{10} to C_{25} were below the limits of detection in Cook Inlet waters and suspended sediments.

The authors concluded that biodegradation of petroleum in Cook Inlet is a significant, low temperature process and that mixing of the waters significantly increased the rate of biodegradation. They recommended that studies should be conducted in the Arctic to determine the physical stability of oil in pack ice and on the surface waters where comparatively little mixing ordinarily occurs.

Port Valdez. Robertson et al. [93,191] determined the hydrocarbon-oxidizing microflora population of the waters of Port Valdez and estimated the rate at which these organisms could be expected to metabolize oil. Comparison of results with those of a previous Cook Inlet study [6] indicated more hydrocarbon-oxidizing microflora in Cook Inlet than in Port Valdez. Three microorganisms were used in the laboratory studies: a hydrocarbon-oxidizing bacterium, a yeast (*Rhodotorula rubra*), and a freshwater algae (*Selenastrum capricornutum*). Microscopic examination showed that these organisms collected around oil droplets in the culture medium. The preference of the bacterium for oil was so extensive that when an oil phase was present, the cells were difficult to collect from the underlying medium. Oil droplets in the yeast culture were densely coated with organisms to the exclusion of nearby media. The algae were in contact with crude oil droplets when medium and oil were shaken together.

Samples were collected at six sites in Port Valdez. Each cubic centimeter of water taken from each of these sites in the spring contained sufficient microorganisms to initiate hydrocarbon degradation. Smaller populations were present in water samples from the same sites taken in the fall. The formation of ^{14}C-labeled carbon dioxide from added ^{14}C-labeled dodecane after *in situ* incubation at a depth of 10 meters in the waters of Port Valdez was used to estimate the rate of biodegradation. The average rate was 0.3 μg/l per day over a range of incubation times.

The sediments of Port Valdez contained only a few microorganisms due to the impermeable bottom and to a shallow aerobic layer. The authors suggested, therefore, that petroleum residues that settled in these sediments would last for a long time.

Port Valdez, Point Barrow, and Fletcher Ice Island. Button [40] reported that microbial hydrocarbon-oxidizers were present at levels of 10^3 to 10^4 cells per liter in the three different and distant systems of Port Valdez, Point Barrow, and Fletcher Ice Island. Hydrocarbon oxidation was initiated by the microorganisms present in 0.1 ml of seawater from Port Valdez and off Point Barrow. One milliliter of winter Arctic Ocean samples (Fletcher Ice Island) was required to initiate the process. Oxidation of ^{14}C-labeled dodecane occurred when the hydrocarbon was added to seawater and, in a sealed container, incubated at a depth of four meters below shore ice near Point Barrow. Radioactive carbon dioxide was released within one day at the three sites. Thus, hydrocarbon oxidation occurred in Arctic Ocean water at -1.7°C. Results suggested that oxidation and cooxidation occurred at low levels (1 μg/l) of dissolved hydrocarbon substrate. The *in situ* oxidation rate of dodecane was 2 to 50 ng per hour.

Button [40] concluded that where oil spills occur in, on, or under ice, processes remain unchanged as long as the oil is in contact with the water. Where the oil is incorporated into the ice by freezing, little biodegradation of the oil would be likely to take place. However, after thawing, the degradative process would again resume in the melted system. Much of the Arctic Ocean ice is thought to have a residence time of about seven years. Thus, oil frozen into the ice would reappear in water in about seven years, assuming that the presence of the oil does not significantly change the migration rate of the ice. Since fractionation and dispersion would be likely during the ice-oil melt process on the sea surface, the bulk of the degradation could be expected to occur either before freezing or after melting.

Prudhoe Bay, Port Valdez, and Arctic Natural Oil Seeps. Atlas [32] and Atlas et al. [192] conducted field and laboratory studies on the effects of oil pollutants in Alaskan coastal waters. The field studies were carried out to assess the interactions of microorganisms and Prudhoe Bay crude oil. The main study site was located at Prudhoe Bay. Other experiments were conducted at Port Valdez, the southern terminus for the Trans-Alaska Pipeline, and at Cape Simpson and Umiat, where there are large natural oil seeps. Crude oil biodegradation was measured in laboratory experiments by monitoring carbon dioxide evolution and loss of weight of the added crude oil.

When Prudhoe Bay crude oil was mixed with brackish water from coastal ponds along Prudhoe Bay several shifts in the microbial populations were noted. When 0.1 ml of petroleum was added to 100 ml pond water, coccoid green algae and diatoms increased in numbers, accompanied by an overall increase in species diversity. Addition of 1.0 ml of petroleum to 100 ml of pond water resulted in a qualitative shift in the algal populations. Coccoid green algae disappeared but the remaining algae did not appear to be inhibited. The protozoan population shifted from amoeboid to flagellated forms at levels of both 0.1 and 1.0 ml oil per 100 ml of water. The bacterial populations increased in numbers by several orders of magnitude, and the organisms appeared to be attached to the oil droplets. The changes in algal and protozoan populations (1) may have been the result of a direct toxic effect on some species from the added oil or its degradation products, or (2) may have been the result of secondary effects caused by increased bacterial growth and a shift in the predator-prey relationships among the various organisms.

Studies on the adaptation of microorganisms to prolonged oil contamination under Arctic conditions were carried out at the site of a natural oil seep at Cape Simpson, Alaska. The microbial flora associated with the water and soil in contact with the oil seepage were markedly different from unaffected adjacent areas. The water and soil in contact with the seepage contained about 10^6 viable fungi per gram of sample, but did not contain any viable bacteria. The fungi consisted of two species of yeasts and one imperfect filamentous fungus.

Miniature, contained-oil slicks that were purposely floated in Prudhoe Bay were subjected to extensive biodegradation under ambient conditions. Some of the slicks were supplemented with an oleophilic fertilizer that contained nitrogen and phosphorus compounds, some were left untreated, and others were poisoned to kill the microorganisms present. Replicate slicks and their underlying water columns were periodically recovered for microbiological analysis. Large increases were found in the numbers of bacteria underlying the unpoisoned slicks over a period of weeks. The largest increase was associated with the fertilized slick. Pseudomonads showed the largest increase in numbers of organisms, but the variety of species present appeared to be unaffected. After five weeks of exposure, the non-biological loss (poisoned slick) was 31%, the natural loss including biodegradation (unfertilized slick) was 60%, and the stimulated loss (fertilized slick) was 80% of the added oil by weight. Prudhoe Bay water is highly stratified with little mixing. The level of nutrients in the water were low compared to the levels required by the organisms grown on laboratory media; nevertheless, the *in situ* tests demonstrated a substantial degree of natural biodegradation of oil slicks. Fertilization of an oil slick with sources of nitrogen and phosphorus enhanced the rate of biodegradation.

The rates of degradation of Prudhoe Bay petroleum by microorganisms determined in the laboratory were less than the rates determined from the field studies. In the laboratory tests, the rate of degradation at 5°C was significant but was much less than that at 25°C. No lag periods occurred before the onset of biodegradation, indicating that the Prudhoe Bay crude oil did not contain a significant amount of a fraction that might inhibit the action of the microorganisms.

The microbial populations of the waters of Prudhoe Bay were greater than those of the waters of Port Valdez and comparable to the levels found in the warmer Raritan Bay, New Jersey, waters [189]. Crude oil-oxidizers in surface waters were 100 per liter at Port Valdez and 700 per liter in Prudhoe Bay. A sizable proportion of the microbiological population was capable of growth at 5°C. The level of the oil-degrading microorganisms was adequate to extensively degrade Prudhoe Bay crude oil in Alaskan waters. The rates of degradation of Prudhoe Bay crude oil were greater for Prudhoe Bay water samples than the rates for Port Valdez water samples. Total microbial populations were about 10^2/ml at Port Valdez and 10^3/ml at Prudhoe Bay.

Comparison of Results from Different Alaskan Waters. Psychrophilic organisms capable of degrading crude oil at ambient temperatures were found in the waters of all study areas. The total numbers of microbial organisms ranged from 10^3 to 10^7 per liter in water from Cook Inlet, Port Valdez, Prudhoe Bay, Arctic Ocean, and the Beaufort Sea. The water and soil at a natural oil seep at Cape Simpson (near Prudhoe Bay) showed 10^6 microorganisms per liter, consisting only of three fungi (two yeasts and one imperfect filamentous fungus). The range in numbers of cells compare with those reported from the North Sea and the Atlantic Ocean.

Seasonal sampling in the south Beaufort Sea showed that maximum bacterial population in the water column was inversely related to phytoplankton bloom, which occurs soon after break-up of the ice. Low nitrate levels in the water column suggested that this nutrient had been depleted by the phytoplankton. The numbers of microorganisms near Port Valdez were higher in the spring than in the fall.

The percentage of oil-oxidizing bacteria in the microbial populations varied from 0.001 to 1.63% in the Beaufort Sea to 10% in Cook Inlet. Concentration of microorganisms in the water decreased with increasing depth, but the relative number of microorganisms capable of growing on hydrocarbons increased.

An assumption has been made that oxidation of petroleum in the Beaufort Sea would not be completed during one open water period [38]. In laboratory tests using an excess of nutrients, the flora of seawater samples required 11 to 12 weeks to degrade the aliphatic fraction of Norman Wells crude oil at 5°C. Thus, oil remaining over winter probably would be entrained in ice and be unavailable for microbial action during that time. Further degradation would occur with the release of the oil into open waters the following spring. The possibility of complete degradation, i.e., the disappearance of the asphaltene fraction, is a matter of conjecture at this time.

A few attempts have been made to determine the rates of biodegradation of petroleum in Alaskan waters. Nevertheless, the combined impact from the many environmental parameters make it impossible to estimate from existing knowledge the amount of time that would be required to degrade petroleum completely. The reports on quantitation invariably have included a statement such as "under our conditions." This qualification is necessary because the relative importance and influence of various parameters is still unknown. For example, even the obviously critical parameter of microorganisms present may only be important qualitatively.

Biodegradation of fertilized and unfertilized crude oil slicks in Prudhoe Bay was studied [32,192]. After five weeks, 80% by weight of the fertilized and 60% of the unfertilized slicks were gone. The rates of loss were faster than the rates determined in corresponding laboratory experiments.

The action of microorganisms on petroleum promoted the development of oil-water emulsions. Emulsification of the crude oil provided more surface areas for microbial action. Dispersion of the oil following emulsification also increased the degradation rate because the microorganisms at the oil-water interface are more active. In fact, any condition that increased dispersion, spreading, or breakup of the petroleum increased the rate of degradation.

POSSIBLE IMPACT OF PETROLEUM ON MICROORGANISMS

The bacterial population of pure seawater normally ranges from ten or less to a few thousand per milliliter, but it is not uncommon to find millions of bacteria per milliliter in oil-polluted seawater or bottom sludge. The growth of bacteria on petroleum may be beneficial to the food web in the sea, because such bacteria are consumed by numerous animal species, such as the copepods and protozoa that graze on the bacteria associated with oil slicks and tar balls.

Surprisingly little information is available about the effects of oil on microbiological communities, including such factors as whether selection occurs in favor of certain genera, what succession of microorganisms may occur, and what potential effects on microorganisms result from the presence of hydrocarbons. Walker and Colwell [43] found that a mixed hydrocarbon substrate limited growth of estuarine and marine microorganisms. Walker et al. [193] studied the effects of No. 2 fuel oil and Louisiana crude oil on a sub-estuary marshland in Chesapeake Bay, Maryland. Both oils limited the total number of viable organisms and appeared to alter the relative amounts of proteolytic, lipolytic, and chitinolytic bacteria in the populations. Cobet and Guard [182] studied effects from 0.2% levels of decane, hexadecane, tetramethylpentadecane, bicyclohexyl, anthracene, trimethylnaphthalene, and a pentane-soluble fraction of Bunker C fuel oil on growth of bacteria isolated from beach sand and from a water column after a spill in San Francisco Bay. Approximately 85% of the bacterial population appeared to be unaffected by decane and the pentane-soluble fraction. The other compounds (including bicyclohexyl) showed no obvious effect on 92 to 98% of the isolates.

According to ZoBell [170] most hydrocarbons and their derivatives are neither bacteriostatic nor bacteriocidal at concentrations ordinarily found in the environment. This is still accepted as a generality. Organisms have been grown on most components of oil. Some components such as toluene, xylene, mesitylene, phenol, and cresol may inhibit growth of some organisms in aqueous media containing 0.01% of the compounds [111,178]; however, other organisms can utilize these compounds as a sole source of energy [108,143,145,178,194].

Calder and Lader [195] studied the effects of dissolved aromatic compounds on *Serratia marinorubra* or *Vibrio parahaemolyticus* in batch culture at 25°C. The aromatic compounds were naphthalene, phenanthrene, pyrene, benzopyrene, 2-methylnaphthalene, 2,6-dimethylnaphthalene, and an oxidation product, 2,3-dihydroxynaphthalene. The dissolved compounds each had a negative effect on at least one of the measurements made, i.e., extent of lag time, growth rate, or maximum cell density. The impact from individual compounds appeared to be related to percent of saturation of the compound rather than to concentration of the compound. The dihydroxynaphthalene had a high degree of solubility and inhibited growth at saturation levels. At similar concentration, dihydroxynaphthalene was more inhibitory to the two organisms than naphthalene.

Atlas and Bartha [189] and Atlas [196] reported that fresh crude oils contain substances that inhibit growth of bacteria. Inhibition varied among different crude oils depending on the relative amounts of high- and low-molecular weight compounds. Temperature was also a critical factor, presumably both for evaporation of inhibitory compounds and for growth of the bacteria that had been isolated from seawater at 28°C. Time for initiation of measurable biological degradation of the crude oils was correlated with the rate of evaporation of the volatile fraction. Results for lag period and temperature effects are known to vary for several reasons, including the petroleum product used, the conditions for introduction of the petroleum, the amount of petroleum-water interface, the methods used for enrichment of the bacterial cultures, the temperatures of incubation, and the methods used for measurement of degradation of the petroleum. The assessment of the possible environmental significance of petroleum-induced lag periods, or inhibition of growth of microorganisms, is not possible at this time.

Morphology

An interesting study of the ultrastructure [25] of *Acinetobacter* sp. grown on alkanes and alkenes demonstrated that hydrocarbons were sequestered in cytoplasm. The same ultrastructure was found in 15 species of microorganisms isolated from a tar ball. Multiple inclusion bodies were readily apparent and characterized an ultrastructure feature that was associated only with hydrocarbon-utilizing bacteria. Development of these inclusion bodies occurred in organisms grown on hexadecene and on an homologous series of alkanes varying in chain length from C_{12} to C_{20}. Analyses by gas chromatography and X-ray diffraction techniques of the inclusion bodies confirmed the presence of pentadecane, hexadecane, hexadecene, heptadecane, or octadecane in the microorganisms grown on the respective hydrocarbons.

Intracytoplasmic membrane structures also were induced in cells grown on hydrocarbons. These membrane structures sometimes were so extensive in the *Acinetobacter* sp. that the cells became four to ten times their normal size. The giant cells were maintained when they were subcultured on hydrocarbon-containing media; they reverted to their normal size when they were subcultured on media without hydrocarbons.

Scott and Finnerty [197] made a comparative study of a diverse group of bacteria and yeasts grown with and without hydrocarbons in the substrate. Ultrastructure examinations showed that all of the microorganisms grown on hydrocarbons contained intracellular electron-transport inclusions and had accumulated unmodified hydrocarbons. Atlas and Heintz [198] found electron-opaque inclusions containing lipid in two species of oil-degrading marine bacteria. Osumi et al. [199] found microbodies in a yeast, *Candida tropicalis*, grown on *n*-alkane; development of the microbodies was correlated with increased catalase activity and oxidation of *n*-alkanes.

Lipids

The lipid composition of psychrophilic microorganisms has been implicated as one among several characteristics that are important to organisms that grow at low temperatures. Thus, changes in lipid composition potentially may have an impact on the activity of the organisms at low temperatures. When bacteria [25,113], yeasts [59,200,201] and filamentous fungi [64,202,203] were grown on alkanes and alkenes, total lipid content of the organisms increased, or the fatty acid composition of the cellular lipids changed, or both. At this point in time we do not know whether these changes are an advantage, a disadvantage, or of no consequence to the organisms.

Intermediary metabolism in cells utilizing alkanes (or other hydrocarbons) is not the same as in cells growing at the expense of glucose or lactic acid; i.e., assimilation of alcohols or fatty acids from alkane oxidation does not proceed by "normal" pathways [91,204,205]. For example, the amounts and kinds of lipids produced by a species of *Nocardia* was different when the organism was grown on glucose than when the organism was grown on *n*-hexadecane. Total lipid content in the cells of microorganisms grown on glucose was 28%; the lipids consisted exclusively of glycerides. The lipid content of the microorganisms grown on hexadecane was 48%; the lipids contained 38% wax, predominantly cetylpalmitate. No wax was found in the lipid of the *Nocardia* sp. grown on hexane or tridecane. When alkanes with 13 to 20 carbon atoms were used as substrates, the fatty acids of the cellular triglycerides and waxes reflected the carbon skeleton of the substrate; e.g., growth on *n*-heptadecane produced lipid having predominantly C_{17} acids with lesser amounts of C_{15} and C_{13}. Cells grown on hexadecane on the other hand, contained only even-numbered fatty acids, C_{12} to C_{18}, with palmitic acid predominating [206]. The data indicated direct incorporation of alkane-derived fatty acids into cell lipids and clearly suggest β-oxidation of fatty acids by *Nocardia* sp. Shorter carbon-length aliphatics (propane and *n*-butane) yielded triglycerides with a "normal" fatty acid composition: the lipids in the cells contained palmitic and stearic acids but no waxes. Patrick and Dugan [207] found that *Acinetobacter* sp. grown on alkanes with an odd number of carbon atoms incorporated lipids that were enriched in odd-chain fatty acids, indicating synthesis or elongation of carbon chains without prior degradation to acetate [208,209].

Olefins were directly incorporated into cellular lipids after oxidation at the methyl end of the molecules [208,210]. Monoterminal oxidation of C_{14} to C_{18} *n*-alkanes was followed by β-oxidation with both the parent fatty acids and the products of β-oxidation incorporated into cellular lipids [113]. Long-chain fatty acids were desaturated but no chain elongation occurred. Cells grown on 2- and 3-methyloctadecane contained the corresponding iso- and anteiso-fatty acids. Cells cultured on 8-heptadecene contained 8- and 9-heptadecenoic acids, 6- and 7-pentadecenoic acids, 9- and 10-methyl heptadecanoic acids, and 7- and 8-methyl pentadecanoic acids.

Chemotaxis

Recent investigations suggest that movement of certain motile microorganisms is not always random. The movement may be influenced by chemicals, which can attract or repel the organisms and thus act as a means of communication. These movements of microorganisms appear to be involved in maintaining balance in heterogeneous microbial communities, because intra- and inter-species responses of bacteria rely upon chemical communication [211]. If compounds are added that inhibit communication, movement becomes random, resulting in probable interruption of normal predator-prey reactions in the microbial community. This interruption could then change the balance of species in the community. Inter-microbial predators are present in all aquatic ecosystems and help to preserve equilibrium of the microflora.

Mitchell [212] found that many marine bacteria prey on other microbial forms such as phytoplankton, bacteria, and fungi. He postulated that if a single collision was necessary for direct predator-prey interaction, the chance frequency of collisions would be too few to result in significant die-off in natural waters. He proposed that bacterial chemotaxis, thus, may act to make predator motion non-random and thereby increase collision frequency. Adler [213,214] described some of the mechanisms of positive chemoreception by bacteria. Chet et al. [215] showed that motile bacterial predators were attracted specifically to exudates from their fungal and diatom prey. The motile bacteria exhibited an apparently beneficial chemical-based response to suitable food sources and their movement was non-random due to this chemotaxis.

Tso and Adler [216] arranged mechanisms of negative chemotaxis and grouped repellents into at least nine classes. Young and Mitchell [217] determined negative chemotaxis with four marine pseudomonads from sublethal concentrations of toxic compounds, including toluene, benzene, and two metal salts of lead and copper. At non-lethal concentrations, viable cells of the pseudomonads detected and avoided areas that contained the toxic compounds; the negative effect of these compounds was stronger than the positive effect from the presence of food. The effective non-lethal concentrations of compounds were high, 0.1% toluene, 0.2 and 0.05% benzene, 0.05% copper sulfate, and 0.001% lead nitrate.

Mitchell et al. [218] reported that hydrocarbons inhibited the ability of marine bacteria to detect living and non-living substrates. They found that 0.6% phenol or toluene and Kuwait crude oil inhibited chemotactic responses by marine bacteria. The bacteria were not immobilized, but their movement was random. Apparently chemoreceptors were blocked, but the effect was reversible. After the bacteria were washed free of hydrocarbons, they were able to detect and move toward substrates. Mitchell [211] also indicated that enzymatic inhibition apparently did not occur. Walsh and Mitchell [219] reviewed what was known about the inhibition of bacterial chemoreception by hydrocarbons and the effect of a broad spectrum of hydrocarbons on bacterial chemotaxis. Those hydrocarbons that affected chemotaxis at concentrations found in the environment may be ecologically significant in that they may alter predator-prey interactions or the rate of biodegradation of organic substrates without affecting gross bacterial physiology. They found that a marine pseudomonad showed a wide variation in chemotactic response in the presence of hydrocarbons. The effect was expressed as an index of chemotaxis inhibition (CI). The CI_{50} was defined as the concentration of the hydrocarbon which reduced the chemotactic response by 50% without affecting gross bacterial physiology. The hydrocarbon concentration ranges which resulted in CI_{50} were greater than would be expected in natural ecosystems. For example, the CI_{50} for Kuwait crude oil was 10 ppm, kerosine 12 ppm, benzene 0.1 ppm, and phenol 120 ppm.

The rate of degradation of organic material and thus the transfer of bound energy in most ecosystems is catalyzed by bacteria. These rates have been assumed to be dependent only upon activity of the bacterial enzymes. However, bacterial populations must be attracted to the organic sources to maintain enzyme concentrations. Bacterial chemotaxis may provide such a mechanism and may thus affect the rate of degradation of organic matter, depending on whether a positive or negative attraction occurs. The fact that the reactions are reversible is important because the results of negative chemotaxis will be temporary and symptoms will disappear when the inhibitory material is removed, or inactivated by dilution or by physical and chemical transformations.

PROSPECTUS

Petroleum released into the arctic and subarctic environment undergoes physico-chemical and microbiological degradative processes similar to those that occur in temperate environments. The actual pathways and rates of these processes in the arctic area and the effects of these processes on arctic marine microbial communities are largely unknown. One of the many questions of concern to the arctic area relate to the effects of year-round low temperature on the rates of degradation of petroleum and the production of intermediate products. What happens to petroleum released in arctic marine regions that are either intermittently or permanently covered with ice? What changes occur in petroleum trapped under, in, or on the permanent ice pack? What happens to petroleum released into the waters or ice of areas seasonally covered with ice? What are the effects of snow cover? Do seasonal exposure conditions affect the pathways and rates of physico-chemical and microbiological degradation of petroleum? What is the nature of the degradation products formed? Clearly, a major research effort is needed to investigate the problems posed by these questions.

Results from studies made during the last few years indicate that microbial degradation by psychrophilic microorganisms at 10°C or lower is the most significant degradative process in the arctic environment. Future studies should be approached from two aspects: the impact of the biochemical activities of microbial communities on petroleum and the impact of petroleum on microbial communities and their activities.

Various intermediate compounds are produced during microbial degradation and utilization of petroleum. Existing information suggests that some metabolites produced by terrestrial animals from petroleum may cause more adverse effects than the parent hydrocarbons. Does the same possibility exist from the metabolites formed by microorganisms and then released into the marine environment? Information is urgently needed on the pathways and rates of microbial degradation of petroleum under arctic conditions, on the identity of the intermediate products of metabolism, on their stability and their rates of accumulation, and on the effects of these products on the marine environment and organisms. In addition, information is needed on the effects of environmental conditions on rates of degradation of both parent hydrocarbons and metabolites to develop energy-flow models to predict the potential impact of the introduction of petroleum into marine arctic and subarctic regions.

We know that petroleum introduced into the environment can change the composition of microbial communities. We do not know the ecological significance of this effect, because we do not understand the roles of these communities nor the interactions among the organisms within the communities. Future studies involving the physico-chemical and the microbial alterations of petroleum in the arctic and subarctic should include related studies on the roles of microbial communities of the region.

The microbial method proposed several years ago to help in the cleanup of spilled oil has not proved to be useful except in certain highly localized areas for several reasons: attempts to develop a standard culture that would degrade the various fractions of petroleum were not successful; questions concerning the possible ecological impact from introduction of a foreign bacterial strain were never answered; experiments indicated that indigenous bacteria in various geographic locations would degrade added petroleum; and nutrients and environmental conditions, rather than the species and numbers of bacteria that were present, appeared to be the limiting factors. Hydrocarbon-oxidizing bacteria, however, can be used to remove traces of hydrocarbons from oil in localized areas. A limited amount of work on possible biological additives to scavenge petroleum has continued. Two approaches are currently underway. One approach is on bacterial genetic changes, i.e., to develop strains of bacteria with different plasmids that will permit induction of enzyme systems to degrade the different kinds of chemical compounds in petroleum. The other approach is to concentrate and isolate stable bacterial degradative enzymes in sufficient quantities to use when a spill occurs. At this time these methods neither supplement nor compete with existing physical cleanup methods. Developments in these fields should be followed because both approaches have potential environmental impacts that could be either adverse or beneficial to marine organisms.

This chapter has dealt primarily with the hydrocarbon components of petroleum because most of our available knowledge is about these compounds. The non-hydrocarbon components have received very little attention. Almost no consideration has been given to understanding the role of other compounds such as organo-metallics and chlorinated compounds that are present in petroleum. A recent publication by Smith et al. [220] illustrates a much needed type of investigation. They determined the effects of a chlorinated hydrocarbon pesticide, heptachlor, on hexadecane metabolism by several fungi. Results varied with different organisms and with different concentrations of the heptachlor. When utilization of hexadecane was greatest, the most degradation of heptachlor occurred. Obviously the complexity of composition of petroleum makes studies of interactions a difficult field. However, careful planning and intelligent choice of potential interactions to investigate can make this a productive approach. Certainly the investigations will be necessary to link laboratory and field investigations.

REFERENCES

1. ZoBell, C.E. (1962). Importance of microorganisms in the sea. In: Proceedings of the Low Temperature Microbiology Symposium, 1961, p. 107-32. Campbell Soup Co., Camden, N.J.
2. Shelton, T.B. and J.V. Hunter (1975). Anaerobic decomposition of oil in bottom sediments. J. Water Pollut. Control Fed. 47:2256-70.
3. ZoBell, C.E. (1973). Microbial degradation of oil: Present status, problems, and perspectives. In: The Microbial Degradation of Oil Pollutants (D.G. Ahearn and S.P. Meyers, eds.), p. 3-16. Publ. No. LSU-SG-73-01. Center for Wetland Resources, Louisiana State University, Baton Rouge, La.
4. Davies, J.A. and D.E. Hughes (1968). The biochemistry and microbiology of crude oil degradation. In: The Biological Effects of Oil Pollution on Littoral Communities (J.D. Carthy and D.R. Arthur, eds.), Suppl. to Field Studies, Vol. 2, p. 139-44. Obtainable from E.W. Classey, Ltd., Hampton, Middx., England.

5. Berridge, S.A., R.A. Dean, R.G. Fallows, and A. Fish (1968). The properties of persistent oils at sea. In: Scientific Aspects of Pollution of the Sea by Oil, Proceedings of a Symposium (P. Hepple, ed.), p. 2-11. Institute of Petroleum, London.
6. Kinney, P.J., D.K. Button, and D.M. Schell (1969). Kinetics of dissipation and biodegradation of crude oil in Alaska's Cook Inlet. In: Proceedings of 1969 Joint Conference on Prevention and Control of Oil Spills, p. 333-40. American Petroleum Institute, Washington, D.C.
7. Ayers, R.C., Jr., H.O. Jahns, and J.L. Glaeser (1974). Oil spills in the Arctic Ocean: Extent of spreading and possibility of large-scale thermal effects. Science 186:843-5.
8. NORCOR Engineering and Research Ltd. (1975). The interaction of crude oil with Arctic Sea ice. Beaufort Sea Tech. Rept. 27, Environment Canada, Victoria, B.C., 206 p.
9. Rosenegger, L.W. (1975). Movement of oil under sea ice. Beaufort Sea Tech. Rept. 28, Environment Canada, Victoria, B.C., 81 p.
10. Walker, E.R. (1975). Oil, ice, and climate in the Beaufort Sea. Beaufort Sea Tech. Rept. 35, Environment Canada, Victoria, B.C., 40 p.
11. Glaeser, J.L. and G.P. Vance (1971). A study of the behavior of oil spills in the Arctic. AD717142. National Technical Information Service, U.S. Dep. of Commerce, Springfield, Va., 60 p.
12. Hoult, D.P., S. Wolfe, S. O'Dea, and J.P. Patureau (1975). Oil in the Arctic. Final Report. CG-D-96-75. National Technical Information Service, U.S. Dep. of Commerce, Springfield, Va.
13. Institute of Petroleum Oil Pollution Analysis Committee (1974). Marine Pollution by Oil, Applied Science Publishers, Barking, Essex, England, p. 14-6.
14. Harrison, W., M.A. Winnik, P.T.Y. Kwong, and D. Mackay (1975). Crude oil spills. Disappearance of aromatic and aliphatic components from small sea-surface slicks. Environ. Sci. Technol. 9:231-4.
15. Sivadier, H.O. and P.G. Mikolaj (1973). Measurement of evaporation rates from oil slicks on the open sea. In: Proceedings of 1973 Joint Conference on Prevention and Control of Oil Spills, p. 475-82. American Petroleum Institute, Washington, D.C.
16. Bohon, R.L. and W.F. Claussen (1951). The solubility of aromatic hydrocarbons in water. J. Am. Chem. Soc. 73: 1571-8.

17. McAuliffe, C. (1966). Solubility in water of paraffin, cycloparaffin, olefin, acetylene, cycloolefin, and aromatic hydrocarbons. J. Phys. Chem. 70:1267-75.
18. Sutton, C. and J.A. Calder (1974). Solubility of higher-molecular-weight *n*-paraffins in distilled water and seawater. Environ. Sci. Technol. 8:654-7.
19. Parker, C.A., M. Freegarde, and C.G. Hatchard (1971). The effect of some chemical and biological factors on the degradation of crude oil at sea. In: Water Pollution by Oil (P. Hepple, ed.), p. 237-44. Institute of Petroleum, London.
20. Boylan, D.B. and B.W. Tripp (1971). Determination of hydrocarbons in seawater extracts of crude oil and crude oil fractions. Nature 230:44-7.
21. Frankenfeld, J.W. (1973). Factors governing the fate of oil at sea; variations in the amounts and types of dissolved or dispersed materials during the weathering process. In: Proceedings of 1973 Joint Conference on Prevention and Control of Oil Spills, p. 485-95. American Petroleum Institute, Washington, D.C.
22. Lysyj, I. and E.C. Russell (1974). Dissolution of petroleum-derived products in water. Water Res. 8:863-8.
23. Wasik, S.P. and R.L. Brown (1973). Determination of hydrocarbon solubility in seawater and the analysis of hydrocarbons in water extracts. In: Proceedings of 1973 Joint Conference on Prevention and Control of Oil Spills, p. 223-7. American Petroleum Institute, Washington, D.C.
24. Burwood, R. and G.C. Speers (1974). Photo-oxidation as a factor in the environmental dispersal of crude oil. Estuarine Coastal Mar. Sci. 2:117-35.
25. Finnerty, W.R., R.S. Kennedy, P. Lockwood, B.O. Spurlock, and R.A. Young (1973). Microbes and petroleum: Perspectives and implications. In: The Microbial Degradation of Oil Pollutants (D.G. Ahearn and S.P. Meyers, eds.), p. 105-25. Publ. No. LSU-SG-73-01. Center for Wetland Resources, Louisiana State University, Baton Rouge, La.
26. Zajic, J.E., B. Supplisson, and B. Volesky (1974). Bacterial degradation and emulsification of No. 6 fuel oil. Environ. Sci. Technol. 8:664-8.
27. Pilpel, N. (1968). Fate of oil in the sea. Endeavor 27:11-3.
28. Davis, S.J. and C.F. Gibbs (1975). The effect of weathering on a crude oil residue exposed at sea. Water Res. 9:275-85.

29. Berridge, S.A., M.T. Thew, and A.G. Loriston-Clarke (1968). The formation and stability of emulsions of water in crude petroleum and similar stocks. In: Scientific Aspects of Pollution of the Sea by Oil, Proceedings of a Symposium (P. Hepple, ed.), p. 35-59. Institute of Petroleum, London.
30. Friede, J.D. (1973). The isolation and chemical and biological properties of microbial emulsifying agents for hydrocarbons. Progress Report. AD 770-630. National Technical Information Service, U.S. Dep. of Commerce, Springfield, Va., 5 p.
31. Guire, P.E., J.D. Friede, and R.K. Gholson (1973). Production and characterization of emulsifying factors from hydrocarbonoclastic yeast and bacteria. In: The Microbial Degradation of Oil Pollutants (D.G. Ahearn and S.P. Meyers, eds.), p. 229-31. Publ. No. LSU-SG-73-01. Center for Wetland Resources, Louisiana State University, Baton Rouge, La.
32. Atlas, R.M. (1973). Fate and effects of oil pollutants in extremely cold marine environments. AD769895. National Technical Information Service, U.S. Dep. of Commerce, Springfield, Va., 33 p.
33. Kinney, P.J., D.K. Button, D.M. Schell, B.R. Robertson, and J. Groves (1970). Quantitative assessment of oil pollution problems in Alaska's Cook Inlet. Univ. Alaska Inst. Mar. Sci. Rept. R 69-16.
34. Hartung, R. and G.W. Klingler (1968). Sedimentation of floating oils. Pap. Mich. Acad. Sci. Arts Lett. LIII: 23-7.
35. Poirier, O.A. and G.A. Thiel (1941). Deposition of free oil by sediments settling in seawater. Bull. Am. Assoc. Pet. Geol. 25:2170-80.
36. Nelson-Smith, A. (1973). Oil Pollution and Marine Ecology. Plenum Press, New York, 260 p.
37. Klein, A.E. and N. Pilpel (1974). The effects of artificial sunlight upon floating oils. Water Res. 8:79-83.
38. Bunch, J.N. and R.C. Harland (1976). Biodegradation of crude petroleum by the indigenous microbial flora of the Beaufort Sea. Beaufort Sea Tech. Rept. 10, Environment Canada, Victoria, B.C., 52 p.
39. Bushnell, L.D. and H.F. Haas (1941). The utilization of certain hydrocarbons by microorganisms. J. Bacteriol. 41:653-73.
40. Button, D.K. (1974). Arctic oil biodegradation. AD-A014-096. National Technical Information Service, U.S. Dep. of Commerce, Springfield, Va., 37 p.
41. Marr, E.K. and R.W. Stone (1961). Bacterial oxidation of benzene. J. Bacteriol. 81:425-30.

42. McKenna, E.J. and M.J. Coon (1970). Enzymatic ω-oxidation. IV. Purification and properties of the ω-hydroxylase of *Pseudomonas olevorans*. J. Biol. Chem. 245:3882-9.
43. Walker, J.D. and R.R. Colwell (1974). Some effects of petroleum on estuarine and marine microorganisms. Can. J. Microbiol. 21:305-13.
44. Walker, J.D. and R.R. Colwell (1976). Enumeration of petroleum-degrading microorganisms. Appl. Environ. Microbiol. 31:198-207.
45. Walker, J.D. and R.R. Colwell (1976). Measuring the potential activity of hydrocarbon-degrading bacteria. Appl. Environ. Microbiol. 31:189-97.
46. Mironov, O.G. and A.A. Lebed (1972). Hydrocarbon-oxidizing microorganisms in the North Atlantic. Hydrobiol. J. 8:71-4.
47. Tagger, S., L. Deveze, and J. LePetit (1976). The conditions for biodegradation of petroleum hydrocarbons at sea. Mar. Pollut. Bull. 7(9):172-4.
48. ZoBell, C.E. (1973). Bacterial degradation of mineral oils at low temperatures. In: The Microbial Degradation of Oil Pollutants (D.G. Ahearn and S.P. Meyers, eds.), p. 153-61. Publ. No. LSU-SG-73-01. Center for Wetland Resources, Louisiana State University, Baton Rouge, La.
49. Morita, R.Y. (1975). Psychrophilic bacteria. Bacteriol. Rev. 39:144-67.
50. Sinclair, N.A. and J.L. Stokes (1965). Obligately psychrophilic yeasts from the polar regions. Can. J. Microbiol. 11:259-69.
51. Kriss, A.E. (1963). Marine Microbiology. Oliver and Boyd, London, 536 p.
52. Bunch, J.N. (1974). Biodegradation of crude petroleum by the indigenous microbial flora of the Beaufort Sea. Interim Rept. Beaufort Sea Proj. Study B5a, Environment Canada, Victoria, B.C.
53. Traxler, R.W. (1973). Bacterial degradation of petroleum materials in low temperature marine environments. In: The Microbial Degradation of Oil Pollutants (D.G. Ahearn and S.P. Meyers, eds.), p. 163-70. Publ. No. LSU-SG-73-01. Center for Wetland Resources, Louisiana State University, Baton Rouge, La.
54. Schwarz, J.R., A.A. Yayanos, and R.R. Colwell (1976). Metabolic activities of the intestinal microflora of a deep-sea invertebrate. Appl. Environ. Microbiol. 31:46-8.
55. Schwarz, J.R. and R.R. Colwell (1976). Microbial activities under deep-ocean conditions. Dev. Ind. Microbiol. 17:299-304.

56. Stokes, J.L. (1971). Influence of temperature on the growth and metabolism of yeasts. In: The Yeasts, Vol. 2: Physiology and Biochemistry of Yeasts (A.H. Rose and J.S. Harrison, eds.), p. 119-34. Academic Press, New York.
57. Ahearn, D.G., S.P. Meyers, and P.G. Standard (1971). The role of yeasts in the decomposition of oils in marine environments. Dev. Ind. Microbiol. 12:126-34.
58. Miller, T.L. and M.J. Johnson (1966). Utilization of normal alkanes by yeasts. Biotechnol. Bioeng. 8:549-65.
59. Klug, M.J. and A.J. Markovetz (1967). Degradation of hydrocarbons by members of the genus *Candida*. J. Bacteriol. 93:1847-52.
60. Turner, W.E. and D.G. Ahearn (1970). Ecology and physiology of yeasts of an asphalt refinery and its watershed. In: Recent Trends in Yeast Research (D.G. Ahearn, ed.) 1:113-23.
61. Komagata, K., T. Nakase, and N. Katsuya (1964). Assimilation of hydrocarbons by yeasts. I. Preliminary screening. J. Gen. Appl. Microbiol. 10:313-21.
62. Cook, W.L., J.K. Massey, and D.G. Ahearn (1973). The degradation of crude oil by yeasts and its effect on *Lesbistes reticulatus*. In: The Microbial Degradation of Oil Pollutants (D.G. Ahearn and S.P. Meyers, eds.), p. 279-82. Publ. No. LSU-SG-73-01. Center for Wetland Resources, Louisiana State University, Baton Rouge, La.
63. Walker, J.D., L. Cofone, Jr., and J.J. Cooney (1973). Microbial petroleum degradation: The role of *Cladosporium resinae*. In: Proceedings of 1973 Joint Conference on Prevention and Control of Oil Spills, p. 821-5. American Petroleum Institute, Washington, D.C.
64. Cooney, J.J. and J.D. Walker (1973). Hydrocarbon utilization by *Cladosporium resinae*. In: The Microbial Degradation of Oil Pollutants (D.G. Ahearn and S.P. Meyers, eds.), p. 25-32. Publ. No. LSU-SG-73-01. Center for Wetland Resources, Louisiana State University, Baton Rouge, La.
65. Colwell, R.R., J.D. Walker, and J.D. Nelson, Jr. (1973). Microbial ecology and the problem of petroleum degradation in Chesapeake Bay. In: The Microbial Degradation of Oil Pollutants (D.G. Ahearn and S.P. Meyers, eds.), p. 185-97. Publ. No. LSU-SG-73-01. Center for Wetland Resources, Louisiana State University, Baton Rouge, La.
66. Ezura, Y., K. Daiku, K. Tajima, T. Kimura, and M. Sakai (1974). Seasonal differences in bacterial counts and heterotrophic bacterial flora in Akkeshi Bay. In: Effect of the Ocean Environment on Microbial Activities (R.R. Colwell and R.Y. Morita, eds.), p. 112-23. University Park Press, Baltimore, Md.

67. Cundell, A.M. and R.W. Traxler (1973). The isolation and characterization of hydrocarbon-utilizing bacteria from Chedabucto Bay, Nova Scotia. In: Proceedings of 1973 Joint Conference on Prevention and Control of Oil Spills, p. 421-6. American Petroleum Institute, Washington, D.C.
68. Mulkins-Phillips, G.J. and J.E. Stewart (1974). Distribution of hydrocarbon-utilizing bacteria in northwestern Atlantic waters and coastal sediments. Can. J. Microbiol. 20:955-62.
69. Sieburth, J. (1967). Seasonal selection of estuarine bacteria by water temperature. J. Exp. Mar. Biol. Ecol. 1:98-121.
70. Traxler, R.W. (1974). Petroleum degradation in low temperature marine and estuarine environments. AD-778-687, National Technical Information Service, U.S. Dep. of Commerce, Springfield, Va., 54 p.
71. Chung, B.H., R.Y. Cannon, and R.C. Smith (1976). Influence of growth temperature on glucose metabolism of a psychrotrophic strain of *Bacillus cereus*. Appl. Environ. Microbiol. 31:39-45.
72. Lynch, W.H., J. MacLeod, and M. Franklin (1975). Effect of temperature on the activity and synthesis of glucose-catabolizing enzymes in *Pseudomonas fluorescens*. Can. J. Microbiol. 21:1560-72.
73. Christophersen, J. (1967). Adaptive temperature responses of microorganisms. In: Molecular Mechanisms of Temperature Adaptation (C.L. Prosser, ed.), p. 327-48. American Association for the Advancement of Science, Washington, D.C.
74. Farrell, J. and A. Rose (1967). Temperature effects on microorganisms. Annu. Rev. Microbiol. 21:101-20.
75. Ingraham, J.L. (1962). Temperature relationships. In: The Bacteria (I.C. Gunsalus and R.Y. Stanier, eds.), Vol. IV, p. 265-96. Academic Press, New York.
76. Ingraham, J.L. and J.L. Stokes (1959). Psychrophilic bacteria. Bacteriol. Rev. 23:97-108.
77. Rose, A.H. (1968). Physiology of microorganisms at low temperatures. J. Appl. Bacteriol. 31:1-11.
78. Stokes, J.L. (1967). Heat-sensitive enzymes and enzyme synthesis in psychrophilic microorganisms. In: Molecular Mechanisms of Temperature Adaptation (C.L. Prosser, ed.), p. 311-23. American Association for the Advancement of Science, Washington, D.C.
79. Morita, R.Y. and L.J. Albright (1965). Cell yields of *Vibrio marinus*, an obligate psychrophile, at low temperature. Can. J. Microbiol. 11:221-7.

80. Morita, R.Y. and S.D. Burton (1970). Occurrence, possible significance, and metabolism of obligate psychrophiles in marine waters. In: Organic Matter in Natural Waters (D.W. Hood, ed.), p. 275-85. Univ. Alaska Inst. Mar. Sci. Publ. No. 1.
81. Geesey, G.G. and R.Y. Morita (1975). Some physiological effects of near-maximum growth temperatures on an obligately psychrophilic marine bacterium. Can. J. Microbiol. 21:811-8.
82. MacLeod, R.A. (1965). The question of the existence of specific marine bacteria. Bacteriol. Rev. 29:9-23.
83. MacLeod, R.A. (1971). Salinity, bacteria, fungi and blue-green algae. In: Marine Ecology (O. Kinne, ed.), Vol. 1, Part 2, p. 689-703. Wiley Interscience Publishers Co., London.
84. Morita, R.Y., L.P. Jones, R.P. Griffiths, and T.E. Staley (1973). Salinity and temperature interactions and their relationship to the microbiology of the estuarine environment. In: Estuarine Microbial Ecology (L.H. Stevenson and R.R. Colwell, eds.), p. 221-32. University of South Carolina Press, Columbia, S.C.
85. Morita, R.Y. (1974). Hydrostatic pressure effects on microorganisms. In: Effect of the Ocean Environment on Microbial Activities (R.R. Colwell and R.Y. Morita, eds.), p. 133-8. University Park Press, Baltimore, Md.
86. Oppenheimer, C.H. and C.E. ZoBell (1952). The growth and viability of sixty-three species of marine bacteria as influenced by hydrostatic pressure. J. Mar. Res. 11: 10-8.
87. ZoBell, C.E. (1946). Action of microorganisms on hydrocarbons. Bacteriol. Rev. 10:1-49.
88. ZoBell, C.E. and R.Y. Morita (1959). Deep-sea bacteria. In: Scientific Results of the Danish Deep-Sea Round the World Expedition 1950-52, Galathea Report, Vol. I, p. 139-54. Copenhagen.
89. ZoBell, C.E. and J. Kim (1972). Effects of deep-sea pressures on microbial enzyme systems. Symp. Soc. Exp. Biol. 26:125-46.
90. Schwarz, J.R., J.D. Walker, and R.R. Colwell (1975). Deep-sea bacteria: Growth and utilization of *n*-hexadecane at *in situ* temperature and pressure. Can. J. Microbiol. 21:682-7.
91. McKenna, E.J. and R.E. Kallio (1965). The biology of hydrocarbons. Annu. Rev. Microbiol. 19:183-208.

92. Floodgate, G.D. (1973). A threnody concerning the biodegradation of oil in natural waters. In: The Microbial Degradation of Oil Pollutants (D.G. Ahearn and S.P. Meyers, eds.), p. 17-24. Publ. No. LSU-SG-73-01. Center for Wetland Resources, Louisiana State University, Baton Rouge, La.
93. Robertson, B.R., S.D. Arhelger, R.A.T. Law, and D.K. Button (1973). Hydrocarbon biodegradation. In: Environmental Studies of Port Valdez (D.W. Hood, W.E. Shiels, and E.J. Kelley, eds.), p. 449-79. Univ. Alaska Inst. Mar. Sci. Occas. Publ. No. 3.
94. Leadbetter, E.R. and J.W. Foster (1959). Oxidation products formed from gaseous alkanes by the bacterium *Pseudomonas methanica*. Arch. Biochem. Biophys. 82:491-2.
95. Horvath, R.S. and M. Alexander (1970). Cometabolism: A technique for the accumulation of biochemical products. Can. J. Microbiol. 16:1131-2.
96. Raymond, R.L., V.W. Jamison, and J.O. Hudson (1971). Hydrocarbon cooxidation in microbial systems. Lipids 6:453-7.
97. Gibson, D.T., J.R. Koch, and R.E. Kallio (1968). Oxidative degradation of aromatic hydrocarbons by microorganisms. I. Enzymatic formation of catechol from benzene. Biochemistry 7:2653-62.
98. Gibson, D.T., J.R. Koch, C.L. Schuld, and R.E. Kallio (1968). Oxidative degradation of aromatic hydrocarbons by microorganisms. II. Metabolism of halogenated aromatic hydrocarbons. Biochemistry 7:3795-802.
99. Horvath, R.S. (1972). Microbial co-metabolism and the degradation of organic compounds in nature. Bacteriol. Rev. 36:146-55.
100. Sethunathan, N. and M.D. Pathak (1971). Development of a diazinon-degrading bacterium in paddy water after repeated applications of diazinon. Can. J. Microbiol. 17:699-702.
101. Horvath, R.S. (1972). Cometabolism of the herbicide, 2,3,6-trichlorobenzoate by natural microbial populations. Bull. Environ. Contam. Toxicol. 7:273-83.
102. Davis, J.B. and R.L. Raymond (1961). Oxidation of alkyl-substituted cyclic hydrocarbons by a *Nocardia* during growth on *n*-alkanes. Appl. Microbiol. 9:383-8.
103. Friede, J., P. Guire, R.K. Gholson, E. Gaudy, and A.F. Gaudy, Jr. (1972). Assessment of biodegradation potential for controlling oil spills on the high seas. AD-759848. National Technical Information Service, U.S. Dep. of Commerce, Springfield, Va., 130 p.

104. Kim, J. and C.E. ZoBell (1974). Occurrence and activities of cell-free enzymes in oceanic environments. In: Effect of the Ocean Environment on Microbial Activities (R.R. Colwell and R.Y. Morita, eds.), p. 368-85. University Park Press, Baltimore, Md.
105. Foster, J.W. (1962). Bacterial oxidation of hydrocarbons. In: Oxygenases (O. Hiyashi, ed.), p. 241-71. Academic Press, New York.
106. Gibson, D.T. (1971). The microbial oxidation of aromatic hydrocarbons. Crit. Rev. Microbiol. 1:199-223.
107. Raymond, R.L. and V.W. Jamison (1971). Biochemical activities of *Nocardia*. Adv. Appl. Microbiol. 14:93-122.
108. Rogoff, M.H. (1961). Oxidation of aromatic compounds by bacteria. Adv. Appl. Microbiol. 3:193-221.
109. Treccani, V. (1964). Microbial degradation of hydrocarbons. Prog. Ind. Microbiol. 4:3-33.
110. Van der Linden, A.C. and G.J.E. Thijsse (1965). The mechanisms of microbial oxidations of petroleum hydrocarbons. Adv. Enzymol. 27:469-546.
111. ZoBell, C.E. (1950). Assimilation of hydrocarbons by microorganisms. Adv. Enzymol. 10:443-86.
112. Abbott, B.J. and W.G. Gledhill (1971). The extracellular accumulation of metabolic products by hydrocarbon-degrading microorganisms. Adv. Appl. Microbiol. 14: 249-388.
113. King, D.H. and J.J. Perry (1975). The origin of fatty acids in the hydrocarbon-utilizing microorganism, *Mycobacterium vaccae*. Can. J. Microbiol. 21:85-9.
114. Chakrabarty, A.M. (1972). Genetic basis of the biodegradation of salicylate in *Pseudomonas*. J. Bacteriol. 112:815-23.
115. Dunn, N.W. and I.C. Gunsalus (1973). Transmissible plasmid coding early enzymes of naphthalene oxidation in *Pseudomonas putida*. J. Bacteriol. 114:974-9.
116. Chakrabarty, A.M., G. Chou, and I.C. Gunsalus (1973). Genetic regulation of octane dissimilation plasmid in *Pseudomonas*. Proc. Natl. Acad. Sci. 70:1137-40.
117. Chakrabarty, A.M. (1974). Dissociation of a degradative plasmid aggregate in *Pseudomonas*. J. Bacteriol. 118: 815-20.
118. Van Eyk, J. and T.J. Bartels (1968). Paraffin oxidation in *Pseudomonas aeruginosa*. I. Induction of paraffin oxidation. J. Bacteriol. 96:706-12.
119. McKenna, E.J. and R.E. Kallio (1964). Hydrocarbon structure: Its effect on bacterial utilization of alkanes. In: Principles and Applications in Aquatic Microbiology (H. Heukelian and N.C. Dondero, eds.), p. 1-14. John Wiley & Sons, New York.

120. Pirnik, M.P., R.M. Atlas, and R. Bartha (1974). Hydrocarbon metabolism by *Brevibacterium erythrogenes*: Normal and branched alkanes. J. Bacteriol. 119:868-78.
121. Walker, J.D., L. Petrakis, and R.R. Colwell (1976). Comparison of the biodegradability of crude and fuel oils. Can. J. Microbiol. 22:598-602.
122. McKenna, E.J. and R.E. Kallio (1971). Microbial metabolism of the isoprenoid alkane, pristane. Proc. Natl. Acad. Sci. 68:1552-4.
123. Westlake, D.W.S., A. Jobson, R. Phillippe, and F.D. Cook (1974). Biodegradability and crude oil composition. Can. J. Microbiol. 20:915-28.
124. May, S.W. and B.J. Abbott (1972). Enzymatic epoxidation. I. Alkene epoxidation by the ω-hydroxylation system of *Pseudomonas oleovorans*. Biochem. Biophys. Res. Commun. 48:1230-4.
125. May, S.W. and B.J. Abbott (1973). Enzymatic epoxidation. II. Comparison between the epoxidation and hydroxylation reactions catalyzed by the ω-hydroxylation system of *Pseudomonas oleovorans*. J. Biol. Chem. 248:1725-30.
126. Abbott, B.J. and C.T. Hou (1973). Oxidation of 1-alkenes to 1,2-epoxyalkanes by *Pseudomonas oleovorans*. Appl. Microbiol. 26:86-91.
127. Schwartz, R.D. (1973). Octene epoxidation by a cold stable alkane-oxidizing isolate of *Pseudomonas oleovorans*. Appl. Microbiol. 25:574-7.
128. Schwartz, R.D. and C.J. McCoy (1976). Enzymatic epoxidation: Synthesis of 7,8-epoxy-1-octene, 1,2-7,8-diepoxyoctane, and 1,2-epoxyoctane by *Pseudomonas oleovorans*. Appl. Environ. Microbiol. 31:78-82.
129. Ooyama, J. and J.W. Foster (1965). Bacterial oxidation of cycloparaffinic hydrocarbons. J. Microbiol. Serol. 31:45-65.
130. Evans, W.C. (1956). Biochemistry of the oxidative metabolism of aromatic compounds by microorganisms. Annu. Repts. Prog. Chem., Chem. Soc. (London) 53:279-94.
131. Axcell, B.C. and P.J. Geary (1973). The metabolism of benzene by bacteria. Purification and some properties of the enzyme *cis*-1,2-dihydroxy cyclohexa-3,5-diene (nicotinamide adenine dinucleotide) oxidoreductase (*cis*-benzene glycol dehydrogenase). Biochem. J. 136:927-34.
132. Reiner, A.M. (1972). Metabolism of aromatic compounds in bacteria. Purification and properties of the catechol-forming enzyme, 3,5-cyclohexadiene-1,2-diol-1-carboxylic acid (NAD^+) oxidoreductase. J. Biol. Chem. 247:4960-5.
133. Barnsley, E.A. (1976). Role and regulation of the *ortho* and *meta* pathways of catechol metabolism in pseudomonads metabolizing naphthalene and salicylate. J. Bacteriol. 125:404-8.

134. Jerina, D.M., J.W. Daly, A.M. Jeffrey, and D.T. Gibson (1971). *Cis*-1,2-dihydroxy-1,2-dihydronaphthalene: A bacterial metabolite from naphthalene. Arch. Biochem. Biophys. 142:394-6.
135. Patel, T.R. and D.T. Gibson (1974). Purification and properties of (+)-*cis*-naphthalene dihydrodiol dehydrogenase of *Pseudomonas putida*. J. Bacteriol. 119:879-88.
136. Gibson, D.T. (1968). Microbial degradation of aromatic compounds. Science 161:1093-7.
137. Gibson, D.T., C.E. Cardini, F.C. Maseles, and R.E. Kallio (1970). Incorporation of oxygen-18 into benzene by *Pseudomonas putida*. Biochemistry 9:1631-5.
138. Gibson, D.T. (1972). Initial reactions in the degradation of aromatic hydrocarbons. In: Degradation of Synthetic Organic Molecules in the Biosphere: Natural, Pesticidal and Various Other Man-Made Compounds. Proceedings of a Conference, San Francisco, Calif., June 12-13, 1971, p. 116-36. National Academy of Sciences, Washington, D.C.
139. Gibson, D.T. and W.K. Yeh (1973). Microbial degradation of aromatic hydrocarbons. In: The Microbial Degradation of Oil Pollutants (D.G. Ahearn and S.P. Meyers, eds.), p. 33-8. Publ. No. LSU-SG-73-01. Center for Wetland Resources, Louisiana State University, Baton Rouge, La.
140. Ribbons, D.W. and P.J. Senior (1970). Enzymic estimation of 2,3-dihydroxybenzoate and 2,3-dihydroxy-*p*-toluate. Anal. Biochem. 36:310-4.
141. Ribbons, D.W. and P.J. Senior (1970). 2,3-Dihydroxybenzoate-3,4-oxygenase from *Pseudomonas fluorescens* - Oxidation of a substrate analog. Arch. Biochem. Biophys. 138:557-65.
142. Sparnins, V.L. and S. Dagley (1975). Alternative routes of aromatic catabolism in *Pseudomonas acidovorans* and *Pseudomonas putida*: Gallic acid as a substrate and inhibitor of dioxygenases. J. Bacteriol. 124:1374-81.
143. Gibson, D.T., M. Hensley, H. Yoshioka, and T.J. Mabry (1970). Formation of (+)-*cis*-2,3-dihydroxy-1-methylcyclohexa-4,6-diene from toluene by *Pseudomonas putida*. Biochemistry 9:1626-30.
144. Davey, J.F. and D.T. Gibson (1974). Bacterial metabolism of *para*- and *meta*-xylene: Oxidation of a methyl substituent. J. Bacteriol. 119:923-9.
145. Gibson, D.T., V. Mahadevan, and J.F. Davey (1974). Bacterial metabolism of *para*- and *meta*-xylene: Oxidation of the aromatic ring. J. Bacteriol. 119:930-6.
146. Catterall, F.A., K. Murray, and P.A. Williams (1971). The configuration of the 1,2-dihydroxy-1,2-dihydronaphthalene formed in the bacterial metabolism of naphthalene. Biochim. Biophys. Acta 237:361-4.

147. Davies, J.I. and W.C. Evans (1964). Oxidative metabolism of naphthalene by soil pseudomonads. Biochem. J. 91: 251-61.
148. Shamsuzzaman, K.M. and E.A. Barnsley (1974). The regulation of naphthalene metabolism in pseudomonads. Biochem. Biophys. Res. Commun. 60:582-9.
149. Shamsuzzaman, K.M. and E.A. Barnsley (1974). The regulation of naphthalene oxygenase in pseudomonads. J. Gen. Microbiol. 83:165-70.
150. Raymond, D.D. (1975). Metabolism of methylnaphthalenes and other related aromatic hydrocarbons by marine bacteria. Diss. Abstr. Int. B. Sci. Eng. 35(10):5014 B., 135 p.
151. Dean-Raymond, D. and R. Bartha (1975). Biodegradation of some polynuclear aromatic petroleum components by marine bacteria. Dev. Ind. Microbiol. 16:97-110.
152. ZoBell, C.E. (1971). Sources and biodegradation of carcinogenic hydrocarbons. In: Proceedings of 1971 Joint Conference on Prevention and Control of Oil Spills, p. 441-51. American Petroleum Institute, Washington, D.C.
153. Poglazova, M.N., G.E. Fedoseeva, A.J. Khesina, M.N. Meissel, and L.M. Shabad (1967). Destruction of benzo-[a]pyrene by soil bacteria. Life Sci. 6:1053-62.
154. Barnsley, E.A. (1975). The bacterial degradation of fluoranthene and benzo[a]pyrene. Can. J. Microbiol. 21:1004-8.
155. Gibson, D.T. (1975). Oxidation of the carcinogens benzo-[a]pyrene and benzo[a]anthracene to dihydrodiols by a bacterium. Science 189:295-7.
156. Yamada, K., Y. Monoda, K. Kodama, S. Nakatani, and T. Akasaki (1968). Microbial conversion of petro-sulfur compounds. Part I. Isolation and identification of dibenzothiophene-utilizing bacteria. Agric. Biol. Chem. 32:840-5.
157. Nakatani, S., T. Akasaki, K. Kodama, Y. Minoda, and K. Yamada (1968). Microbial conversion of petro-sulfur compounds. Part II. Culture conditions of dibenzothiophene-utilizing bacteria. Agric. Biol. Chem. 32:1205-11.
158. Kodama, K., S. Nakatani, K. Umehara, K. Shimizu, Y. Minoda, and K. Yamada (1970). Microbial conversion of petro-sulfur compounds. Part III. Isolation and identification of products from dibenzothiophene. Agric. Biol. Chem. 34:1320-4.
159. Laborde, A. and D.T. Gibson (1975). Bacterial oxidation of dibenzothiophene. Abstr. Annu. Meet. Am. Soc. Microbiol. Q45.
160. Hou, C.T. and A.I. Laskin (1976). Microbial conversion of dibenzothiophene. Dev. Ind. Microbiol. 17:351-62.

161. McCarty, P.L. (1972). Energetics of organic matter degradation. In: Water Pollution Microbiology (R. Mitchell, ed.), p. 91-118. John Wiley & Sons, New York.
162. McCarty, P.L. (1965). Thermodynamics of biological synthesis and growth. In: Second International Conference on Water Pollution Research, p. 169-99. Pergamon Press, New York.
163. McCarty, P.L. (1971). Energetics and bacterial growth. In: Organic Compounds in Aquatic Environments (S.J. Faust, ed.), p. 495-531. Marcel Dekker, Inc., New York.
164. Gunsalus, I.C. and C.W. Shuster (1961). Energy-yielding metabolism in bacteria. In: The Bacteria (I.C. Gunsalus and R.Y. Stanier, eds.), Vol. II, Metabolism, p. 1-58. Academic Press, New York.
165. Fasoli, U. and W. Numann (1973). A proposal for the application of Monod's mathematical model to the biodegradation of mineral oil in natural waters. Water Res. 7:409-18.
166. Vaccaro, R.F. and H.W. Jannasch (1966). Studies on heterotrophic activity in seawater based on glucose assimilation. Limnol. Oceanogr. 11:596-607.
167. Mill, J. and D.G. Hendry (1976). Estimation procedures for persistence of organic compounds in the environment. Presented at the 172nd National Meeting, American Chemical Society, August 29-September 3, 1976. In: Preprints of Div. Environ. Chem., p. 172-5.
168. Johnston, R. (1970). The decomposition of crude oil residues in sand columns. J. Mar. Biol. Assoc. U.K. 50:925-37.
169. Hunt, P.G., F.R. Koutz, R.P. Murrmann, and T.G. Martin (1973). Microbial degradation of petroleum in continental shelf sediments. AD 772698. National Technical Information Service, U.S. Dep. of Commerce, Springfield, Va., 16 p.
170. ZoBell, C.E. (1964). The occurrence, effects and fate of oil polluting the sea. Adv. Water Pollut. Res. 3:85-118.
171. Bailey, N.J.L., A.M. Jobson, and M.A. Rogers (1973). Bacterial degradation of crude oil: Comparison of field and experimental data. Chem. Geol. 11:203-21.
172. Floodgate, G.D. (1972). Biodegradation of hydrocarbons in the sea. In: Water Pollution Microbiology (R. Mitchell, ed.), p. 153-71. John Wiley & Sons, New York.
173. ZoBell, C.E. and J.F. Prokop (1966). Microbial oxidation of mineral oils in Barataria Bay bottom deposits. Z. Allg. Mikrobiol. 6:143-62.
174. Mulkins-Phillips, G.J. and J.E. Stewart (1974). Effect of environmental parameters on bacterial degradation of Bunker C oil, crude oils and hydrocarbons. Appl. Microbiol. 28:915-22.

175. Scarratt, D.J. and V. Zitko (1972). Bunker C oil in sediments and benthic animals from shallow depths in Chedabucto Bay, N.S. J. Fish. Res. Board Can. 29: 1347-50.
176. Gordon, D.C., Jr. and P.A. Michalik (1971). Concentration of Bunker C fuel oil in the waters of Chedabucto Bay, April 1971. J. Fish. Res. Board Can. 28:1912-4.
177. Rashid, M.A. (1974). Degradation of Bunker C oil under different coastal environments of Chedabucto Bay, Nova Scotia. Estuarine Coastal Mar. Sci. 2:137-44.
178. ZoBell, C.E. (1969). Microbial modification of crude oil in the sea. In: Proceedings of 1969 Joint Conference on Prevention and Control of Oil Spills, p. 317-26. American Petroleum Institute, Washington, D.C.
179. Atlas, R.M. and R. Bartha (1972). Biodegradation of petroleum in seawater at low temperatures. Can. J. Microbiol. 18:1851-5.
180. Pierce, R.H., Jr., A.M. Cundell, and R.W. Traxler (1975). Persistence and biodegradation of spilled residual fuel oil on an estuarine beach. Appl. Microbiol. 29:646-52.
181. Cundell, A.M. and R.W. Traxler (1976). Psychrophilic hydrocarbon-degrading bacteria from Narragansett Bay, Rhode Island, U.S.A. Mater. Org. (Berl.) 11:1-17.
182. Blumer, M. and J. Sass (1972). Oil pollution: Persistence and degradation of spilled fuel oil. Science 176:1120-2.
183. Cobet, A.B. and H.E. Guard (1973). Effect of a bunker fuel on the beach bacterial flora. In: Proceedings of 1973 Joint Conference on Prevention and Control of Oil Spills, p. 815-9. American Petroleum Institute, Washington, D.C.
184. Walker, J.D., R.R. Colwell, and L. Petrakis (1975). Microbial petroleum degradation: Application of computerized mass spectrometry. Can. J. Microbiol. 21:1760-7.
185. Walker, J.D. and R.R. Colwell (1974). Microbial degradation of model petroleum at low temperatures. Microbiol. Ecol. 1:63-95.
186. Soli, G. (1973). Marine hydrocarbonoclastic bacteria: Types and range of oil degradation. In: The Microbial Degradation of Oil Pollutants (D.G. Ahearn and S.P. Meyers, eds.), p. 141-6. Publ. No. LSU-SG-73-01. Center for Wetland Resources, Louisiana State University, Baton Rouge, La.
187. Mechalas, B.J., T.J. Meyers, and R.L. Kolpack (1973). Microbial decomposition patterns using crude oil. In: The Microbial Degradation of Oil Pollutants (D.G. Ahearn and S.P. Meyers, eds.), p. 67-79. Publ. No. LSU-SG-73-01. Center for Wetland Resources, Louisiana State University, Baton Rouge, La.

188. Lee, R.F. and C. Ryan (1976). Biodegradation of petroleum hydrocarbons by marine microbes. In: Biodeterioration of Materials (Sharpley and Kaplan, eds.), Vol. 3. Applied Science Publishers Ltd., Essex, England. In press.
189. Atlas, R.M. and R. Bartha (1973). Abundance, distribution and oil biodegradation potential of microorganisms in Raritan Bay. Environ. Pollut. 4:291-300.
190. Grainger, E.H. (1975). Biological productivity in the southern Beaufort Sea: The physical-chemical environment and the plankton. Beaufort Sea Tech. Rept. 12a, Environment Canada, Victoria, B.C., 82 p.
191. Robertson, B., S. Arhelger, P.J. Kinney, and D.K. Button (1973). Hydrocarbon biodegradation in Alaskan waters. In: The Microbial Degradation of Oil Pollutants (D.G. Ahearn and S.P. Meyers, eds.), p. 171-84. Publ. No. LSU-SG-73-01. Center for Wetland Resources, Louisiana State University, Baton Rouge, La.
192. Atlas, R.M., E.A. Schofield, F.A. Morelli, and R.E. Cameron (1974). Interactions of microorganisms and petroleum pollutants in the Arctic. Abstr. Annu. Meet. Am. Soc. Microbiol, p. 64.
193. Walker, J.D., P.A. Seesman, and R.R. Colwell (1974). Effects of petroleum on estuarine bacteria. Mar. Pollut. Bull. 5(12):186-8.
194. Claus, D. and N. Walker (1964). The decomposition of toluene by soil bacteria. J. Gen. Microbiol. 36:107-22.
195. Calder, J.A. and J.H. Lader (1976). Effect of dissolved aromatic hydrocarbons on the growth of marine bacteria in batch culture. Appl. Environ. Microbiol. 32:95-101.
196. Atlas, R.M. (1975). Effects of temperature and crude oil composition on petroleum biodegradation. Appl. Microbiol. 30:396-403.
197. Scott, C.C.L. and W.R. Finnerty (1976). A comparative analysis of the ultrastructure of hydrocarbon-oxidizing micro-organisms. J. Gen. Microbiol. 94:342-50.
198. Atlas, R.M. and C.E. Heintz (1973). Ultrastructure of two species of oil-degrading marine bacteria. Can. J. Microbiol. 19:43-5.
199. Osumi, M., F. Fukuzumi, Y. Teranishi, A. Tanaka, and S. Fukui (1975). Development of microbodies in *Candida tropicalis* during incubation in a *n*-alkane medium. Arch. Microbiol. 103:1-11.
200. Gill, C.O. and C. Ratledge (1973). Regulation of *de nova* fatty acid biosynthesis in the *n*-alkane utilizing yeast, *Candida* 107. J. Gen. Microbiol. 78:337-47.

201. Ratledge, C. (1973). Lipid biosynthesis in hydrocarbon-utilizing organisms. In: The Microbial Degradation of Oil Pollutants (D.G. Ahearn and S.P. Meyers, eds.), p. 313. Publ. No. LSU-SG-73-01. Center for Wetland Resources, Louisiana State University, Baton Rouge, La.
202. Cerniglia, C.E. and J.J. Perry (1974). Effect of substrate on the fatty acid composition of hydrocarbon-utilizing filamentous fungi. J. Bacteriol. 118:844-7.
203. Boyer, J.M. and M.A. Pisano (1975). Effect of growth on hexadecane on the lipids of *Paecilomyces persicinus*. Dev. Ind. Microbiol. 16:391-400.
204. Lukins, H.B. and J.W. Foster (1963). Methyl ketone metabolism in hydrocarbon-utilizing mycobacteria. J. Bacteriol. 85:1074-87.
205. Raymond, R.L. and J.B. Davis (1960). *N*-alkane utilization and lipid formation by a *Nocardia*. Appl. Microbiol. 8:329-34.
206. Davis, J.B. (1964). Microbial incorporation of fatty acids derived from *n*-alkanes into glycerides and waxes. Appl. Microbiol. 12:210-4.
207. Patrick, M.A. and P.R. Dugan (1974). Influence of hydrocarbons and derivatives on the polar lipid fatty acids of an acinetobacter isolate. J. Bacteriol. 119:76-81.
208. Dunlap, K.R. and J.J. Perry (1967). Effect of substrate on the fatty acids of hydrocarbon-utilizing microorganisms. J. Bacteriol. 94:1919-21.
209. Makula, R. and W.R. Finnerty (1968). Microbial assimilation of hydrocarbons. I. Fatty acids derived from normal alkanes. J. Bacteriol. 95:2102-7.
210. Makula, R. and W.R. Finnerty (1968). Microbial assimilation of hydrocarbons. II. Fatty acids derived from 1-alkenes. J. Bacteriol. 95:2108-11.
211. Mitchell, R. (1972). Ecological control of microbial imbalances. In: Water Pollution Microbiology (R. Mitchell, ed.), p. 273-88. Wiley-Interscience, New York.
212. Mitchell, R. (1968). Factors affecting the decline of non-marine microorganisms in seawater. Water Res. 2: 535-9.
213. Adler, J. (1966). Chemotaxis in bacteria. Science 153:708-16.
214. Adler, J. (1969). Chemoreceptors in bacteria. Science 166:1588-97.
215. Chet, I., S. Fogel, and R. Mitchell (1971). Chemical detection of microbial prey by bacterial predators. J. Bacteriol. 106:863-7.
216. Tso, W. and J. Adler (1974). Negative chemotaxis in *Escherichia coli*. J. Bacteriol. 118:560-76.

217. Young, L.Y. and R. Mitchell (1973). Negative chemotaxis of marine bacteria to toxic chemicals. Appl. Microbiol. 25:972-5.
218. Mitchell, R., S. Fogel, and I. Chet (1972). Bacterial chemoreception: An important ecological phenomenon inhibited by hydrocarbons. Water Res. 6:1137-40.
219. Walsh, F. and R. Mitchell (1973). Inhibition of bacterial chemoreception by hydrocarbons. In: The Microbial Degradation of Oil Pollutants (D.G. Ahearn and S.P. Meyers, eds.), p. 275-8. Publ. No. LSU-SG-73-01. Center for Wetland Resources, Louisiana State University, Baton Rouge, La.
220. Smith, N.G., A.W. Bourquin, S.A. Crow, and D.G. Ahearn (1976). Effect of heptachlor on hexadecane utilization by selected fungi. Dev. Ind. Microbiol. 17:331-6.

List of Abbreviations

Å	angstrom (10^{-10} meter)
API	American Petroleum Institute
ASTM	American Society for Testing and Materials
Atl.	Atlantic
bbl	barrel (42 U.S. gallons)
BOD	biochemical oxygen demand
^{14}C	radioactive carbon-14 labeled compound
μCi	microcurie
CI	chemotaxis inhibition
COD	chemical oxygen demand
CPI	carbon preference index (see reference 28, Chapter 1)
cSt	centistoke
DBT	dibenzothiophene
E_h	oxidation-reduction potential
Fs	spreading coefficient
F.O.	fuel oil
FS	fluorescence spectrometry
GC(GLC)	gas chromatography (gas-liquid chromatography)
GC/GR	gas chromatography and gravimetry
GC/MS	gas chromatography and mass spectrometry
GR	gravimetry
^{3}H	radioactive tritium-labeled compound
HC	hydrocarbon
IP	Institute of Petroleum (Great Britain)
IR	infrared spectroscopy
LOT	"Load on Top": procedure for recovery of oil residues from storage tanks of oil tankers (see reference 9, Chapter 2)
LPG	liquefied petroleum gas
MS	mass spectrometry
NAS	National Academy of Sciences
NIO	National Institute of Oceanography (Great Britain), now Institute of Oceanographic Sciences
NORCOR	NORCOR Engineering and Research Ltd.
NSFO	Navy Special fuel oil
NSO	nitrogen-, sulfur-, and oxygen-containing organic compounds (polar compounds)
NW	northwest
OEP	odd-even (carbon number) predominance (see reference 30, Chapter 1)
OM	oil material

OWD	oil-in-water dispersion
Pac.	Pacific
PCB	polychlorinated biphenyl
pE	redox potential
pH	hydrogen-ion concentration
Phyt.	phytane(see Fig. 2, Chapter 1)
Pris.	pristane (see Fig. 2, Chapter 1)
St	stoke
SW	seawater
TLC	thin-layer chromatography
UCM	unresolved complex mixture envelope
UV	ultraviolet spectroscopy
WSF	water-soluble fraction

List of Symbols

%	percent (parts per hundred)
‰	salinity (parts per thousand)
$\cong$	approximately equal
$<$	less than
$>$	greater than
$\leqslant$	less than or equal to
$\geqslant$	greater than or equal to
η	absolute or dynamic viscosity in centipoise
p	probability
ρ	density in grams per cubic centimeter
r	sample correlation coefficient
μ	micro
υ	kinematic viscosity in centipoise

Geographical Name Index

G

H

I

K

L

M

N

O

P

Q

R

S

T

U

V

W

Y

Subject Index

A

B

D

E

F

G

T

U

V

A
B 7
C 8
D 9
E 0
F 1
G 2
H 3
I 4
J 5

W

X

Y

Z